Applied Biochemistry and Bioengineering

VOLUME 4

Advisory Board

Applied Biochemistry and Bioengineering

Edited by

Lemuel B. Wingard, Jr.
Department of Pharmacology
University of Pittsburgh School of Medicine
Pittsburgh, Pennsylvania

Leon Goldstein
Department of Biochemistry
Tel Aviv University
Tel Aviv, Israel

Ephraim Katchalski-Katzir
Department of Biophysics
The Weizmann Institute of Science
Rehovot, Israel

VOLUME 4
IMMOBILIZED MICROBIAL CELLS

Edited by

Ichiro Chibata
Research and Development Headquarters
Tanabe Seiyaku Co., Ltd.
Yodogawa-ku, Osaka, Japan

Lemuel B. Wingard, Jr.
Department of Pharmacology
University of Pittsburgh School of Medicine
Pittsburgh, Pennsylvania

1983

ACADEMIC PRESS
A Subsidiary of Harcourt Brace Jovanovich, Publishers
NEW YORK LONDON
PARIS SAN DIEGO SAN FRANCISCO SÃO PAULO SYDNEY TOKYO TORONTO

ACADEMIC PRESS, INC.
111 Fifth Avenue, New York, New York 10003

United Kingdom Edition published by
ACADEMIC PRESS, INC. (LONDON) LTD.
24/28 Oval Road, London NW1 7DX

Library of Congress Catalog Card Number: 76-9161

ISBN 0-12-041104-0

PRINTED IN THE UNITED STATES OF AMERICA

83 84 85 86   9 8 7 6 5 4 3 2 1

Contents

Contributors

Numbers in parentheses indicate the pages on which the authors' contributions begin.

Joaquim M. S. Cabral (189), Laboratório de Engenharia Bioquímica, Departamento de Engenharia Química, Instituto Superior Técnico, Universidade Técnica de Lisboa, 1000 Lisbon, Portugal

Ichiro Chibata (1), Research and Development Headquarters, Tanabe Seiyaku Co., Ltd., Yodogawa-ku, Osaka 532, Japan

Saburo Fukui (153), Laboratory of Industrial Biochemistry, Department of Industrial Chemistry, Kyoto University, Kyoto 606, Japan

S. B. Karkare (311), Department of Chemical and Biochemical Engineering, Rutgers University, New Brunswick, New Jersey 08903

Isao Karube (281), Research Laboratory of Resources Utilization, Tokyo Institute of Technology, Nagatsuta-cho, Midori-ku, Yokohama 227, Japan

John F. Kennedy (189), Research Laboratory for the Chemistry of Bioactive Carbohydrates and Proteins, Department of Chemistry, University of Birmingham, Birmingham B15 2TT, England

Joachim Klein (11), Institute of Technical Chemistry, Technical University of Braunschweig, D-3300 Braunschweig, Federal Republic of Germany

Pekka Linko (53), Laboratory of Biochemistry and Food Technology, Department of Chemistry, Helsinki University of Technology, SF-02150, Espoo 15, Finland

Yu-Yen Linko (53), Laboratory of Biochemistry and Food Technology, Department of Chemistry, Helsinki University of Technology, SF-02150, Espoo 15, Finland

Hideo Ochiai (153), Laboratory of Biochemistry, College of Agriculture, Shimane University, Shimane 690, Japan

Shuichi Suzuki (281), Research Laboratory of Resources Utilization, Tokyo Institute of Technology, Nagatsuta-cho, Midori-ku, Yokohama 227, Japan

Atsuo Tanaka (153), Laboratory of Industrial Biochemistry, Department of Industrial Chemistry, Kyoto University, Kyoto 606, Japan

Tetsuya Tosa (1), Research Laboratory of Applied Biochemistry, Tanabe Seiyaku Co., Ltd., Yodogawa-ku, Osaka 532, Japan

K. Venkatasubramanian (311), Department of Chemical and Biochemical Engineering, Rutgers University, New Brunswick, New Jersey 08903, and H. J. Heinz Company, World Headquarters, Pittsburgh, Pennsylvania 15230

W. R. Vieth (311), Department of Chemical and Biochemical Engineering, Rutgers University, Piscataway, New Jersey 08854

Fritz Wagner (11), Institute of Biochemistry and Biotechnology, Technical University of Braunschweig, Braunschweig, Federal Republic of Germany

Preface

The immobilization of enzymes on solid supports has evolved from a stage of research curiosity to one of industrial applications between the 1950s and the late 1960s or early 1970s. These early industrial uses of single enzymes immobilized on solid supports stimulated a variety of research projects to ascertain the practicality of immobilizing multiple enzymes to carry out sequences of reactions. One alternative to isolating and immobilizing multiple enzymes is to see if the enzymes can be left intact in the parent microbial cells and simply immobilize the cells. Experiments begun in the late 1960s have shown that practical applications can be developed using immobilized microbial cells instead of immobilized enzymes. Since then, many studies have been carried out to explore the possibilities for use of immobilized dormant or growing microbial cells as well as organelles, and plant and animal cells.

These immobilized biocatalysts have been and are being applied in a wide variety of fields of high importance in biotechnology from industrial processes to medical research, and also in many areas of related science and engineering. With the advent of genetic engineering developments, the production of microbial cells suitable for immobilization and tailored for specific end uses is becoming a reality.

This volume is aimed at providing an overview of the methods of immobilization, applications, and ways of utilizing immobilized microbial cells and subcellular organelles and chloroplasts as biocatalysts. Later volumes in this series will deal specifically with plant and animal cells.

Contents of Previous Volumes

VOLUME 3

Applied Biochemistry and Bioengineering

VOLUME 4

Immobilized Cells: Historical Background

Ichiro Chibata and Tetsuya Tosa

Research Laboratory of Applied Biochemistry
Tanabe Seiyaku Co. Ltd.
Osaka, Japan

I. IMPROVEMENT OF ENZYME CHARACTERISTICS: IMMOBILIZATION OF ENZYMES

Enzymes are protein biocatalysts that participate in the many chemical reactions that occur in living organisms. Enzymes differ from ordinary chemical catalysts in their unique ability to catalyze reactions under very mild conditions—such as in neutral, aqueous solution at ordinary temperature and pressure—and in their very high specificity.

The external utilization of enzymes by humans was carried out before the concept of an enzyme was developed. In other words, the utilization of enzymes began with human history and has gradually expanded into a variety of fields, such as brewing, food production, textiles, tanning, and medicine. Furthermore, recent developments in the field of biochemistry and subsequent clarification of the mechanisms of enzyme reactions, the development of new enzyme sources, and especially the progress is applied microbiology and genetic engineering have all markedly accelerated the utilization of enzymes.

APPLIED BIOCHEMISTRY AND BIOENGINEERING
Volume 4

ISBN 0-12-041104-0

Although enzymes have many advantages as catalysts, they are essentially produced by an organism for its own requirements and not for external use by humans. When we talk about the catalytic activity of enzymes, we say enzymes are efficient and advantageous as catalysts. However, they are not always ideal catalysts for practical applications. In some cases, the aforementioned advantages turn out to be disadvantages for catalysis. For example, enzymes are generally unstable and cannot be used in organic solvents or at elevated temperatures.

Conventionally, enzyme reactions have been carried out in batch processes by incubating a mixture of substrate and soluble enzyme. In this case, it is technically very difficult to recover the active enzyme from the reaction mixture for reuse. Accordingly, the enzyme and other contaminated proteins are removed by denaturation using pH adjustment or heat treatment, to isolate the product from the reaction mixture. This is not an economical use of enzymes, because the remaining active enzyme is thrown away after each batch reaction.

Two approaches can be used to eliminate the disadvantages inherent in ordinary chemical catalysts and enzymes, and for obtaining more superior catalysts for application purposes—that is, highly active and stable catalysts having appropriate specificity. One approach uses recently developed techniques of organic synthesis and polymer chemistry to synthesize catalysts having enzyme-like activities. These catalysts are sometimes called "synzymes." The other approach involves modification of enzymes produced by organisms. The immobilization of enzymes is included in this latter approach.

If active and stable water-insoluble enzymes (i.e., immobilized enzymes) are prepared, most of the aforementioned disadvantages are eliminated and the following advantages are expected:

1. Stability of enzymes is improved.
2. A catalyst can be tailor-made for specific use.
3. Enzymes can be reused.
4. Continuous operation becomes practical.
5. Reactions require less space.
6. Better control of reaction is possible.
7. Higher purity and yield of products may be obtained.
8. Resources can be conserved and pollution minimized.

In 1916, Nelson and Griffin reported that invertase extracted from yeast was adsorbed on charcoal, and the adsorbed enzyme showed the same activity as the native enzyme. Thirty-two years later Sumner (1948) found that urease from jack bean became water-insoluble on standing in 30% alcohol and sodium chloride for 1–2 days at room temperature, and the water-insoluble urease showed enzyme activity. It has therefore been known for some time that enzymes in water-insoluble form show catalytic activity. However, the early reports merely observed the phenomenon and did not actually immobilize enzymes to improve their properties for efficient application. The first attempt to immobilize an enzyme for application purposes was made in 1953, when Grubhofer and Schleith immobilized such enzymes as carboxypeptidase, diastase, pepsin, and ribonuclease by using diazotized polyaminopolystyrene resin. Prior to this, Micheel and Ewers (1949) had carried out the immobilization of a physiologically active protein. Several years later, Campbell *et al.* (1951) prepared an immobilized antigen by binding albumin to a diazonium derivative of *p*-aminobenzylcellulose. Subsequently, a number of articles on the preparation and application of immobilized antigens and antibodies were published. These reports on immobilized physiologically active proteins were considered to be the forerunners of the studies on immobilized enzymes. Following Grubhofer's investigation, fewer than 10 articles were published on immobilized enzymes in the 1950s. In the 1960s, many articles on immobilized enzymes appeared. In particular, Katzir Katchalski and co-workers at the Weizmann Institute of Science in Israel carried out extensive studies on new immobilization techniques and on the enzymatic, physical, and chemical properties of immobilized enzymes.

In addition, since the early 1960s, Tosa *et al.* (1966) have been investigating immobilized enzymes with the goal of utilizing them for continuous industrial production. In 1969, we succeeded in the industrialization of the continuous optical resolution of DL-amino acids using immobilized aminoacylase (Chibata *et al.*, 1972) (Table I). This was the world's first industrial application of an immobilized enzyme.

In the late 1960s, studies on immobilized enzymes also developed rapidly in the United States, Europe, and Japan; and reports on immobilized enzymes increased markedly. Since the end of the 1960s the term *enzyme engineering* has been used as a science and technology aimed at the efficient utilization of enzymes. In 1971, the first Enzyme Engineering Conference was held at Henniker, New Hampshire; the predominant theme of this conference was immobilization of enzymes. At this conference it was proposed that immobilized enzymes be defined as "enzymes physically confined or localized in a certain defined region of space with retention of their catalytic activities, and which can be used repeatedly and continuously." Accordingly, enzymes modified to water-insoluble form by proper techniques satisfied this definition of immobilized enzymes. Furthermore, when enzyme reactions using substrates of high molecular weight were carried out in a reactor equipped with a semipermeable ultrafiltration membrane, a reaction product of low molecular weight was removed continuously through the membrane without leakage of enzyme from the reactor. This also seemed to be a kind of immobilized enzyme system. Before that time, various terms such as "water-insoluble enzyme," "trapped enzyme," "fixed enzyme," and "matrix-supported enzyme" had been used.

A classification of immobilized enzymes was also proposed at the 1971 conference. Enzymes were classified as native or modified. Immobilized enzymes belong in the modified category, along with chemically modified soluble enzymes and biologically (i.e., genetically) modified enzymes. For practical use as catalysts, enzymes in the following three forms can be considered: (1) soluble, (2) soluble immobilized, and (3) insoluble immobilized. Hence, for forms (2) and (3), the term "immobilized enzyme" is more suitable than "insoluble enzyme."

The main topics at this biannual conference have continued to be immobilized enzymes. Work on immobilized enzymes has been very active in Japan since the end of the 1960s, and at present Japan is one of the leading countries in this field.

Applications of immobilized enzymes have been expanded into new fields besides synthetic chemical reactions, for example, chemical and clinical analysis, medicine, food processing, and elucidation of reaction mechanisms. Especially in the late 1960s, immobilization of physiologically active substances, including enzymes, had been successfully carried out by Porath and co-workers at Uppsala University in Sweden. These immobilization techniques were developed as part of specific isolation procedures called "affinity chromatography" by Anfinsen at the National Institute of Health and by Cuatrecasas at The Johns Hopkins University, both in the United States.

II. IMMOBILIZATION OF MICROBIAL CELLS

Although enzymes are produced by all organisms—animals, plants, and microorganisms—enzymes from microbial sources are the most suitable for industrial purposes for the following reasons: (1) the production cost is low, (2) the conditions for production are not restricted by location and season, (3) the time required for production is short, and (4) mass production is possible. Microbial enzymes can be classified into two groups: extracellular (i.e., excreted from the cells into the broth) and intracellular (i.e., retained in the cells during cultivation). In order to utilize intracellular enzymes, it is necessary to extract them from the microbial cells. However, these extracted enzymes are generally unstable and often undesirable for practical use as immobilized enzymes. Also, many useful chemical substances have to be produced by fermentation in order to utilize the catalytic activities of multienzyme systems in the microorganisms.

With the aim of eliminating the necessity for extracting the enzyme from microbial cells and utilizing the multienzyme systems of microbial cells, direct immobilization of whole microbial cells was therefore attempted. Continuous enzymatic reaction with immobilized microbial cells was investigated (Chibata *et al.*, 1974; Sato *et al.*, 1975), and continuous production of L-aspartic acid using immobilized microbial cells was successfully industrialized. This is considered to be the first industrial application of immobilized microbial cells. In subsequent work, the industrial production of L-malic acid from fumaric acid and of L-alanine from L-aspartic acid were started in 1974 (Yamamoto *et al.*, 1976) and in 1982 (Yamamoto *et al.*, 1980; Takamatsu *et al.*, 1981), respectively, using immobilized microbial cells.

At present, it is reported that seven immobilized enzyme–microbial cell systems (Table I) have been industrialized. The continuous production of high-fructose syrup by glucose isomerase is becoming one of the major fields of application of immobilized systems.

With regard to immobilized microbial cells, many problems remain,

TABLE I
CURRENT APPLICATIONS OF IMMOBILIZED ENZYMES AND IMMOBILIZED MICROBIAL CELLS IN INDUSTRY

Immobilized enzymes and microorganisms	Application	Operating since
Aminoacylase	Optical resolution of DL-amino acid	1969
Glucose isomerase	Isomerization of glucose to fructose	1973
Penicillin amidase	Production of 6-APA	1973
Escherichia coli (aspartase)	Production of L-aspartic acid	1973
Brevibacterium ammoniagenes (fumarase)	Production of L-malic acid	1974
β-Galactosidase	Hydrolysis of lactose	1977
Pseudomonas dacunhae (L-Aspartate β-decarboxylase)	Production of L-alanine	1982

among them the limitation of permeability of the substrate and product through cellular membranes, and the occurrence of side reactions. However, if these problems can be solved, the future of immobilized microbial cell systems appears very promising, as the enzyme systems within the microorganism become more efficiently utilized.

Immobilized microbial cells can be defined by substituting the word enzymes for microbial cells in the definition given for immobilized enzymes, that is, "microbial cells physically confined or localized in a certain defined region of space with retention of their catalytic activities, and which can be used repeatedly and continuously." The immobilized microbial cells can be growing, resting, or dead; but the enzyme activities are kept in the active state. When cells are in the growing state, it is sometimes difficult to distinguish immobilized systems as previously defined from certain kinds of conventional continuous-fermentation processes. These "immobilized growing cells," or "immobilized living cells," are described in the next section.

From our experience with one industrialized immobilized enzyme system and three immobilized microbial cell systems, we think that reactions by immobilized microbial cells are advantageous in the following areas:

1. When enzymes are intracellular
2. When enzymes extracted from cells are unstable during and after immobilization
3. When the microorganism contains no interfering enzymes, or when any interfering enzymes are readily inactivated or removed
4. When the substrates and products are not high molecular weight compounds

In these cases, the following advantages of immobilized microbial cells may be expected.

1. Processes for extraction and/or purification of enzyme are not necessary.
2. Yield of enzyme activity on immobilization is high.
3. Operational stability is generally high.
4. Cost of enzyme is low.
5. Application for multistep enzyme reaction may be possible.

Another aspect to be considered is the volume of liquid to be processed. For the unit production of a desired compound, the required volume of fermentation broth is much smaller in the case of a continuous method using immobilized cells as compared to conventional batch fermentation. Thus, the continuous process using immobilized cells is very advantageous also from the viewpoint of reducing plant pollution problems.

One of the problems encountered when using immobilized microbial cell systems is contamination by bacteria. The use of thermophilic and halophilic bacteria may allow regulation of the reaction conditions so that few contaminating bacteria will be able to survive.

III. IMMOBILIZATION OF MICROBIAL CELLS IN LIVING OR GROWING STATE

The industrialized reactions listed in Table I are primarily catalyzed by a single enzyme: the immobilized cells are dead, but the enzyme is both active and stable. However, many useful compounds, especially ones produced by fermentation methods, are usually formed by multistep reactions catalyzed with many kinds of enzymes in living microbial cells. Also, these reactions often require generation of ATP and other coenzymes such as NAD, NADP, and coenzyme A. If immobilized cells are kept in the living state, they may be applicable for carrying out these multienzyme reactions.

As is well known, the trickle-filter vinegar fermentation process developed in the beginning of the last century was based on films of living microorganisms. This process may be regarded as a kind of immobilized living cell system. Except for this trickle-filter system, none of the processes utilizing immobilized living cells has been industrialized. However, in the late 1970s several investigators started studies on immobilized living cells for the production of useful compounds such as alcohols, organic acids, amino acids, antibiotics, and enzymes, and for the decomposition of poisonous chemicals. The details are described by Kennedy and Cabral (Chapter 5, this volume).

IV. IMMOBILIZATION OF SUBCELLULAR PARTICLES AND OF PLANT AND ANIMAL CELLS

In the mid-1970s, several articles were published on the immobilization of subcellular particles, such as chloroplasts, microbodies, peroxisomes, and mitochondria. Arkles and Brinigar (1975) reported the immobilization of rat liver mitochondria by adsorption on alkylsilanized glass bead. This is considered to be the first article on immobilization of subcellular particles. Two years later Tanaka *et al.* (1977) immobilized yeast peroxisomes containing alcohol oxidase, catalase, and D-amino acid oxidase into matrices of photo-cross-linkable resin. Afterwards, Kastle *et al.* (1978) entrapped microsomes obtained from rat liver and used them as an extracorporeal drug detoxifier. Yagi and Ochiai (1978) immobilized chloroplasts to prepare a chloroplast electrode and to generate a photocurrent. These immobilized subcellular preparations are very interesting topics for both academic and application studies.

In addition to studies on the immobilization of microbial whole cells and subcellular particles, articles on immobilization of plant and animal cells have very recently been published. Immobilized plant cells are considered to be advantageous catalysts for production of medicines originating from plants. Immobilized animal cells will be practical for biological sensors, extracorporeal shunt systems, and the production of useful biomaterials.

In the case of plant cells, Lambert *et al.* (1979) immobilized algae (*Anabaena cylindrica*) by adsorbing it onto glass beads and then examining the preparation for the evolution of hydrogen gas. Brodelius *et al.* (1979) immobilized cells of *Morinda, Catharanthus,* and *Digitalis* by the calcium alginate method, and used them for transformation of natural products such as anthraquinones, ajmalicine isomers, and digoxin. In the case of animal cells, Rechnitz *et al.* (1979) immobilized a thin slice of porcine kidney at the surface of a membrane electrode that sensed ammonia gas by means of a gas-permeable membrane, and used the device for determination of L-amino acids. Furthermore, Ikariyama *et al.* (1979) immobilized the posterior silk glands of the silkworm in polyacrylamide gel, and this immobilized organ produced silk protein in the presence of amino acids and energy sources. Nilsson and Mosbach (1980) immobilized animal cell cultures to microcarriers such as gelatin beads and chitosan beads. Works relating to immobilization of plant and animal cells are still very few, but it is expected that studies of this type will increase in the near future.

V. CONCLUSION

As stated before, the techniques that started with immobilized enzymes have been adapted to immobilized microbial cells, to immobilized living microbial cells, and further, to immobilized plant and animal cells. These immobilized systems are applied in a variety of fields, and they play a very important role in the field of biotechnology.

Genetic engineering as well as enzyme engineering have recently come to be viewed as two of the most promising techniques in biotechnology. The two technologies, in which immobilized biocatalysts play the main role, are not competitive technologies;

rather, each area complements the other. Genetic engineering, to be an efficient production technology, should be combined with fermentation technology, enzyme engineering, and further with the technologies of isolation and purification processes. Were a novel microorganism having the desired characteristics to be produced by genetic engineering and utilized via immobilized enzymes or immobilized living microbial cells, this would be a very promising production technology.

Therefore, we are convinced that if cooperation is accelerated among scientists and engineers in a variety of fields related to biotechnology (e.g., genetic engineering, fermentation technology, enzyme engineering, and separation-process technology), then biotechnology will contribute to the future welfare of humankind.

REFERENCES

Arkles, B., and Brinigar, W. S. (1975). *J. Biol. Chem.* **250,** 8856.

Brodelius, P., Deus, B., Mosbach, K., and Zenk, M. H. (1979). *FEBS Lett.* **103,** 93.

Campbell, D. H., Luescher, E., and Lerman, L. S. (1951). *Proc. Natl. Acad. Sci. U.S.A.* **37,** 575.

Chibata, I., Tosa, T., Sato, T., Mori, T., and Matuo, Y. (1972). *Ferment. Technol. Today, Proc. Int. Ferment. Symp., 4th, 1972* p. 383.

Chibata, I., Tosa, T., and Sato, T. (1974). *Appl. Microbiol.* **27,** 878.

Grubhofer, N., and Schleith, L. (1953). *Naturwissenschaften* **40,** 508.

Ikariyama, Y., Aizawa, M., and Suzuki, S. (1979). *J. Solid-Phase Biochem.* **4,** 69.

Kastle, P. R., Baricos, W. H., Chambers, R. P., and Cohen, W. (1978). *Enzyme Eng.* **4,** 199.

Lambert, G. R., Daday, A., and Smith, G. D. (1979). *FEBS Lett.* **101,** 125.

Micheel, F., and Ewers, J. (1949). *Makromol. Chem.* **3,** 200.

Nelson, J. M., and Griffin, E. G. (1916). *J. Am. Chem. Soc.* **38,** 1109.

Nilsson, K., and Mosbach, K. (1980). *FEBS Lett.* **118,** 145.

Rechnitz, G. A., Arnold, M. A., and Meyerhoff, M. E. (1979). *Nature (London)* **278,** 466.

Sato, T., Mori, T., Tosa, T., Chibata, I., Furui, M., Yamashita, K., and Sumi, A. (1975). *Biotechnol. Bioeng.* **17,** 1797.

Sumner, J. B. (1948). *Science* **108,** 410.

Takamatsu, S., Yamamoto, K., Tosa, T., and Chibata, I. (1981). *J. Ferment. Technol.* **59,** 489.

Tanaka, A., Yasuhara, S., Osumi, M., and Fukui, S. (1977). *Eur. J. Biochem.* **80,** 193.

Tosa, T., Mori, T., Fuse, N., and Chibata, I. (1966). *Enzymologia* **31,** 214.

Yagi, T., and Ochiai, H. (1978). *Adv. Hydrogen Energy* **3,** 1293.

Yamamoto, K., Tosa, T., Yamashita, K., and Chibata, I. (1976). *Eur. J. Appl. Microbiol.* **3,** 169.

Yamamoto, K., Tosa, T., and Chibata, I. (1980). *Biotechnol. Bioeng.* **22,** 2045.

Methods for the Immobilization of Microbial Cells

Joachim Klein

Institute of Technical Chemistry
Technical University of Braunschweig
Braunschweig, Federal Republic of Germany

Fritz Wagner

Institute of Biochemistry and Biotechnology
Technical University of Braunschweig
Braunschweig, Federal Republic of Germany

APPLIED BIOCHEMISTRY AND BIOENGINEERING
Volume 4

ISBN 0-12-041104-0

I. INTRODUCTION

Other chapters in this volume give a historical description of various immobilization methods. Therefore, we have omitted the historical aspects and have concentrated on selected topics that represent relevant problem fields originating from (1) the immobilization procedure, (2) cell biology, and (3) reaction conditions or engineering application. Within each topic we have the chance to compare various immobilization procedures and to evaluate their strengths and drawbacks. Although starting from different viewpoints, we hope to be able finally to develop a fair picture of all principally important contributions to the field. A number of recent review articles are available (Jack and Zajic, 1977; Durand and Navarro, 1978; Klein and Wagner, 1978, 1979; Ash, 1979; Venkatasubramanian and Vieth, 1979; Vishnoi, 1980; Chibata and Tosa, 1981; Bucke and Wiseman, 1981; Kolot, 1981a, b) covering many more subjects and details than can be considered in this rather comprehensive chapter. Special references are restricted to those articles that contribute to the development of immobilization methodology itself and not to the numerous fields of application.

II. PRINCIPAL STRATEGIES OF IMMOBILIZATION

A. Definition of Immobilization

Before discussing the detailed aspects of whole-cell immobilization, it seems necessary to structure the whole field with respect to some strategic lines. The best starting point might be a definition, which is based on the definition of immobilized enzymes given at the first Enzyme Engineering Conference in 1971. Immobilized cells are "physically confined or localized in a certain defined region of space with retention of their catalytic activity and—if possible or even necessary—their viability and which can be used repeatedly and continuously." This definition covers three different aspects, which should be looked upon more closely with regard to specific features of microbial cell immobilization as compared to enzyme immobilization.

The first aspect, dealing with "confinement or localization," is of geometrical nature and in effect requires heterogenization by formation of a macroscopic catalytically active solid phase dispersed in or in contact with a liquid reactant medium free of catalyst. This compartmentalization of the cell mass usually means that the cell density in the solid phase becomes very high. In comparison to a free-cell suspension reactor, considerable savings in reactor volume become possible. Cell loadings of 0.7 g wet cells/ml catalyst have been reported for epoxy carriers (Klein and Eng, 1979a), and even higher values are possible if a partial drying process is applied to calcium alginate-immobilized cells (Klein and Wagner, 1978). On a cell number basis, cell densities in carrageenan of 10^{10} cells/ml have been achieved (Chibata, 1979).

Therefore, in the solid phase for a very large number of cells, relative to each other in a fixed position and with the exception of some membrane configurations (Drioli *et al.*, 1982; Inoles *et al.*, 1982), transport of reactants to and from the cells is governed by diffusion only. As a consequence of diffusional transport resistance, gradients of concentration or pH in the particle may be established. As a result, the environmental situation of the immobilized species may become different from that of the liquid bulk phase. On the macroscale very different reactor configurations are possible, such as packed-bed (Lilly, 1978) columns or suspensions of granular particles, fixed tubular membranes (Vieth and Venkatsubramanian, 1979), or fixed fibers (Marconi, 1978). Furthermore, a complete variation of residence time distribution between plug flow and well-stirred tank characteristics is possible.

This definition excludes reactor configurations where the cells are still freely suspended as isolated entities but retained in a reaction space by a separation unit, as in membrane filtration (Webster *et al.*, 1981) or centrifugation. In such reactors the catalyst species are mobilized in convective-flow regions and homogeneously distributed in the liquid, and all cells are equally accessible to the reactant medium. As to the reactor type, it is possible to use only one well-stirred vessel or a series of them (cascade). With respect to "confinement and localization," cell immobilization is not different from enzyme immobilization.

The second aspect of the definition is the "retention of enzymatic activity (and for cells, especially) their viability." Retention of enzymatic activity need not be complete but only a fraction that is high enough to be of practical interest. Retention of enzymatic activity of whole cells well above 50% is not uncommon, but values below 25% may not be acceptable. A specific problem for whole cells in contrast to isolated enzymes is the retention of their *viability.* It has long been questionable whether it is realistic to expect such retention, because only indirect evidence for immobilized cell viability has been available. However, following the first direct observations on cell growth under reincubation with nutrients in a matrix (Klein *et al.*, 1976, 1978;

Somerville *et al.*, 1977), more and more direct data on cell growth inside of a carrier have been reported (Klein and Wagner, 1979; Chibata, 1979; Ohlson *et al.*, 1979), so that the effect itself is undoubtedly accepted today.

In one immobilization experiment, the percentage of retention of enzyme activity and viable cells will be different figures. Depending on the level of complexity of the enzyme-catalyzed reaction, viability of the cells may not be required at all (Chibata and Tosa, 1977). It may be a fringe benefit in connection with the catalyst operational stability (Klein *et al.*, 1979b). Or it may be a prerequisite for catalytic action in a resting cell (Klein and Schara, 1981) or a growing cell situation, as in ethanol production (Wada *et al.*, 1979, 1981; Linko and Linko, 1981).

The third aspect of the definition is the repeated use in a continuous fashion. These possibilities are usually related to easy separation of the catalyst from the reaction medium without loss of activity. As with immobilized enzymes, heterogenization of the cells by appropriate methods of immobilization solves this problem. It is generally of interest, however, to know whether the immobilization contributes to the stability of the catalytic activity of the enzymatic system itself, and thus to the *extent* of repeated use. In this respect there may be differences for dissolved or immobilized *enzymes*. In principle, however, soluble enzymes can show rather high stability and reuse factors (Wandrey, 1979), and special immobilization techniques (e.g., multiple point attachment) must be applied to improve the stability significantly (Mosbach, 1982).

The enzymatic activity of free suspended *cells* shows generally a rather fast decay of 1 day or less. In such a case only two or three repeated batches may be possible. However, immobilization of cells usually gives rise to a substantial stabilization of enzymatic activity, and therefore in most cases immobilization becomes a necessary prerequisite for the extended repeated use of cells. The comparison of such stability figures of free and immobilized cells has been made elsewhere (Klein and Wagner, 1978), and further evidence has been reported since then (Kokubu *et al.*, 1981). The half-life under operational conditions is of the order of 20–50 days without applying special stabilization techniques. It should at least be mentioned (Chibata, 1979) that the operational stability can be extended to a year if a special chemical treatment is applied to the cells after immobilization.

Another problem of definition is the borderline between immobilized *enzymes* and *cells* when the cell structure has been destroyed after immobilization. Our definition of immobilization of *cells* applies to all processes where the complete biomass, obtained from fermentation, is used in the immobilization procedure. Any posttreatment to denature the cells by the application of heat or chemicals, which may finally lead to a state where the active enzyme is no longer covered by the cell wall, does not change the situation wherein whole cells have been subject to immobilization.

B. Methods of Immobilization

In realizing the crucial importance and significance of immobilization for obtaining long-life whole-cell catalysts, we have to ask what principal ways can be used to reach this goal. In such a discussion it has become routine to adopt the following categories from immobilized enzyme work:

1. Physical cross-linking (flocculation)
2. Covalent cross-linking
3. Adsorption on insoluble matrices
4. Covalent binding to insoluble matrices
5. Physical entrapment in porous materials
6. Encapsulation

These methods can be put into a scheme in which the *type of bonding* of the cells in the catalyst particle is the dominating parameter (Fig. 1). This scheme, however, is oriented toward the *state* of the immobilized cell catalyst and not so much toward the route along which the catalyst has been prepared.

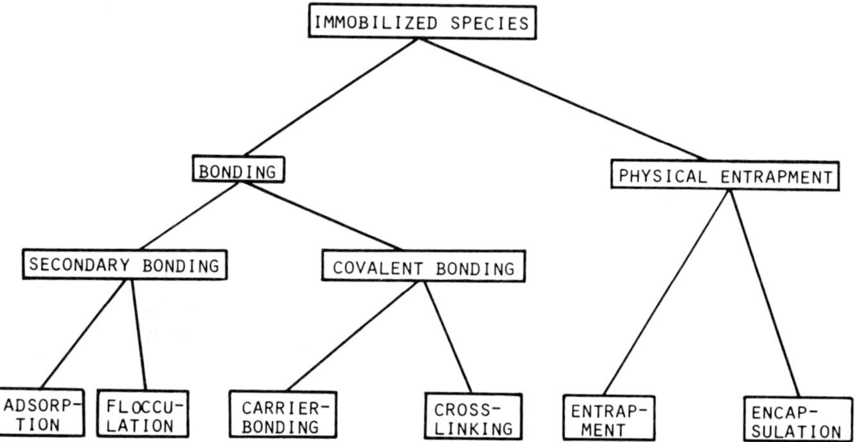

Fig. 1. Classification of immobilized enzymes or cells with respect to the type of bonding in the catalyst particle.

In this sense four different categories can be defined as follows:

1. Carrier-free immobilization
2. Immobilization of a given biomass *onto* a preformed carrier
3. Immobilization of a given biomass in the course of carrier preparation
4. Immobilization by growth of immobilized cell

Some explanations and comments on each of these groups may be given:

1. Carrier-Free Immobilization

This means essentially a compartmentalization of cells by adsorptive or covalent cross-linking of cells with each other. Adsorptive processes—such as flocculation or pelletization of cells in the course of their fermentation or by secondary processes under appropriate parameter variations (ionic strength, pH, temperature)—can be very simple and effective means of preparing catalytically active particles in accordance with the aforementioned definition (Section II,A) (McGinis, 1975; Ash, 1979). We would like to add to this group also methods using only a small amount of flocculating aids (e.g., polymer); in such cases direct cell interaction is still the dominating feature of particle stability (Lee and Long, 1974; Tsumura and Kasumi, 1976; Bungard *et al.*, 1979). The second possibility is direct chemical cross-linking of cells with bivalent coupling reagents such as glutaraldehyde (Lartique and Weetall, 1976; Prescott and Dunn, 1959; Chibata *et al.*, 1974). Glutaraldehyde cross-linking can also be applied following a flocculation step, possibly enhanced by flocculation aids such as polyelectrolytes (Ahn and Byun, 1979).

The main advantages of these methods are that very high cell densities can be obtained and that the adsorption processes are carried out under physiologically mild conditions. The main disadvantage is the poor mechanical stability under compression and shear. Furthermore, if the cell activity is high, the considerable packing density of the cells may lead to severe problems in transport limitation, for example, in the supply of oxygen for those cells in the center of the core. Finally, it should be kept in mind that the adsorptive processes require specific properties of the cell surface as a prerequisite for self-attachment and agglomeration. These conditions will be limited to only a few classes of microorganisms.

2. Immobilization onto a Preformed Carrier

This is the typical strategy in the field of enzyme immobilization, especially by covalent bonding. In such a process of catalyst preparation the carrier matrix can be produced without limitation in chemical or physical conditions. The flexibility of carrier materials and preparation methods is therefore very high, and the optimization of mechanical stability and pore structure can be carried out without constraints originating from physiological parameters. For whole-cell immobilization the problem arises of how to bring the rather large cell entities to adhere to the outer and inner surface of the matrix in a reasonable capacity and stability against washout. However, the technology of adsorbed cells has been in use for vinegar production since 1823, and the first article on immobilized cells made use of the adsorption technique (Hattori and Furusaka, 1960).

The matrix must contain pores larger than cell diameters to allow for cell penetration to the inner surface. This obviously sets limits on the surface area available for cell attachment capacity and mechanical stability. Attachment of cells can be achieved by adsorption or covalent bonding, with a high priority to adsorptive processes. Usually porous particulate carriers are applied, and immobilization is achieved by immersing the matrix particles in a cell suspension. A more recent approach is the application of hollow-fiber membranes, where the cells can be forced into the macroporous support structure by convective flow of the suspension liquid through the membrane (Inoles *et al.*, 1982). The immobilization of whole cells by surface attachment is a main point of discussion in Section III, with regard to the state of the art in resting cell systems. Growth of cells after immobilization adds a new dimension, especially to surface adsorption.

3. Immobilization in the Course of Carrier Preparation

This is by far the most common approach in whole-cell immobilization. This technique in principle consists of two methods, namely entrapment and encapsulation. However, the latter is only of marginal importance (Mohan and Li, 1975; Fukushima *et al.*, 1976; Miyoshi *et al.*, 1977; Ghose and Chand, 1978; Ado *et al.*, 1980; Sakimae and Onishi, 1980).

Due to the size of whole cells, it is rather simple to prepare networks of such porosity that complete cell retention is guaranteed, and transport processes for substrates and products are fast enough to obtain a high efficiency of catalytic activity. The main problem to be solved is to match the conditions of carrier preparation to the physiological requirements for the retention of enzymatic activity and cell viability as well. However, following the pioneering work of Mosbach and Mosbach (1966), it took only a few years before the first industrial applications of entrapped cells were reported (Chibata *et al.*, 1974). Immobilization by entrapment, due to its simplicity and flexibility, has now become by far the most extensively used immobilization method. A detailed discussion of the respective approaches is the subject of Section IV.

4. Growth of Immobilized Cells within the Biocatalyst

This method is attracting increasing interest. The reason for preparing a biocatalyst with whole cells in a living state may be to increase the cell concentration within the beads. This is done by incubation of freshly prepared biocatalyst under suitable growth conditions or by reactivation of the biocatalyst after a certain degree of deactivation for both one-step and multistep enzyme systems. Finally, biocatalysts with living cells allow the utilization of multistep enzyme systems with the participating coenzymes under either stationary or growing conditions. In this context, it should be made clear that the use of biocatalysts with living cells underlies the same problems as with free cells in respect to undesired side reactions.

a. *In situ* preparation of biocatalysts by growth of immobilized cells under controlled conditions: The most commonly used method in this field is the physical entrapment of cells in ionic networks. In the first step, precultured cells with or without harvesting are mixed with a linear polyelectrolyte under sterile conditions. Then the mixture is added dropwise to a solution of counterions with gentle stirring. Gel beads with homogeneous partition of a small number of cells are obtained. From experimental data (Wada *et al.*, 1979; Chua *et al.*, 1980; Wagner *et al.*, 1982), a cell loading in the range of 0.1–1.0 mg wet cell mass/ml biocatalyst can be obtained. In the second step, the gel beads containing a small number of entrapped living cells are incubated in a given nutrient medium. It has been shown that the growth rate of the entrapped cells is the same or even better than that of the free cells in submerged culture (Wada *et al.*, 1979). Also, the cell yield based on the carbon and energy source of the entrapped cells is identical with that of the free cells. From these results we can calculate the cell loading of living cells in the final biocatalyst. However, as a consequence of the diffusion limitations of the nutrients during the exponential growth phase, gradients of the cell mass in a bead are observed (Wada *et al.*, 1979; Wagner *et al.*, 1982). Homogeneous partitioning of the cell mass in the biocatalyst can be expected during the stationary growth phase. An alternative is to accept a biocatalyst with either heterogeneous or homogeneous cell mass partitioning, which in the first place depends on the kinetic parameters for a given enzymatic process.

This immobilization technique was found to be applicable to many single-step and multistep reactions. With κ-carrageenan as porous support, the production of L-aspartic acid from fumarate, L-isoleucine from D-threonine, L-sorbose from D-sorbitol, and ethanol from glucose using heterogeneous immobilized cell systems of *Escherichia coli, Serratia marcescens, Acetobacter suboxydans*, and *Sacharomyces cerevisiae*, respectively, were investigated (Wada *et al.*, 1979, 1980, 1981). In addition, after growth of *Enterobacter aerogenes* in κ-carrageenan, the production of 2,3-butanediol from glucose (Chua *et al.*, 1980), and with *Brevibacterium fuscum* immobilized in the same manner, the oxidation of dehydrocholic acid to 12-keto-chenodeoxycholic acid (Sawada *et al.*, 1981) were studied. The application of biocatalysts prepared by growth of immobilized cells also was demonstrated by using *Pseudomonas denitrificans* in calcium alginate for denitrification of water (Nilsson *et al.*, 1980) and *S. cerevisiae* for the production of ethanol from glucose (Larsson and Mosbach, 1979; Larsson *et al.*, 1982). The preparation of L-tryptophan from indole and L-serine with entrapped *E. coli* in chitosan was activated by growth (Wagner *et al.*, 1982).

b. *In situ* preparation of biocatalysts by germination of immobilized spores: Immobilized spores have been prepared by physical entrapment in various insoluble matrices, such as photo-cross-linked resins (Sonomoto *et al.*, 1981; Tanaka *et al.*, 1982), calcium alginate (Ohlson *et al.*, 1980; Häggström and Molin, 1980; Krouwel *et al.*, 1980; Sonomoto *et al.*, 1981), κ-carrageenan (Sonomoto *et al.*, 1981), and polyacrylamide (Ohlson *et al.*, 1980). The advantages of this method come from several aspects. Mycelial cells of sporulating microorganisms are fragile, and the immobilization of active and vegetative mycelia of such microorganisms is difficult. Spores are generally more resistant to physical and chemical stress than are vegetative cells. Therefore, with immobilized spores of a given microorganism a wider range of chemicals and methods can be used without damage of the cell during the immobilization technique. Spores of anaerobic microorganisms are not sensitive to oxygen and can be handled without precaution. Finally, after the preparation of the immobilized spores, this biocatalyst can be washed with an aseptic agent to kill contaminated vegetative microorganisms while the spores survive (Krouwel *et al.*, 1981).

Germination of entrapped spores under suitable growth conditions within the aforementioned networks has been shown with different microbiological systems. After germination of immobilized spores of *Curvularia lunata*, the entrapped mycelium thus obtained showed excellent activity in a reaction medium to hydroxylate cortexolone (Reichstein's substance S) to hydrocortisone (Ohlson *et al.*, 1980;

Sonomoto *et al.*, 1981; Tanaka *et al.*, 1982). In these experiments the activity of the biocatalysts was comparable to that of the free mycelium, and the hydroxylation system could be reactivated by incubating the entrapped mycelium under growth conditions. The use of biocatalysts after germination of immobilized spores is also applicable in anerobic processes involving a number of consecutive reactions requiring the regeneration of cofactors. This was shown with calcium alginate-immobilized cells of *Clostridium acetobutylicum* for the production of butanol–acetone–ethanol (Häggström and Molin, 1980) and with calcium alginate-immobilized cells of *Clostridium butylicum* producing butanol and isopropanol (Krouwel *et al.*, 1980). In the last case the average productivity for continuous processing was approximately four times higher than the productivity of a classical batch fermentation with free cells.

c. Use of immobilized cells under growth conditions: Many useful compounds produced by biotechnological reactions are usually formed by multistep reactions catalyzed by a number of enzymes within the cells. These reactions often require the regeneration of ATP and other coenzymes. One of the advantages of the use of immobilized cells in the living state is the possibility of restoring or even increasing the *in situ* enzymatic activities by nutrient addition (see reviews by Klein and Wagner, 1979; Mosbach, 1982). Under limiting nutrient supply in a continuous process, the entrapped cell mass in the biocatalyst and the productivity of this biocatalyst can be held at steady state. Such processes are discussed in Section VII under operational stability.

In this section it should be mentioned briefly that additional benefits can be obtained with immobilized cells under growth conditions. For instance, during continuous production of ethanol in high concentrations using immobilized viable yeast cells under growth conditions, a selection of the growing yeast cells with decreased substrate and product inhibition was observed (Larsson and Mosbach, 1979; Wada *et al.*, 1981). It might be concluded, therefore, that immobilized living cells in a reversible network are an interesting source for selection or adaptation of organisms under chemostat conditions.

III. CARRIERS FROM DIFFERENT ORIGINS

A. General Considerations

The overwhelming majority of published articles in the field of microbial cell immobilization represents only a very narrow selection of all the methods and substances available. The most widely used method is entrapment, and the most often chosen substances are polyacrylamide, calcium alginate, and carrageenan. The literature by contrast offers a large variety of alternatives, which usually are discussed by comparing different immobilization methods. Especially with a technical application in mind, it seems worthwhile to look at factors related to the origin of the carrier material as well.

From the chemist's point of view we have to distinguish between inorganic and organic materials. In each group there are naturally occurring bulk materials. There also are purified naturally occurring substances obtained by processing the aforementioned bulk materials. Finally, there are synthetic materials of different levels of sophistication. It is not unfair to assume that the cost factor per unit volume of carrier material will increase in the same order. In the group of inorganic substances, typical examples of bulk materials are bricks (Navarro *et al.*, 1976; Corrieu *et al.*, 1976) and sand particles (Anonymous, 1976). More advanced materials are ceramics or metal hydroxides (Weetall *et al.*, 1974; Kennedy *et al.*, 1976; Messing *et al.*, 1979; Marcipar *et al.*, 1980). Finally, controlled-pore glass is a synthetic material of well-defined structure (Navarro and Durand, 1977; Messing *et al.*, 1979).

In the group of organic materials, wood chips (Fetzer, 1930; Moo-Young *et al.*, 1980) and anthracite (Scott and Hancher, 1976) can be mentioned as natural bulk materials. Polymers of natural origin are collagen (Vieth *et al.*, 1973), cellulose (Johnson and Ciegler, 1969; Linko *et al.*, 1978), carrageenan (Chibata, 1979; Tosa *et al.*, 1979), alginates (Klein and Wagner, 1978), and albumin (Petre *et al.*, 1978). In the subgroup of synthetic organic materials, a huge variety exist, ranging from simple structures like poly(vinyl chloride) (PVC) (Corrieu *et al.*, 1976) and polypropylene (Cochet *et al.*, 1979) to the more complex polyacrylamide (Mosbach and Mosbach, 1966), ion-exchange resins (Hattori and Furusaka, 1960; Seyan and Kirwan, 1979), epoxides (Klein and Eng, 1979a), and polyurethanes (Fukui *et al.*, 1980b).

A number of factors have to be considered in the selection of a carrier material if a large-scale application is planned: (*a*) the material has to be available on a large scale at a low price, (*b*) the process of immobilization has to be simple and effective with regard to retention of enzymatic activity, (*c*) the capacity and efficiency of the immobilized cells has to be high, and (*d*) reactor design with respect to the mechanical handling of the biocatalyst has to be simple. With regard to factor (*a*), the bulk materials of inorganic and organic nature are certainly advantageous. For factor (*b*) from the point of view of simple immobilization, the method of adsorption deserves priority. The fact that in practice more emphasis has been given to entrapment into more complex carriers of organic origin is related to factors (*c*) and (*d*). This is because a much higher cell capacity can be immobilized; and the handling in reaction engineering (with regard to catalyst stability, cell washout, fluidized-bed application, etc.) is more flexible with entrapment.

The method of entrapment is a domain of organic polymers, where the only question that remains is whether naturally occurring or synthetic polymers are to be preferred. Whereas the development started with a synthetic polymer, polyacrylamide (PAAm) (Mosbach and Mosbach, 1966; Chibata and Tosa, 1977), the emphasis has shifted to naturally based polymers like alginate or carrageenan. A similar substitution has occurred in industrial applications (Chibata, 1979). It is especially important to keep in mind the application of immobilized cells in industry. However, synthetic carriers such as epoxides and polyurethanes may well be used if the biocatalyst is applied in synthetic chemistry and if special requirements of chemical and mechanical stability of the carrier have to be met.

An evaluation of the suitability of inorganic carriers compared to organic ones is restricted to those methods where the cells are attached to a preformed carrier in an adsorptive or chemically bonded fashion. We therefore have to discuss the relevance of attachment techniques as compared to others in cell immobilization and the advantages of different materials therein.

B. Carriers for Cell Adsorption

The general term "attachment" includes the methods of adsorption and covalent bonding. We discuss mainly the more important adsorption method, leaving the aspect of covalent bonding to the end of this section.

The adsorption method has two advantages, namely simplicity and physiological conditions. A preformed carrier is mixed with the cell suspension; and the cells adhere to the surface in a more or less complete way. Because the carrier is inert and no additional chemicals are involved in the process, the immobilization is carried out under the same physiological conditions in which the cells are kept in suspension. It is therefore not surprising that the first immobilization of viable cells was achieved with an adsorptive process (Hattori and Furusaka, 1960). The successful application of adsorption techniques in ethanol production with living yeast cells (Navarro and Durand, 1977; Moo-Young *et al.*, 1980) and the applications in steroid conversion (Atrat and Groh, 1981) give ample evidence for the physiological advantages. The critical points of the adsorption method, in contrast, are bonding capacity and bonding strength. To obtain substantial accumulation of biomass at all, the carrier materials should have pore diameters that are large in comparison to the cell diameter. In the case of bacteria and yeast, a factor of 4 to 5 has been reported (Messing *et al.*, 1979). With increasing pore size, however, the surface area decreases; thus some optimum pore size or optimum surface area can be expected. Such experimental data are available (Messing *et al.*, 1979; Navarro and Durand, 1977).

With respect to the absolute number of cells adsorbed per unit mass of carrier, comparison of organic and inorganic materials is difficult because the densities are not usually reported. Cell loading on a volumetric basis would be the most interesting figure. On a dry-weight basis, values between 248 and 2 mg/g are reported for different materials from wood chips to porous silica (Durand and Navarro, 1978). On the basis of number of cells, values of 10^9 cells/g for *E. coli* and 10^7–10^8 cells/g for yeast have been determined for porous glass carriers (Messing *et al.*, 1979). This should be compared to 6×10^9 cells/ml for yeast cells entrapped in carrageenan (Chibata, 1979), showing the much higher loading capacity of the latter method. Considering the fact that controlled-pore ceramics or glass have a very narrow pore-size distribution, which can be optimized for each microbial species, the adsorption capacity of real bulk materials like sand, bricks, technical ceramics, or PVC chips will definitely be significantly smaller.

Ion-exchange resins as porous polymer beads also have rather low bonding capacities of the order of 10 mg/g (Durand and Navarro, 1978). Entrapment of *E. coli* in calcium alginate gels or epoxid resins gives cell loadings of 0.3–1.0 g wet cells/ml catalyst (Klein and Wagner, 1979). One might argue, however, that for a fair comparison the very high loading values of the entrapment methods should be corrected for two effects: (*a*) activity loss due to toxic effects during immobilization and (*b*) activity loss due to transport limitations. For effect (*a*), especially in alginate and carrageenan entrapment the yield of enzymatic activity and cell viability is between 80 and 100%. In (*b*), with very active cells, transport limitation may indeed play an important role. But if the catalyst particles are small enough (<1 mm), catalytic efficiency can be kept between 30 and 50%. Therefore, even in cases where the efficiency factor for an adsorptive catalyst preparation is close to 100%, the entrapment catalyst will be superior in its effective volumetric activity due to the much higher loading capacity. From the standpoint of adsorption alone, wood chips seem to show the highest adsorption capacity (Moo-Young *et al.*, 1980).

The second critical point mentioned is the bonding strength. It is easy to imagine that the adsorption of a rather large entity like a whole cell to a rough surface will not allow the development of strong interaction forces. A very detailed analysis of the thermodynamic requirements with regard to the surface energy of the cells on the one hand and the carrier on the other has been given (Gerson and Zajic, 1979). These problems obviously have attracted substantial interest in soil microbiology and in processes like bacterial leaching. In the area of biocatalysis, however, only some experimental data are available (Ash, 1979). There is some information about the influence of pH (Navarro and Durand, 1977; Marcipar *et al.*, 1980), ionic strength (Vijayalakshmi *et al.*, 1979), and support composition (Marcipar *et al.*, 1980) for inorganic supports. Organic supports, especially in connection with metal cations as chelating agents, have also been studied (Vijayalakshmi *et al.*, 1979; Kolot, 1981a,b; Kennedy *et al.*, 1980). Beyond van der Waals forces, electrostatic interactions can be of importance. This can be expected in cases where the negatively charged bacterial or yeast cells are adsorbed on positively charged anion-exchange resins (Seyan and Kirwan, 1979). However, the same cells also adhere to negatively charged inorganic carriers such as glass or ceramic. Partial covalent bonding must be postulated to account for such observations. The data available are far from being conclusive enough to allow one to predict selection of a carrier for a specific adsorption problem. Furthermore, there are no significant trends toward a higher priority of inorganic or organic materials. More experimental data on surface properties of cells and carriers have to be determined, where a bridging to other fields like soil microbiology or biomedical application of polymers may be helpful (Tanzawa *et al.*, 1980; Sakurai *et al.*, 1980).

In concluding this section, two additional aspects should be discussed briefly. First, as mentioned earlier, cell attachment may not only be caused by adsorption, but also by covalent bonding. In this context, the bonding of cells to methyl hydroxides like zirconium or titanium has been studied extensively (Kennedy, 1979a,b). Also, the surface modification of silica or cellulose with organic functional groups is a step in this direction. Second, carrier materials discussed in this section are, practically speaking, either pure organic or pure inorganic materials; and there are only a few examples of hybrid structures. One example is the reinforcement of an organic gel with inorganic filler (Zueva *et al.*, 1980). In contrast, surface modification of silica with organic molecules (Kolot, 1981a,b) to modify adsorptivity or polarity can be considered in this context as well. A third example is the incorporation of chelating cations in polysaccharide carriers. It will be interesting to observe whether more developments of this hybrid type will appear in the future.

IV. THE CHEMISTRY OF ENTRAPMENT IN POLYMERIC CARRIERS

Physical entrapment in porous polymeric carriers is undoubtedly the most widely used immobilization technique for whole cells. The size of immobilized species make it rather simple to prepare porous networks that combine complete cell retention with high porosity for substrate and product transport. The chemistry of such a network formation, which inherently takes place in the presence of the cells to be immobilized (see Section II,B,3), can be classified along the following three different lines:

1. Precursor size
   Monomer
   Oligomer
   Polymer
2. Cross linking reaction
   Covalent polymerization
   Covalent polycondensation
   Ionic bonding
   Secondary bonding (hydrogen bonding, polar or hydrophobic interaction)
3. Reaction conditions
   Solvent
   Temperature
   pH
   Catalyst

We select the first topic as a route to evaluate the "pros and cons" of the various methods of entrapment, discussing the other topics from case to case. In doing so we must keep in mind that the following performance criteria have to be considered: (*a*) chemical network stability, (*b*) mechanical network stability, (*c*) variation and control of catalyst shape, (*d*) network porosity, (*e*) cell loading capacity, (*f*) yield of immobilized activity, and (*g*) viability of immobilized cells.

A. Polymeric Networks from Monomeric Precursors

Starting with monomeric precursors there is obviously only one useful way to obtain a polymeric network, namely by polymerization.

Polycondensation, although possible in principle, should preferably be used with oligomeric precursors and would lose its advantages if used in the monomeric form. The variability of polymerization reactions exists with respect to the choice of (*a*) monomer (or monomers), (*b*) cross-linking, (*c*) initiation reaction, and (*d*) technology of polymerization. It was in this group of reactions that the pioneering work of Mosbach and Mosbach (1966) opened the area of polymer entrapment of whole cells. In retrospect the obvious choice was the polyacrylamide system, which is so familiar to any biochemist from electrophoresis. The challenge that remained was to show that the enzymatic activity of whole cells could be preserved. In any case a system was developed that led relatively soon to the first successful industrial application (Chibata *et al.*, 1974), and up to now is used in many scientific laboratories (e.g., Kim and Ryu, 1980; Kokubu *et al.*, 1981; Koshcheenko *et al.*, 1981a).

The monomers to be applied in this approach are primarily water-soluble. Besides acrylamide (AAm), consideration has been given to methacrylamide (MAAm), hydroxyethylmethacrylate (HEMA), vinylpyrrolidon, 2-vinylpyridine, acrylic acid, among others (Kumakura *et al.*, 1978, 1979; Klein and Schara, 1980). Copolymerization of two monomers has also been described, with special emphasis on the incorporation of charged groups (Schara, 1977). Monomer concentrations in the range of 5–20% are used, where higher values than 20% would increase toxicity problems and decrease porosity of the gel without significant improvement of the mechanical strength. The choice of the monomer controls the rate of polymerization and the directly related rate of heat evolution. Furthermore, the polarity (hydrophilicity) of the matrix and the toxicity in the interaction of monomers with cells can be varied. An illustrative example for these different properties is the comparison of AAm and MAAm. AAm polymerizes much faster, and more care has to be taken to avoid enzyme denaturation by heat evolution. The lower polymerization rate of MAAm furthermore makes possible an optimal timing of mixing the cells with the polymerizing solution. Polymethacrylamide (PMAAm) is less hydrophilic than PAAm (high molecular weight linear PMAAm samples are even insoluble in water); so the PMAAm matrix gives improved mechanical stability compared to PAAm. A drawback is the higher toxicity of MAAm, leading to a significantly lower yield of immobilized activity (50% compared to 80%) (Klein and Schara, 1980, 1981).

For completeness it should be mentioned, that water-insoluble, glass-forming monomers such as methylmethacrylate have been tested as well to obtain carriers of higher mechanical stability and hydrophobicity. Activity yield in earlier experiments was low (Klein *et al.*, 1979a), but recently, on the basis of new polymerization technology, high activities were reported (Yoshii *et al.*, 1981).

Cross-linking in free-radical polymerization is achieved by the addition of divinyl components. In aqueous systems the selection is very limited, and practically in all cases methylenbisacrylamide has been applied. Certainly a drawback of this compound is its limited water solubility ($C_{max} \approx 20\%$) so that variation in monomer-cross-linker ratio at a total monomer level of 20% is limited as well. PAAm gels therefore always have to be considered as homogeneous, microporous gels with all consequences for limited diffusivity or cell growth. A higher fraction of cross-linker could be advantageous in obtaining macroreticular networks. In this sense alternative cross-linking substances would be welcome.

Initiation is typically achieved by water-soluble free-radical initiators such as $(NH_4)S_2O_8$, in combination with tetramethylethylendiamine (TEMED) as coinitiator. pH and temperature can be selected to control polymerization and gelation rates with respect to the optimal yield of enzymatic activity. Especially a rather low polymerization temperature between 4 and 10°C has proved to be advantageous in retention of high catalytic activity (Klein and Schara, 1980; Koshcheenko *et al.*, 1981b).

A less conventional cross-linking and polymerization system has been described, in which γ radiation is applied to obtain polymeric networks from mono-, di-, or trimethyacrylate monomers. An advantage of this approach may be the fact that very low temperatures (−24°C) can be used (Kumakura *et al.*, 1978, 1979; Y. Tanaka *et al.*, 1980; Yoshii *et al.*, 1981).

Polymerization technology is usually rather simple when the polymerizing solution is mixed with the cell suspension and bulk polymerization is applied. The flat polymeric block is then pressed through a sieve plate to obtain smaller but irregularly shaped particles (Chibata *et al.*, 1974). A more elegant method is suspension polymerization, which requires the selection of an appropriate suspension agent and surfactant to stabilize the heterogeneous mixture. Although in earlier cases hydrocarbons have been used, dibutylphthalate is a much better nontoxic suspending liquid. If the process is well designed, spherical particles of controlled size in the range of 0.1–2 mm diameter can be prepared, and no further treatment other than washing is necessary before using the catalyst particles in a reaction (Klein *et al.*, 1979a; Klein and Schara, 1980). A special advantage of the suspension method is the much better temperature control as compared to the bulk method. Therefore, activity yields in PAAm close to 100% with viable cells could be obtained (Schara, 1979).

As mentioned before, the mechanical stability of the PAAm gel itself is not very high, though column packing is possible (Chibata *et al.*, 1974). Some improvement can be achieved by incorporating inactive, mechanically stable particles as filler in the polymerizing solution. Organic polymers or inorganic materials such as ceramics, glass, or silicates have been tested with some success (Klein and Schara, 1980; Zueva *et al.*, 1980). The volume fraction of this filler, which does not participate in the network-forming reaction, cannot be very large without inhibiting the formation of a continuous network. The same effect holds for the microbial cells to be immobilized, so that cell loadings of *E. coli* of more than 20% wet weight are not possible.

In light of the aforementioned preparation conditions, the following performance profile should be understood:

1. The polymeric network, formed in its backbone by covalent C—C bonds, is *chemically stable*, so there is no limitation with regard to application in very different reaction media. Matrix properties may change, however, if under certain pH conditions chemical modifications (e.g., saponification) of the vinylpolymer side groups occur. Thus an initially neutral matrix may become ionic by saponification of amide groups to carboxylate groups at basic pH.
2. The mechanical stability of these polymers is rather limited, because networks of gel type rather than macroreticular resins are obtained. However, industrial application in column reactors has been possible. Less hydrophilic monomers or inert fillers can improve the situation, but only in a limited range.
3. Granular particles (from bulk polymerization) and spherical beads (from suspension polymerization) can be obtained.
4. Under typical polymerization conditions the average degree of polymerization between two cross-links is in the order of 10. Therefore, in a gel-type matrix the network meshes are rather small.
5. Cell loading capacity is limited (e.g., in the case of *E. coli* cells) to about 20% on a wet-weight basis. For larger cells (e.g., yeast), even smaller limiting values will exist.
6. If the polymerization conditions of AAm are carefully controlled, an activity yield of close to 100% can be realized. Heat evolution in bulk polymerization, however, can give rise to severe deactivation.
7. Growth experiments after immobilization showed that *Candida tropicalis* and *E. coli* cells could even be immobilized in the viable state.

In giving an overall rating, polymerization of monomers, especially of AAm, is a fast and simple immobilization technique, which may well be used in laboratory tests, if low mechanical stability and low cell loading are not inhibitive. Industrial applications, however, are not likely anymore.

B. Polymeric Networks from Oligomeric Precursors

Preparing a polymer from oligomeric rather than from monomeric precursors has three advantages: (1) the number of molecules in the system (and thus in contact with cells) is smaller, (2) their size (which may control transport through the cell wall) is larger, and (3) the heat of reaction is reduced according to the degree of polymerization of the precursor. All three factors are advantageous in reducing toxic influences on the cells. By contrast, typical condensation reactions require the presence of rather reactive functional groups, for example, amino, epoxy, or isocyanate, which may react unfavorably with the cellular material. Also, reaction conditions (temperature, pH, solvents) are not typically physiological. As will be seen, introduction of polymerizable end groups can eliminate this problem; however, the shape and stability of such networks is limited. So, we will discuss typical polycondensation reactions first.

1. Epoxy Resins

Epoxy resins are formed by the reaction of bifunctional epoxy oligomers with appropriate polyfunctional amino compounds. Typical reaction conditions in technical applications are higher temperatures, higher pH (>10), and the absence of water. The product usually is a glassy nonporous polymeric block. Such a system for cell immobilization can be successfully applied by the following means: (*a*) selecting a water-dispersible epoxy precursor, (*b*) selecting an appropriate polyaminoamide for curing at room temperature and moderate pH (≤9), and (*c*) using ionotropic gelation as an intermediate step to stabilize the bead shape of the resin particle during the curing time of about 24 h and to introduce porosity by redissolution of the alginate from the interpenetrating network. Irreversible shrinking of the ionotropic matrix by partial drying is a very flexible possibility to control particle size and cell packing density (Klein and Wagner, 1978; Klein and Eng, 1979a). Despite the rather mild conditions of the network formation, some irreversible inactivation of enzymes cannot be avoided. It was shown recently that the basic amino component is the stronger toxic reagent, and the yield of immobilized adducts can be improved if the polycondensation reaction is allowed to proceed and the cells are added just before gelation (Klein and Kressdorf, 1982).

The performance profile of such an epoxid preparation is as follows:

1. The chemical stability of the covalent polymeric network is excellent.
2. The matrix has excellent mechanical properties with regard to column-packing and stirred systems, even at high cell loading (up to 70% bio wet mass).
3. Preferably spherical beads of variable size can be prepared, controlled by the ionotropic gelation process.
4. Electron micrographs show rather uniform distribution of isolated cells, embedded in a microporous polymeric structure. Thus diffusion limitations are to be expected.
5. Cell loading capacity can be rather high, and values up to 70% (*E. coli*) or 30% (bakers' yeast) on a wet-weight basis can well be obtained. Shrinking of the ionotropic gel by partial drying is the key step to obtaining such cell densities.
6. Yield of immobilized activity is moderate; 40% has been a typical value for penicillin acylase in *E. coli* cells (Eng, 1980).
7. Growth experiments of immobilized cells show the viability of *E. coli* and yeast cells (Eng, 1980; Klein and Kressdorf, 1982).

2. Polyurethane Networks from Polycondensation

Polyurethane elastomers are a second condensation system successfully applied in whole-cell immobilization. Polyfunctional oligomers with isocyanate end groups can be reacted with hydroxyl or amino groups containing compounds to form a polyurethane network. An especially suited reaction system is obtained when polyisocyanates are used that require only water for the cross-linking reaction. *In situ* formation of CO_2 gas gives an open macroporous structure—polyurethane (PU) foam—under volumetric expansion (Fukui *et al.*, 1980a; A. Tanaka *et al.*, 1980; Klein and Kluge, 1981). The reaction with water is usually very fast; therefore, the cells have to be mixed with the system right from the beginning. The foam expansion can be reduced, if the polyurethane formation is performed under elevated pressure. Foam formation can be totally suppressed, if gelation is performed at low temperatures, such that a transparent homogeneous gel is obtained (Klein and Kluge, 1981). Variation of the hydrophobicity of the polyether blocks in the isocyanate precursors allows adjustment of the matrix for special solubility conditions (Fukui *et al.*, 1980a; A. Tanaka *et al.*, 1980).

The following performance profile can be given:

1. The covalent polymeric network is very stable under exceedingly variable reaction conditions.
2. The matrix is highly elastic and especially the PU foam is very compressible. The matrix seems especially suited for stirred systems; however, column packing is possible as well if the flow rate is low enough.
3. Blocks, flat sheets (membranes), granular particles, and beads can be obtained. In this sense PU is a very versatile system.
4. The foam has open macropores, whereas the matrix itself is rather dense. The gel-type preparations will have very small pores.
5. The cell loading capacity is low in the beginning because only a limited amount of water can be tolerated. Volume expansion by foaming reduces the cell density further.
6. The yield of immobilized activity is moderate to low where the reason for enzyme deactivation is not at all clear.
7. So far, no viable cells have been found after immobilization.

3. Photo-cross-linking Oligomers

Finally, the aforementioned polymerizing oligomer system should be discussed. Polyglycol oligomers are functionalized with polymerizable vinyl end groups. Under the addition of a photosensitizer and illumination with UV light for several minutes, flat network sheets can be obtained. The oligomer chain length can be controlled to determine the network porosity; the chemical composition of the polyglycol precursor determines the hydrophilicity of the matrix (Omata *et al.*, 1979b, 1981; Yamane *et al.*, 1979; Fukui *et al.*, 1980a).

The following performance profile can be given:

1. The covalent polymeric network is chemically stable.
2. Mechanical stability is limited.
3. Only more or less extended flat sheets can be prepared.
4. Gel-type networks of homogeneous structure are obtained, with the advantage of exact network density control by setting the precursor size.
5. Cell loading capacity is low.
6. Yield of immobilized enzymatic activity is moderate.
7. Viability of cells is preserved to some extent.

An overall rating of this group of network formation is difficult. The advantages are the generally excellent chemical and mechanical stability of the matrix in very different geometrical shapes. The system can still be improved with regard to enzyme activity and cell viability. It is our impression that the potential of polycondensation reactions for whole-cell immobilization is by far not exhausted, and many other interesting systems may be presented in the future with a good chance for practical application.

C. Polymeric Networks from Long-Chain Prepolymers

The variety of systems developed in this group is so large that a subcatagorization with respect to different mechanisms of network formation is advisable. These groups are defined as follows and discussed subsequently.

1. *Precipitation:* Network formation without chemical reaction, due to *phase separation* by mixing a polymer solution with a precipitant
2. *Gelation:* Network formation without chemical reaction, due to *phase transition* by parameter change (e.g., temperature, pH)
3. *Ionotropic gelation:* Network formation with chemical reaction (ion exchange) due to ionic cross-linking of polyionic chains with multivalent counterions
4. *Covalent cross-linking:* Network formation with chemical reaction due to chemical cross-linking of functionalized polymers themselves or by cross-linking functionalized polymers with low molecular weight bifunctional reagents

1. Precipitation

There was very early industrial interest in this technique on the basis of fiber formation from cellulose triacetate solution (Dinelli, 1972). Cellulose and cellulose triacetate have also been extensively studied in another laboratory (Y. Y. Linko *et al.*, 1978; P. Linko *et al.*, 1980). Typical solvents used were dimethyl sulfoxide (DMSO), acetone, formamide, or mixtures of DMSO and *N*-ethylpyridinium chloride. Also, synthetic polymers such as polystyrene (from DMF or THF solution) have been tested (Hackel *et al.*, 1975). The main drawback of this method is the solvent–nonsolvent system. Because the matrix should be stable in aqueous media, water has to function as a precipitant; moreover, an organic but water-miscible organic liquid has to serve as the solvent. Most cells will not tolerate these solvents, so applications will be limited to stable enzymes in nonliving cells. Precipitation by electrolytic sedimentation under isothermal conditions is at least in principle an interesting alternative to avoid this difficulty (Li *et al.*, 1980).

2. Gelation

Temperature change is a well-known control variable for gelation by phase transition. Besides gelatin, mainly agar has been used in 1–4% solutions (Chibata, 1978; Suzuki *et al.*, 1980; Totsuoka and Hara, 1981; Brodelius, 1982). This very simple immobilization method suffers from the fact that rather soft, mechanically unstable gels are obtained. Furthermore elevated temperatures have to be used for polymer dissolution and mixing with cells.

A most significant contribution to the field of whole-cell immobilization was the introduction of the κ-carrageenan systems (Chibata, 1978). This is a heteropolysaccharide, containing unit structures of β-D-galactose sulfate and 3,6-anhydro-α-D galactose. This polymer is soluble in water at temperatures between 40 and 60°C, and gelation occurs by cooling to room temperature. Block, bead, and membrane structures can be prepared. Not only temperature change but also contact with monovalent ions such as K^+, Rb^+, Cs^+, and NH_4^+, leads to gelation (Chibata, 1979; Tosa *et al.*, 1979; Nishida *et al.*, 1979). The conditions for gelation of κ-carrageenan, however, include many more systems, which certainly extend the use of this polymer beyond those methods that are connected with the proper definition of gelation. However, they are mentioned in this context because the mechanism is not always clear. Such gelation media are multivalent cations, aliphatic or aromatic diamines, and amino acid derivatives (Chibata, 1979). There are two main lines for the application of κ-carrageenan gels. The first is the area of monoenzyme-catalyzed conversions, where dead cells can be used. In this case posttreatment with glutaraldehyde and/or hexamethylendiamine stabilizes the enzyme function (half-life up to 680 days). The second is the area of immobilized living cells, where growth of cells in the matrix is an essential feature. Very high cell densities of up to 6×10^{10} cells/ml have been obtained (Chibata, 1979).

Another system to be mentioned under the topic of gelation is collagen. This polymer is usually applied after cross-linking (see Section IV,C,4), but simple evaporation of water can also be used for gelation from a microdispersed suspension (Constantinides *et al.*, 1981).

Summarizing the fields of precipitation and gelation, the following performance profile can be given:

1. Under usual conditions the networks are chemically stable in aqueous media. In the case of thermal gelation the temperature range of application is limited.
2. The mechanical stability of the cellulose-based precipitates is good, especially in fiber form. The precipitations from gelation are rather soft, where the compression stability of κ-carrageenan is comparable to PAAm. However, column packing and fluidization are possible.
3. Formation of different structures such as blocks, fibers, beads, and membranes is possible. Application of collagen is limited to membrane structure.
4. Homogeneous gels are formed, where the pore size depends on the polymer content.
5. Loading capacity of cells very much depends on the choice of system. In precipitation as well as in gelation, rather high values can be obtained. Extremely high values ($>10^{10}$ cells/ml) have been found in κ-carrageenan, not by direct immobilization but by growth of immobilized cells.
6. The systems in this group are characterized by very different toxicity levels. In precipitation, the application of organic solvents is critical; in precipitation and gelation the exposure of cells to the higher temperatures causes problems. So the yield very much depends on the choice of organisms.
7. The aspects given in (6) also hold for cell viability. Precipitation systems are unlikely to be successful; however, in gelation systems very high yields of viable cells have been found. This applies for agar and especially κ-carrageenan.

3. Ionotropic Gelation

Two laboratories mainly contributed to the development of this method (Hackel *et al.*, 1975; Kierstan and Bucke, 1977; Klein and Wagner, 1978). From this group the cross-linking of alginates with Ca^{2+} is by far the most popular method. For example, a 3% sodium alginate solution containing the suspended cells is dropped into a 2% $CaCl_2$ solution; size-controlled spherical beads are formed in less than an hour. The flexibility of this method is illustrated by the variability of the following parameters: (1) Alginates of different molecular weight and chemical composition (fractions of mannuronic and guluronic subunits) can be chosen, so that alginate solutions of concentrations between 0.5 and 10% can be used. Depending on the chemical structure, the minimum requirement for Ca^{2+} addition to the reaction solution for gel stabilization is different ($Ca^{2+} : Na^+$ ratio between 0.2 and

0.05 mol/mol) (Klein and Manecke, 1982). (2) The $CaCl_2$ concentration in the precipitation bath can be varied between 0.05 and 2%. (3) Working temperature can be chosen between 0 and 80°C. (4) Beads in the size range from 0.1 to 5 mm diameter can be prepared with rather uniform size distribution. (5) Cell loading by direct entrapment can be high (up to 30 g wet cells/ml catalyst).

A special feature of the Ca^{2+}–alginate system is the controlled decrease of particle size by partial drying, where the particles do not reswell under reequilibration with aqueous media. The advantage of this process step is twofold: the mechanical stability increases significantly without loss of porosity and the cell loading can be brought up to very high levels (e.g., 1 g wet cells/ml catalyst) (Klein and Wagner, 1978). A survey of the variety of polymer–counterion systems is given in Fig. 2. Special emphasis should be given to the fact that positively

POLYELECTROLYTES		MULTIVALENT COUNTERIONS
POLYANIONS		
$-COO^-$	ALGINATE	Ca^{2+}, Al^{3+}, Zn^{2+}, Co^{2+}, Ba^{2+}, Fe^{2+}, Fe^{3+}.....
	CARBOXYMETHYL-CELLULOSE	Ca^{2+}, Al^{3+}.....
	CARBOXY-GUARGUM	Ca^{2+}, Al^{3+}.....
	COPOLY-STYRENE-MALEIC ACID	Al^{3+}.....
$-PO_3^{2-}$	PHOSPHO-GUARGUM	Ca^{2+}, Al^{3+}.....
POLYCATIONS		
$-NH_3^+$	CHITOSAN	TRIPOLYPHOSPHATE PYROPHOSPHATE TETRAPOLYPHOSPHATE OCTAPOLYPHOSPHATE HEXAMETAPHOSPHATE $[Fe(CN)_6]^{4-}$, $[Fe(CN)_6]^{3-}$ POLY-ALDEHYDO-CARBONIC ACID POLY-1-HYDROXY-1-SULFONATE-PROPENE-2

Fig. 2. Gel-forming polyions and counterions for ionotropic gelation.

charged polymers like chitosan can also be applied. This eliminates the main drawback of the alginate gels, namely their instability in phosphate buffer solutions (Vorlop and Klein, 1981). In connection with calcium alginate gels, some special developments should at least be mentioned: the coimmobilization of magnetite or ferrite to prepare magnetic beads (Larsson and Mosbach, 1979), the bonding of an enzyme to alginate prior to gel formation with $CaCl_2$ (Haegerdal and Mosbach, 1980), and the incorporation of calcium alginate into a polyurethane foam (Makiguchi *et al.*, 1980).

The performance profile can be given as follows:

1. Network formation by ionic bonding is a reversible process. Therefore, special care must be given to the composition of the reaction medium with regard to the polysalt matrix. Typical problems are pH stability and interaction of the low molecular weight counterions with solution components via complex formation or precipitation.
2. The mechanical stability of the particles is good to excellent with regard to column packing and normal agitation (fluidized bed). Problems may arise under very high shear rates in mechanically stirred vessels.
3. The formation of regular beads, including controlled variation of particle size, is easy. Any other geometry (e.g., fiber or sheet) can be produced as well.
4. Primarily macroporous networks are formed, as shown by electron microscopy as well as by diffusion and porosity measurements. The macroporous structure is preserved during the shrinking of beads under partial drying.
5. Loading capacity can be extremely high (up to 1 g wet weight cells/g catalyst) without losing mechanical stability. The main problem of increased loading is catalyst efficiency due to diffusional resistance.
6. The yield of immobilized enzymatic activity is generally between 80 and 100% (under reaction-controlled conditions).
7. The immobilization method is very mild so that viable cells of very different origin and stability levels can be effectively entrapped.

4. Covalent Cross-linking

As mentioned already, the covalent cross-linking of polymer precursors has two aspects, namely cross-linking of polymers themselves (case 1) or addition of a cross-linking reagent (case 2). A typical example of case 1 is the modification of poly(vinyl alcohol) (PVA) by esterification with 3-mercaptopropionic acid, in such a way that pendant thiol groups are incorporated along the chain. In the presence of oxygen the thiol groups are oxidized; and covalent S—S cross-links from chain to chain are formed (Klein and Manecke, 1982). Another possibility is the esterification of PVA with acrylic acid to obtain pendant vinyl groups, which can be polymerized by photoinitiation (Klein and Manecke, 1982). In case 2, glutaraldehyde is most commonly used as the cross-linking agent. Typical polymers applied in this approach are collagen (Vieth *et al.*, 1973), albumin (Petre *et al.*, 1978; Barbotini and Thomasset, 1980; Drioli *et al.*, 1982), polyethylenimine (Gestrelius, 1980; Klein and Manecke, 1982), and gelatin (Bachman *et al.*, 1981). Other cross-linking agents are water-soluble polyacrolein bisulfite addition compounds or diamines (Klein and Manecke, 1982; Chibata, 1979; Pollack *et al.*, 1978).

Especially in comparison to the ionotropic gels, the performance is as follows:

1. The network is chemically stable and irreversibly formed, so the problem of redissolution under reaction conditions is eliminated.
2. Mechanical stability of the preparations is very different and depends on the substitution efficiency in case 1 preparations and on the cross-linker in case 2.
3. Usually only flat sheets (membranes) or irregularly shaped granular materials can be prepared. A technology for bead preparation is at least difficult.
4. Microporous, more homogeneous gels are formed.
5. Cell loading capacity is limited to lower values, because at higher loadings the formation of a continuous network is inhibited.
6. Depending on the chemistry of the reaction, a substantial loss of enzymatic activity has to be envisaged.
7. Immobilization of viable cells will be difficult task, especially in cases where a cross-linking reagent such as glutaraldehyde is used. In this respect the case 1 preparations are advantageous.

5. Summary

Summarizing the subject of network formation from polymeric precursors, the immense variability of substances and methods becomes obvious. Naturally occurring polymers have an especially important role, whereas those from chemical systems are certainly interesting as well. The methods vary from simple physical processes to more complex chemical reaction schemes, and it is clear that the problem of toxicity (e.g., cell viability) very much depends on the network-forming conditions. The most widely used immobilization procedures are those utilizing alginate, carrageenan, and albumin, demonstrating dominance of this technique today. Future developments very likely will be devoted to the covalent stabilization of networks, primarily prepared by "noncovalent" processes (Birnbaum *et al.*, 1981), thus combining the advantages of low toxicity, control of shape, and high cell loading with the advantage of an irreversibly cross-linking network. Combination of two polymers in one matrix and sequential application of two immobilization procedures may be considered in this respect.

V. NETWORKS FOR HYDROPHILIC–HYDROPHOBIC REACTIONS

Reactions catalyzed by immobilized whole cells have so far been restricted commonly to the aqueous phase or to an aqueous solution containing a water-soluble organic solvent. In the case of the biotransformation of hydrophobic or water-insoluble compounds, it is desirable to carry out biotransformations in an appropriate organic solvent system, if the enzymatic activity is maintained in such a reaction system. Recently, Fukui's group developed methods to make gel-entrapped biocatalysts of the desired hydrophobicity that are applicable for transformation of hydrophobic substances in organic solvent systems.

Synthetic photo-cross-linkable resin prepolymers or urethane prepolymers of either hydrophilic or hydrophobic character were taken for the immobilization of whole cells of *Nocardia rhodocrous*. The rate of conversion of 3β-hydroxy-Δ^5-steroids to the corresponding 3-keto-Δ^4-steroids in a water-saturated mixture of benzene and *n*-heptane (1 : 1 by volume) was studied (Omata *et al.*, 1979b). Depending on the polarity of the substrates cholesterol, β-sitosterol, stigmasterol, and dehydroepiandrosterone, the activity of the immobilized cells was found to correspond to the partition coefficients of substrates between the matrices and the external solvent system. From studies on the effect of the polarity of substrates and solvents and the hydrophilic–hydrophobic balance between the matrices and the reaction system for steroid transformations (Omata *et al.*, 1979, 1980; Tanaka *et al.*, 1979; Yamane *et al.*, 1979; Fukui *et al.*, 1980a; Fukui and Tanaka, 1981), it was concluded that the activity of entrapped cells correlated closely to partition of substrates between biocatalysts and external solvents. The application of hydrophobic matrices is preferable for bioconversion of lipophilic compounds. The stability of the multistep enzyme system in the whole cells was improved by these immobilization methods. In addition, different reaction routes for product formation could be selected by using resin prepolymers of appropriate hydrophilic–hydrophobic balance for immobilization of whole cells (Fukui *et al.*, 1980b).

The bioconversion of lipophilic substrates under hydrophobic conditions was successfully demonstrated by stereoselective hydrolysis of DL-methyl esters to L-menthol in organic solvents using *Rhodotorula minuta* entrapped within hydrophobic photo-cross-linked gels or polyurethane gels (Omata *et al.*, 1981; Fukui and Tanaka, 1981). With freeze-dried *Corynebacterium simplex* cells immobilized in collagen and tanned with glutaraldehyde, the dehydrogenation of hydrocortisone to prednisolone was performed successfully in the presence of the external cofactor menadione in 15% ethanol (Constantinides, 1980). Another example of bioconversion of the more polar progesterone to 11α-hydroxyprogesterone comes from the work of Leguy *et al.* (1982). The highest rate of progesterone transformed was obtained with *Aspergillus phoenicis* immobilized in calcium alginate, but as far as selectivity is concerned, κ-carrageenan or polyurethane as matrices gave better results. In contrast, continuous side-chain cleavage of the water-soluble cholesterol derivative 4-cholesten-3-(*O*-carboxymethyl)oxime to 4-androsten-17-on-3-(*O*-carboxymethyl)oxime with *Mycobacterium phlei* immobilized in polyacrylamide was carried out successfully (Atrat *et al.*, 1981). Synthesis of the water-insoluble adenine arabinoside from uracil arabinoside and adenine by transglycosidation was catalyzed by entrapped whole cells of *Enterobacter aerogenes* in photo-cross-linkable resin prepolymers or urethane prepolymers in a 40% aqueous solution of dimethyl sulfoxide (Yokozeki *et al.*, 1982; Fukui and Tanaka, 1981). Under this condition the synthesis of adenine arabinoside was not significantly affected by the hydrophilic or hydrophobic nature of the biocatalysts.

VI. INTERACTION OF MICROBIAL CELL–SUPPORT

A. Electrostatic Interactions

The use of ion-exchange resins as a method for the immobilization of whole cells has been long established and well documented. Despite these applications, little attempt has been made to study the strength of the electrostatic interaction between the ion-exchange resin and the surface of cells. Recently, various anionic exchange resins were tested for their ability to adsorb cells of *Saccharomyces cerevisiae*, which have a net negative charge on their surface (Dangulis *et al.*, 1981). The anion-exchange resins varied greatly in their ability to bind the cells. The most effective resin was XE-352, which retained 130–140 mg cells/g resin. The adsorption onto this resin was found to be relatively insensitive to environmental conditions in the case of ethanol production from glucose. The bonding strength and accessibility of the surface of the various adsorbents to cells was not estimated. Immobilization of *S. cerevisiae* by inclusion in silica hydrogel led to a biocatalyst without significantly changing the biological activity (Rouxhet *et al.*, 1981). It was pointed out that entrapment in silica does not involve the formation of specific chemical bonds between the cell and the silica hydrogel matrix. Retention by surface interactions is involved in the adsorption of *Zymomonas mobilis* to borosilicate glass (Arcuri *et al.*, 1980). Under growth conditions the number of cells associated with the matrix increased rapidly, which indicates weak forces (van der Waals, hydrogen bonding, electrostatic) involved in the cell–matrix interactions.

A more detailed analysis of the interaction of various strains of *Acetobacter* sp. and hydrous titanium(IV) oxide or hydrous titanium(IV) chelated cellulose was described (Kennedy *et al.*, 1980). It has been shown that hydrous titanium(IV) oxide or a chelate or titanium with cellulose can flocculate and/or immobilize cells of cellulose- or non-cellulose-producing strains of *Actobacter* and that these biocatalysts can be successfully used in the continuous oxidation of an aqueous solution of ethanol to acetic acid. The mechanism proposed for polymer-induced immobilization assumes that hydrous titanium(IV) oxide is a cross-linking agent between cellulose molecules. This inorganic polymer can act as a cationic polyelectrolyte, which could interact with the negatively charged cell surface by way of a complex formation and/or proton transfer.

B. Modification of Entrapped Cells in Ionic Networks by Partial Covalent Bonding

In order to obtain immobilized whole cells in ionic networks with suitable stability under application conditions, various methods for stabilization of the biocatalysts by covalent cross-linking via polyfunctional reagents were investigated. Treatment of *E. coli* cells immobilized in κ-carrageenan and locust bean gum with glutaraldehyde or glutaraldehyde and hexamethylenediamine stabilized their aspartase activities (Nishida *et al.*, 1979). Similar results were obtained by glutaraldehyde treatment of κ-carrageenan-immobilized *Pseudomonas dacunhae* having L-aspartase β-decarboxylase activity (Yamamoto *et al.*, 1980). Enhanced stabilization was obtained in the case of immobilized *E. coli* cells by incubation of the κ-carrageenan gel with hexamethylenediamine before treatment with glutaraldehyde (Nishida *et al.*, 1979). The effect may be a reduction of aspartase protein mobility within the gel and/or decreased enzyme leakage from the gel by cross-linking among the chemical reagents, cell wall, and parts of the cell proteins. Stabilization of L-aspartase β-decarboxylase activity of *Ps. dacunhae* immobilized with κ-carrageenan was successfully improved further by treatment of the immobilized cells with glutaraldehyde and L-lysine. In this case leakage of cells or enzyme protein is also prevented, but in addition the free carboxyl group of L-lysine fixed into the gel optimizes the microenvironmental pH for enzyme reaction. Another example for effective interaction between cells and the network was found by addition of polyethyleneimine to the immobilization procedure using *Brevibacterium flavum* and κ-carrageenan. The introduction of polyethyleneimine in the immobilized system increased significantly the heat stability of fumarase in the cells in comparison to *B. flavum* immobilized only with κ-carrageenan (Tosa *et al.*, 1982). The industrial production of L-malic acid was changed in 1980 to this improved method. The toxicity of reagents such as glutaraldehyde obviously limits the common applicability of these methods, especially when growing cells are necessary for multienzyme reactions (Jirku *et al.*, 1980a,b; Hartmeier, 1981).

Mosbach's group has developed three methods for covalent stabilization of alginate gels for the entrapment of living cells (Birnbaum *et al.*, 1981). The goal of this study was to overcome the instability of calcium alginate gel in the presence of various complexing anions (e.g., phosphate) which is an essential substrate for growth of microorganisms. In the first procedure, the immobilized cells in calcium alginate beads were preincubated with polyethyleneimine and then covalently cross-linked with glutaraldehyde. In the second method, the alginate sol was partially oxidized with meta-periodate to form aldehyde groups. This modified alginate sol was mixed with nonmodified alginate and cells. Finally, gelation in calcium chloride and cross-linkage with polyethyleneimine was carried out. In the third method, free carboxyl groups in the alginate sol were activated by esterification with *N*-hydroxysuccinimide and a carbodiimide derivative. Cells were added to the activated alginate sol and gelated in calcium chloride; the beads formed were treated with polyethyleneimine. In these three methods *S. cerevisiae* was entrapped, and the productivity of ethanol from glucose was tested. Highly active and stable biocatalysts were obtained with a relative activity of 70–95% of the nonmodified biocatalysts. In the cross-linked preparations living cells still existed, because the modified beads could be activated under growth conditions. Furthermore, it was mentioned that the immobilized cells also were involved in the cross-linking reaction (Birnbaum *et al.*, 1981). In comparison to the first method, similar results were obtained in the alginate system with *S. cerevisiae* by cross-linking with polyethyleneimine (Veliky and Williams, 1981).

C. Toxicity

During the preparation of immobilized whole cells in polymeric porous networks, a certain decrease in relative activity and number of viable cells quite generally can be attributed to the toxicity of the reagents (Klein and Wagner, 1979). We can assume that monomers such as acrylamide or glutaraldehyde, in view of their permeation through the cell wall and their reactivity, have a higher toxicity than oligomers or prepolymers. Recently it was demonstrated that the viability of *Arthrobacter globiformis* and the relative activity of steroid Δ^1-dehydrogenase depend strongly on reaction time and temperature during the polymerization of acrylamide to the polyacrylamide network. A comparison of the 11α-hydroxylase activities of the filamentous fungus *Rhizopus nigricans* immobilized in polyacrylamide, alginate, and agar gels showed that no activity was detected in the case of cells immobilized in polyacrylamide. In contrast to acrylamide, the polymeric alginate and agar preparations were both able to hydroxylate progesterone. Besides the toxic affect of the acrylamide monomer on the cells, the network structure of the polyacrylamide may exert a physical effect on the relatively large and fragile hyphae of *R. nigricans* (Maddox *et al.*, 1981).

The use of immobilized whole cells in polyacrylamide gel requires some distinction between the application of a single enzyme system and a multifunctional enzyme system. As we have seen in the multifunctional reaction of Δ^1-dehydrogenase and 11β-hydroxylase, the monomeric acrylamide shows a high toxicity to the biological system. In such systems it might be much better to prepare the polyacrylamide network from a prepolymerized acrylamide. Such a method has been developed for the immobilization of whole cells of *Streptomyces clavuligerus* entrapped in polyacrylamide for the synthesis of cephalosporin. The production of this β-lactam antibiotic was chosen as a representative model for a biosynthetic activity because it is catalyzed by a multienzyme sequence. The cells were immobilized starting from water-soluble polyacrylamide chains, partially substituted with acrylhydrazide groups, and then cross-linked by stoichiometric amounts of dialdehydes in the presence of cells (Freeman and Aharonowitz, 1981). These immobilized cells produced cephalosporin with yields similar to those of free resting cells. When the same cells were immobilized by direct polymerization of acrylamide monomers, the yield was significantly lower. Going back to the single enzyme systems, the toxicity of acrylamide monomers can increase the relative activity in certain cases. Entrapment of *Caldariella acidophila* by polymerization of acrylamide results in a sharp increase of β-galactosidase activity as compared with intact free cells. Such an increase is probably affected by the acrylamide monomers permeabilizing the cell membrane (De Rosa *et al.*, 1980a).

VII. OPERATIONAL STABILITY

The overall macroscopic parameter to quantify the operational stability of a biocatalyst is its half-life. This is the time for an activity loss of 50% of its initial valve. In the literature, half-lives ($t_{1/2}$) from a few days to 2 years have been reported. In general, the immobilized cells have a much higher operational stability than the free suspended ones. The operational stability is influenced by a great number of parameters; and it will be helpful to distinguish between two types of processes with immobilized cells: single-enzyme systems and multistep enzyme systems. Quite different operational stability was observed in one specific process dependent on the cell species and immobilization methods. Both variables can influence the diffusion processes and the microenvironmental conditions.

A. Single-Enzyme Reactions

The operational stability was enhanced by variation of the permeability of the immobilized cells (De Rosa *et al.*, 1980b; Couderc and Baratti, 1980) and by contact with chemicals (Morikawa *et al.*, 1980a; Takata *et al.*, 1980). Successful progress was made by covalent cross-linking of immobilized cells (Nishida *et al.*, 1979; Yamamoto *et al.*, 1980; Tosa *et al.*, 1982). The increase in operational stability may be effected by cross-linkage of the cell wall and matrix and/or by cross-linking of the enzyme to the cell compartments.

B. Multistep Enzyme Reactions

Improving the operational stability of immobilized whole cells for use with immobilized living cells under growth conditions mainly involves multistep enzyme systems together with coenzymes. One such approach is the continuous production of ethanol from carbohydrates. High operational stability was found with immobilized yeast cells under limiting growth conditions (Wada *et al.*, 1980; Larsson and Mosbach, 1979; Cho and Choi, 1981; Williams and Munnecke, 1981; Dangulis *et al.*, 1981; Sitton *et al.*, 1981; Linko *et al.*, 1981; Linko and Linko, 1981; Hahn-Hägerdahl and Mattiason, 1982; McGhee *et al.*, 1982; Chieng *et al.*, 1982; Day, 1982; Fukushima and Hanai, 1982) and with immobilized *Zymomonas mobilis* (Arcuri *et al.*, 1980; Margaritis *et al.*, 1981; Linko and Linko, 1981). The production of compounds such as dihydroxyacetone (Nabe *et al.*, 1979), 2,3-butanediol (Chua *et al.*, 1980), glutathione (Murata *et al.*, 1980), L-lactic acid (Stenroos *et al.*, 1982), erythromycin (Montoya *et al.*, 1982), and bacitracin (Morikawa *et al.*, 1980a) with immobilized cells involving multistep enzyme systems and coenzymes have been described. Such processes utilize discontinuous or continuous regeneration in the presence of suitable nutrients to get acceptable operational stability. Finally, an interesting improvement in operational stability can be obtained by coimmobilization of an inorganic catalyst to the immobilized whole cells. The formation of α keto acids from D-α-amino acids with immobilized cells of *Trigonopsis variabilis* containing D-amino acid oxidase and coimmobilized manganese oxide was developed (Brodelius *et al.*, 1981). The whole system could be stabilized by catalytic decomposition of hydrogen peroxide, which is produced by the D-amino acid oxidase.

VIII. ENGINEERING ASPECTS

A discussion of the methods of immobilization would be incomplete without considering some aspects related to their industrial application. This mainly means giving reference to those methods that have been developed for the quantitative description and characterization of biocatalyst behavior. Most of these aspects have been discussed in detail in a work published by Buchholz (1979).

The first aspects to consider in a technical application are the shape and size of the catalyst and the problems involved in the scale-up of the immobilization procedure. As has been shown by industrial experience, this scale-up problem can be handled successfully for granular PAAm particles (Chibata *et al.*, 1974), spherical κ-carrageenan beads (Chibata, 1979), fibers (Marconi, 1978), and cylindrical pellets (Bungard *et al.*, 1979). The membrane-type catalyst, a very early development in the whole field, has been developed only on a pilot scale

(Vieth and Venkatasubramanian, 1979). There is no doubt that large-scale production of spherical PAAm beads (Klein and Schara, 1980) and spherical calcium alginate beads (Klein and Wagner, 1978; Cheetham *et al.*, 1979) is possible.

As has been demonstrated by Chibata (1979), the polysaccharide carriers can also be fabricated into laminar membrane sheets. Membrane structures have recently attracted increased interest (De Rosa *et al.*, 1980b). To the extent that information is publicly available, only packed-column tower reactors have been used industrially so far. Many other reactor configurations are possible, a fluidized bed being a good candidate.

The second aspect to be considered is the characterization of the mechanical properties of the biocatalyst. The parameter of primary interest has been the compression behavior as related to fixed-bed column applications. A method has been described that gives quantitative information on the elastic and plastic deformation behavior of a single bead under compression (Klein and Washausen, 1979a; Klein *et al.*, 1980). Also, there is the direct measurement of pressure drop along the column. There is a quantitative correlation between single-particle behavior and compression behavior of a packed bed under forced fluid flow conditions, as has been shown for different carrier materials (Klein and Kluge, 1979, 1981). Similar methods of column-compression behavior have been developed in other laboratories (Bungard *et al.*, 1979; Cheetham *et al.*, 1979). Special emphasis has been given to the viscoelastic behavior of calcium alginate columns (Cheetham, 1979).

A second mechanical parameter is the determination of surface abrasion in a stirred-tank reactor, simulating a suspension in a fermenter-type reactor (Klein and Eng, 1979b; Klein *et al.*, 1980). By such methods it is possible to compare different carriers and to find limiting values of cell loading for a given carrier type (Klein and Eng, 1979b; Klein and Kressdorf, 1982). Another stability test is especially related to immobilization by adsorption, where the critical fluid velocity (shear rate) for the transition from cell retention to cell washout is determined (Kolot, 1981a,b).

With the increased importance of gas-forming reactions such as CO_2 production in ethanol fermentation, the tensile stress developed in a carrier (which finally may lead to particle disruption) becomes a problem (Krouwel and Kossen, 1980). A method to evaluate such tensile stress situations has been described (Krouwel, 1982).

The third engineering aspect is the kinetic performance of the biocatalyst. As is well established in the field of heterogeneous catalysis, external and internal mass transfer can play an important role in the overall reaction behavior. It is therefore highly advisable to consider transport phenomena in each practical catalyst development. A size-dependent or cell capacity-dependent reaction rate is a clear but indirect indication that transport steps are becoming rate controlling. In this context it is desirable to obtain direct information on the yield of immobilized activity (Eng, 1980; Klein *et al.*, 1981) and on the transport coefficients—that is, the effective diffusivity of the rate-controlling substances (Klein and Washausen, 1979b; Klein and Manecke, 1982). With such data at hand it becomes possible to calculate the catalytic efficiency from the Thiele modulus and to correlate the results with experimental data (Klein *et al.*, 1979b, 1980, 1981; Klein and Manecke, 1982). Comparable model calculations have also been developed in other laboratories (Roels and Van Tillberg, 1979; Boersma *et al.*, 1979). Such calculations of catalytic efficiency are of practical importance not only in the evaluation of a given catalyst but as a strategic tool in the optimization of particle size and cell loading for a given reaction (Klein and Manecke, 1982). Finally, the relevance of transport phenomena for the understanding and modeling of time-dependent activity decay under operational conditions should be mentioned (Klein *et al.*, 1981; Klein and Vorlop, 1982).

If carefully reviewed, scanning electron microscopy can be an important tool in the morphological characterization of a biocatalyst. One aspect is the structure (e.g., porosity) of the carrier material. More important and less covered by artifacts is the cell density and cell distribution within the carrier. Usually a homogeneous cell density is assumed in an entrapment carrier. However, due to transport-limited growth, asymmetric radial cell densities with higher values in the surface zones have been found in reaction systems, where continued cell growth is required (Chibata, 1979; Wada *et al.*, 1979, 1980; Siess and Divies, 1981). Such information is important in connection with correct model calculations of catalytic efficiency.

REFERENCES

Ado, Y., Kimura, K., and Samejima, H. (1980). *Enzyme Eng.* **5,** 295–304.
Ahn, B., and Byun, S.-M. (1979). *Hanguk Sikp'um Kwahakhoe Chi* **11**(3), 192–199.
Anonymous (1976). *Chem. Eng.* (*N.Y.*) **83,** 87.
Arcuri, E. J., Worden, R. M., and Shumate, S. E. (1980). *Biotechnol. Lett.* **2,** 499.
Ash, D. G. (1979). *Spec. Publ. Soc. Gen. Microbiol.* **2,** 57–86.
Atrat, P., and Groh, H. (1981). *Z. Allg. Mikrobiol.* **21**(1), 3–6.
Atrat, P., Hüller, E., and Hörhold, C. (1981). *Eur. J. Appl. Microbiol. Biotechnol.* **12,** 157.
Bachman, S., Gabricka, L., and Gasyna, Z. (1981). *Starch* **33**(2), 63–66.
Barbotini, I. N., and Thomasset, B. (1980). *Biochimie* **62**(5–6), 359–365.

Birnbaum, S., Pendleton, R., Larsson, P. O., and Mosbach, K. (1981). *Biotechnol. Lett.* **3/8,** 393–400.

Boersma, J. G., Vellenga, K., DeWilt, H. G. J., and Joosten, G. E. H. (1979). *Biotechnol. Bioeng.* **21**(10), 1711–1724.

Brodelius, P. (1982). *Enzyme Eng.* **6,** 203–204.

Brodelius, P., Nilsson, K., and Mosbach, K. (1981). *Appl. Biochem. Biotechnol.* **6,** 293.

Buchholz, K., ed. (1979). "DECHEMA/Monographien," Vol. 84. Verlag Chemie, Weinheim.

Bucke, C., and Wiseman, A. (1981). *Chem. Ind. (London)* No. 7, pp. 234–240.

Bungard, S. J., Reagan, R., Rodgers, P. J., and Wyncoll, K. R. (1979). *ACS Symp. Ser.* **106,** 139–146.

Cheetham, P. S. (1979). *Enzyme Microb. Technol.* **1**(3), 183–188.

Cheetham, P. S., Blunt, K., and Bucke, C. (1979). *Biotechnol. Bioeng.* **21**(12), 2155–2168.

Chibata, I. (1978). *Enzyme Eng.* **4,** 335–337.

Chibata, I. (1979). *Kem.—Kemi* **12,** 705–714.

Chibata, I., and Tosa, T. (1977). *Adv. Appl. Microbiol.* **22,** 1–27.

Chibata, I., and Tosa, T. (1981). *Annu. Rev. Biophys. Bioeng.* **10,** 197–216.

Chibata, I., Tosa, T., and Sato, T. (1974). *Appl. Microbiol.* **27,** 878–885.

Chieng, L. C., Hsiao, H. Y., Flickinger, M. C., Chen, L. F., and Tsao, G. T. (1982). *Enzyme Microb. Technol.* **4,** 93.

Cho, G. H., and Choi, C. Y. (1981). *Biotechnol. Lett.* **3,** 667.

Chua, I. W., Erarslan, A., Kinoshita, S., and Taguchi, H. (1980). *J. Ferment. Technol.* **58,** 123.

Cochet, N., Marcipar, A., and Hebeault, J. M. (1979). *Cell. Immobilisees Colloq. 1979* pp. 211–217.

Constantinides, A. (1980). *Biotechnol. Bioeng.* **22,** 119.

Constantinides, A., Bhatia, D., and Vieth, W. K. (1981). *Biotechnol. Bioeng.* **23**(4), 899–916.

Corrieu, G., Blachere, A., Ramirez, A., Navarro, J. M., Durand, G., Duteurte, B., and Moll, M. (1976). *Proc. Int. Ferment. Symp., 5th,* p. 298.

Couderc, R., and Baratti, J. (1980). *Biotechnol. Bioeng.* **22,** 1155.

Dangulis, A. J., Brown, N. M., Cluett, W. R., and Dunlop, D. B. (1981). *Biotechnol. Lett.* **3,** 651.

Day, D. F. (1982). *Enzyme Eng.* **6,** 343–345.

De Rosa, M., Gambacorta, A., and Nicolaus, B. (1980a). *Biotechnol. Lett.* **2,** 29.

De Rosa, M., Gambocorta, A., Esposito, E., and Drioli, E. (1980b). *Biochimie* **62**(8–9), 517–522.

Dinelli, D. (1972). *Process Biochem.* **7,** 9–12.

Drioli, E., Dorio, G., de Rosa, M., Gambacorta, A., and Nicilaus, B. (1982). *Enzyme Eng.* **6,** 209–210.

Durand, G., and Navarro, J. M. (1978). *Process Biochem.* **13,** 14–23.

Eng, H. (1980). Ph.D. Dissertation, T.U. Braunschweig.

Fetzer, W. R. (1930). *Food. Ind.* **2,** 130.

Freeman, A., and Aharonowitz, Y. (1981). *Biotechnol. Bioeng.* **23,** 2747.

Fukui, S., and Tanaka, A. (1981). *Adv. Biotechnol.* [*Proc. Int. Ferment. Symp.*], *6th, 1980* Vol. 3, pp. 343–348.

Fukui, S., Omata, T., Yamane, T., and Tanaka, A. (1980a). *Enzyme Eng.* **5,** 347–353.

Fukui, S., Ahmed, S. A., Omata, T., and Tanaka, A. (1980b). *Eur. J. Appl. Microbiol. Biotechnol.* **10,** 289.

Fukui, S., Sonomoto, K., Itoh, N., and Tanaka, A. (1980c). *Biochimie* **62**(5–6), 381–386.

Fukushima, M., Fujuii, T., and Morishita, M. (1976). Japan Patent 70,884.
Fukushina, S., and Hanai, S. (1982). *Enzyme Eng.* **6,** 347–348.
Gerson, D. F., and Zajic, J. E. (1979). *ACS Symp. Ser.* **106,** 29–57.
Gestrelius, S. (1980). *Enzyme Eng.* **5,** 439–442.
Ghose, T. K., and Chand, S. (1978). *J. Ferment. Technol.* **56**(4), 315.
Hackel, U., Klein, J., Megnet, R., and Wagner, F. (1975). *Eur. J. Appl. Microbiol.* **1,** 291.
Haegerdal, B., and Mosbach, K. (1980). *Food Process Eng.* [*Proc. Int. Congr.*], *2nd, 1979* Vol. 2, pp. 129–132.
Häggström, L., and Molin, N. (1980). *Biotechnol. Lett.* **2,** 241.
Hahn-Hägerdahl, B., and Mattiason, B. (1982). *Eur. J. Appl. Microbiol. Biotechnol.* **14,** 140.
Hartmeier, W. (1981). *Adv. Biotechnol.* [*Proc. Int. Ferment. Symp.*], *6th, 1980* Vol. 3, pp. 377–382.
Hattori, T., and Furusaka, J. (1960). *J. Biochem.* (*Tokyo*) **48,** 831.
Inoles, D. S., Smith, W. J., Taylor, D. R., Cohen, S. N., Michaels, A. S., and Robertson, C. R. (1982). ACS/IEC Winter Symposium Boulder, Colorado, 1982.
Jack, T. R., and Zajic, J. E. (1977). *Adv. Biochem. Eng.* **5,** 126–145.
Jirku, V., Turkova, J., Veruovic, B., and Kubanek, V. (1980a). *Biotechnol. Lett.* **2,** 451.
Jirku, V., Turkova, J., and Krumphanzl, V. (1980b). *Biotechnol. Lett.* **2,** 509.
Johnson, D. E., and Ciegler, A. (1969). *Arch. Biochem. Biophys.* **130,** 384.
Kennedy, J. F. (1979a). *Chem. Soc. Rev.* **8**(2), 221–257.
Kennedy, J. F. (1979b). *ACS Symp. Ser.* **106,** 119–131.
Kennedy, J. F., Barker, S. A., and Humphrey, S. (1976). *Nature* (*London*) **261,** 242.
Kennedy, J. F., Humphreys, J. D., Barker, S. A., and Greenshilds, R. N. (1980). *Enzyme Microb. Technol.* **2**(3), 209–216.
Kierstan, M., and Bucke, C. (1977). *Biotechnol. Bioeng.* **19,** 387.
Kim, Y. S., and Ryu, D. D. Y. (1980). *Arch. Pharmacol. Res.* **3**(1), 7–12.
Klein, J., and Eng, H. (1979a). *Biotechnol. Lett.* **1,** 171–176.
Klein, J., and Eng, H. (1979b). *DECHEMA-Monogr.* **84,** 292–299.
Klein, J., and Kluge, M. (1979). *DECHEMA-Monogr.* **84,** 285–291.
Klein, J., and Kluge, M. (1981). *Biotechnol. Lett.* **3/2,** 65–70.
Klein, J., and Kressdorf, B. (1982). *Biotechnol. Lett.* **4**(6), 375–380.
Klein, J., and Manecke, G. (1982). *Enzyme Eng.* **6,** 181–189.
Klein, J., and Schara, P. (1980). *J. Solid-Phase Biochem.* **5**(2), 61–78.
Klein, J., and Schara, P. (1981). *Appl. Biochem. Biotechnol.* **6,** 91–117.
Klein, J., and Vorlop, K. D. (1982). *ACS Symp. Ser.* (in press).
Klein, J., and Wagner, F. (1978). *DECHEMA-Monogr.* **82,** 142–164.
Klein, J., and Wagner, F. (1979). *DECHEMA-Monogr.* **84,** 265–325.
Klein, J., and Washausen, P. (1979a). *DECHEMA-Monogr.* **84,** 277–284.
Klein, J., and Washausen, P. (1979b). *DECHEMA-Monogr.* **84,** 300–314.
Klein, J., Hackel, U., Schara, P., and Washausen, P. (1976). *Proc. Int. Ferment. Symp., 5th,* 295.
Klein, J., Hackel, U., Schara, P., Washausen, P., Wagner, F., and Martin, C. J. A. (1978). *Enzyme Eng.* **4,** 339–341.
Klein, J., Hackel, U., Schara, P., and Eng, H. (1979a). *Angew. Makromol. Chem.* **76/77,** 329–350.
Klein, J., Hackel, U., and Wagner, F. (1979b). *ACS Symp. Ser.* **106,** 101–118.
Klein, J., Washausen, P., Kluge, M., and Eng, H. (1980). *Enzyme Eng.* **5,** 359–362.
Klein, J., Eng, H., Vorlop, K. D., and Wagner, F. (1981). *Commun. Eur. Congr. Biotechnol., 2nd,* p. 118.

Kokubu, T., Karube, I., and Suzuki, S. (1981). *Biotechnol. Bioeng.* **23**(1), 29–39.

Kolot, F. B. (1981a). *Process Biochem.* **16**, Part I, 2–9.

Kolot, F. B. (1981b). *Process Biochem.* **16**, Part II, 30–33, 46.

Koshcheenko, K. A., Sukhodolskaya, G. V., Tyurin, V. S., and Skryabin, G. K. S. (1981a). *Biotechnol. Bioeng. Symp.* **12**, 161.

Koshcheenko, K. A., Avramova, T. L., and Sukhodol'skaya, G. V. (1981b). *Izv. Akad. Nauk SSSR, Ser. Biol.* No. 2, pp. 174–180.

Krouwel, P. G. (1982). Thesis Dr. Technol. Sci., Delft University Press, Delft, The Netherlands.

Krouwel, P. G., and Kossen, N. W. F. (1980). *Biotechnol. Bioeng.* **22**(3), 681–687.

Krouwel, P. G., van der Laan, W. F. M., and Kossen, N. W. F. (1980). *Biotechnol. Lett.* **2**, 253.

Krouwel, P. G., van der Laan, W. F. M., and Kossen, N. W. F. (1981). *Biotechnol. Lett.* **3**, 158.

Kumakura, M., Yoshida, M., and Kaetsu, I. (1978). *J. Solid-Phase Biochem.* **3**(3), 175–183.

Kumakura, M., Yoshida, M., and Kaetsu, I. (1979). *Biotechnol. Bioeng.* **21**(4), 679–688.

Larsson, P. O., and Mosbach, K. (1979). *Biotechnol. Lett.* **1**(2), 501–506.

Larsson, P. O., Birnbaum, S., and Mosbach, K. (1982). *Adv. Biotechnol.* [*Proc. Int. Ferment. Symp.*], *6th, 1980* Vol. 1, pp. 717–720.

Lartique, D. J., and Weetall, H. (1976). U.S. Patent 3,939,041.

Lee, G. K., and Long, M. E. (1974). U.S. Patent 3,821,086.

Legoy, M. D., Ergan, F., Dhulster, P., Kim, N. N., and Gellf, G. (1982). *Enzyme Eng.* **6** 129–130.

Li, T. H., Wang, T. K., Li, C.-T., Ku, T.-C., Chia, C.-Y., and Li, H.-L. (1980). *Shengwu Huaxue Yu Shengwu Wu li Jinzhan* **34**, 79–81.

Lilly, M. D. (1978). *DECHEMA-Monogr.* **82**, 165–180.

Linko, P., Poutanen, K., Wechstrom, L., and Linko, Y. Y. (1980). *Biochimie* **62**(5–6), 387–394.

Linko, Y. Y., and Linko, P. (1981). *Biotechnol. Lett.* **3**(1), 121–126.

Linko, Y. Y., Pohjola, L., Viokasi, R., and Linko, P. (1978). *Enzyme Eng.* **4**, 345–347.

Linko, Y. Y., Jalanka, H., and Linko, P. (1981). *Biotechnol. Lett.* **3**, 263.

McGhee, J. E., Julian, G. S., Detroy, R. W., and Bothast, R. J. (1982). *Biotechnol. Bioeng.* **24**, 1155.

McGinis, R. (1975). *Sugar J.* **38**, 8.

Maddox, I. S., Dunnill, P., and Lilly, M. D. (1981). *Biotechnol. Bioeng.* **23**, 345.

Makiguchi, N., Arita, M., and Asai, Y. (1981). *J. Ferment. Technol.* **58**(2), 167–169.

Marcipar, A., Cochet, N., Brackenridge, L., and Lebeault, J. M. (1980). *Biotechnol. Lett.* **1**(2), 65.

Marconi, W. (1978). *DECHEMA-Monogr.* **82**, 88–141.

Margaritis, A., Bajpai, P. K., and Wallace, J. B. (1981). *Biotechnol. Lett.* **3**, 613.

Messing, R. A., Oppermann, R. A., and Kolot, F. B. (1979). *ACS Symp. Ser.* **106**, 13–28.

Miyoshi, T., Ishimatsu, Y., and Kimura, S. (1977). Japanese Patent 120,185.

Mohan, R. R., and Li, N. N. (1975). *Biotechnol. Bioeng.* **17**, 1137.

Moo-Young, M., Lamptey, J., and Robinson, C. W. (1980). *Biotechnol. Lett.* **2**(12), 541–548.

Morikawa, Y., Karube, I., and Suzuki, S. (1980a). *Biotechnol. Bioeng.* **22**, 1015.

Morikawa, Y., Karube, I., and Suzuki, S. (1980b). *Eur. J. Appl. Microbiol. Biotechnol.* **10**, 23.

Mosbach, K. (1982). *J. Chem. Technol. Biotechnol.* **32**, 179.

Mosbach, K., and Mosbach, R. (1966). *Acta Chem. Scand.* **20,** 2807–2810.

Murata, K., Tani, K., Kato, J., and Chibata, I. (1980). *Eur. J. Appl. Microbiol. Biotechnol.* **10,** 11.

Nabe, K., Izuo, N., Yamada, S., and Chibata, I. (1979). *Appl. Environ. Microbiol.* **38,** 1056.

Navarro, J. M., and Durand, G. (1977). *Eur. J. Appl. Microbiol.* **4,** 243.

Navarro, J. M., Durand, G., Duteurtre, B., Moll, M., and Corrieu, G. (1976). *Ind. Aliment. Agric.* **93,** 695–703.

Nilsson, I., Ohlson, S., Häggström, L., Molin, N., and Mosbach, K. (1980). *Eur. J. Appl. Microbiol. Biotechnol.* **10,** 261.

Nishida, Y., Sato, T., and Chibata, I. (1979). *Enzyme Microb. Technol.* **1**(2), 95–99.

Ohlson, S., Larsson, P. O., and Mosbach, K. (1979). *Eur. J. Appl. Microbiol. Biotechnol.* **7,** 103–110.

Ohlson, S., Flygare, S., Larsson, P. O., and Mosbach, K. (1980). *Eur. J. Appl. Microbiol. Biotechnol.* **10,** 1.

Omata, T., Tanaka, T., Yamane, T., and Fukui, S. (1979a). *Eur. J. Appl. Microbiol. Biotechnol.* **6**(3), 207–215.

Omata, T., Jida, T., Tanaka, J., and Fukui, S. (1979b). *Eur. J. Appl. Microbiol. Biotechnol.* **8,** 143.

Omata, T., Tanaka, A., and Fukui, S. (1980). *J. Ferment. Technol.* **58,** 339.

Omata, T., Iwamoto, N., Kimura, T., Tanaka, A., and Fukui, S. (1981). *Eur. J. Appl. Microbiol. Biotechnol.* **11**(4), 199–204.

Petre, D., Noel, C., and Thomas, D. (1978). *Biotechnol. Bioeng.* **20,** 127–134.

Pollack, A., Baughan, R. L., Adalsteinsson, Ö., and Whitesides, C. M. (1978). *J. Am. Chem. Soc.* **100**(1), 302–304.

Prescott, S. C., and Dunn, C. G. (1959). "Industrial Microbiology." McGraw-Hill, New York.

Roels, J. A., and Van Tilberg, R. (1979). *ACS Symp. Ser.* **106,** 147–172.

Rouxhet, P. G., van Haecht, J. L., Dideler, J., Gerard, P., and Briquet, M. (1981). *Enzyme Microb. Technol.* **3,** 49.

Sakimae, A., and Onishi, H. (1980). Ger. Offen. 2,930,985.

Sakurai, Y., Akaike, T., Kataoka, K., and Okano, T. (1980). *In* "Biomedical Polymers" (E. P. Goldberg and A. Nakajima, eds.), pp. 335–379. Academic Press, New York.

Sawada, H., Kinoshita, S., Yoshida, T., and Taguchi, H. (1981). *J. Ferment. Technol.* **59,** 111.

Schara, P. (1977). Ph.D. Dissertation, Technical Univ., Braunschweig.

Scott, C. D., and Hancher, G. W. (1976). *Biotechnol. Bioeng.* **18,** 1393.

Seyan, E., and Kirwan, D. (1979). *Biotechnol. Bioeng.* **21**(2), 271–281.

Siess, M. H., and Divies, S. (1981). *Eur. J. Appl. Microbiol. Biotechnol.* **12**(1), 10–15.

Sitton, O. C., Magruder, G. C., and Gaddy, J. L. (1981). *Adv. Biotechnol.* [*Proc. Int. Ferment. Symp.*], *6th, 1980* Vol. 2, pp. 231–237.

Somerville, A. J., Mason, J. R., and Ruffell, R. N. (1977). *Eur. J. Appl. Microb.* **4,** 75–85.

Sonomoto, K., Hog, M., Tanaka, A., and Fukui, S. (1981). *J. Ferment. Technol.* **59,** 465.

Stenroos, S. L., Linko, Y. Y., Linko, P., Harju, M., and Heikonen, M. (1982). *Enzyme Eng.* **6,** 299–301.

Suzuki, S., Karube, T., Kuriyama, S., Suzuki, N., Shirogami, T., and Takamura, T. (1980). *Biochimie* **62**(5–6), 355–358.

Takata, I., Yamamoto, K., Tosa, T., and Chibata, I. (1980). *Enzyme Microb. Technol.* **2,** 30.

Tanaka, A., Jin, I. N., Kawamoto, S., and Fukui, S. (1979). *Eur. J. Appl. Microbiol. Biotechnol.* **7,** 351.
Tanaka, A., Hagi, N., Gellf, G., and Fukui, S. (1980). *Agric. Biol. Chem.* **44**(10), 2399–2405.
Tanaka, A., Sonomoto, K., Hog, M., Usui, N., Nomura, K., and Fukui, S. (1982). *Enzyme Eng.* **6,** 131–133.
Tanaka, Y., Haruta, T., Suzuki, Y., Hayashi, T., and Kawashima, K. (1980). *Shokuhin Sogo Kenkyusho Kenkyu Hokoku* **36,** 78–83.
Tanzawa, H., Nagaoka, S., Suzuki, J., Kobayashi, S., Masubuchi, Y., and Kikuchi, T. (1980). *In* "Biomedical Polymers" (E. P. Goldberg and A. Nakajima, eds.), pp. 189–210. Academic Press, New York.
Tosa, T., Sato, T., Mori, T., Yamamoto, K., Tahata, I., Nishida, Y., and Chibata, I. (1979). *Biotechnol. Bioeng.* **21**(10), 1697–1709.
Tosa, T., Takata, I., and Chibata, I. (1982). *Enzyme Eng.* **6,** 237–238.
Totsuoka, A., and Hara, S. (1981). *Hakko Kogaku Kaishi* **59**(3), 231–237.
Tsumura, N., and Kasumi, T. (1976). *Proc. Int. Ferment. Symp., 5th,* p. 291.
Veliky, I. A., and Williams, R. E. (1981). *Biotechnol. Lett.* **3,** 275.
Venkatasubramanian, K., and Vieth, W. R. (1979). *Prog. Ind. Microbiol.* **15,** 61–68.
Vieth, W. R., and Venkatasubramanian, K. (1979). *ACS Symp. Ser.* **106,** 1–12.
Vieth, W. R., Wang, S. S., and Sainti, S. (1973). *Biotechnol. Bioeng.* **15,** 565.
Vijayalakshmi, M., Marcipar, A., Segard, E., and Broun, G. B. (1979). *Ann. N.Y. Acad. Sci.* **326,** 249–254.
Vishnoi, P. S. (1980). *J. Sci. Ind. Res.* **39**(10), 578–580.
Vorlop, K. D., and Klein, J. (1981). *Biotechnol. Lett.* **3,** 9.
Wada, M., Kato, J., and Chibata, I. (1979). *Eur. J. Appl. Microbiol. Biotechnol.* **8**(4), 241–247.
Wada, M., Kato, J., and Chibata, I. (1980). *Eur. J. Appl. Microbiol. Biotechnol.* **10**(4), 275–287.
Wada, M., Kato, I., and Chibata, I. (1981). *Eur. J. Appl. Microbiol. Biotechnol.* **11**(2), 167–171.
Wagner, F., Lang, S., Bang, W. G., Vorlop, K. D., and Klein, J. (1982). *Enzyme Eng.* **6,** 251–257.
Wandrey, C. (1979). *Ann. N.Y. Acad. Sci.* **326,** 87–96.
Webster, I. A., Shuler, M. L., and Rony, P. R. (1979). *Biotechnol. Bioeng.* **21**(10), 1725–1748.
Weetall, H. H., Havewala, N. B., Pitcher, W. H., Jr., Defar, C. C., Vann, W. P., and Yaverbaum, S. (1974). *Biotechnol. Bioeng.* **16,** 689–696.
Williams, D., and Munnecke, D. M. (1981). *Biotechnol. Bioeng.* **23,** 1813.
Yamamoto, K., Tosa, T., and Chibata, I., (1980). *Biotechnol. Bioeng.* **22,** 2045.
Yamane, T., Nakatani, H., Omata, T., Sada, E., Tanaka, A., and Fukui, S. (1979). *Biotechnol. Bioeng.* **21**(11), 2133–2145.
Yokozeki, Y. S., Utagawa, T., Takinami, K., Hirose, Y., Tanaka, A., Sonomoto, K., and Fukui, S. (1982). *Eur. J. Appl. Microbiol. Biotechnol.* **14,** 225–231.
Yoshii, F., Fujimura, T., and Kaetsu, I. (1981). *Biotechnol. Bioeng.* **23**(4), 833–841.
Zueva, N. N., Shcherbakova, U. N., Yakovleva, V., Nikitin, Yu. S., Ausyuk, L. V., Chau Thi Tuyet Mai, and Berezin, I. V. (1980). *Prikl. Biokkim. Mikrobiol.* **16**(6), 918–924.

Applications of Immobilized Microbial Cells

Pekka Linko and Yu-Yen Linko

Laboratory of Biochemistry and Food Technology
Department of Chemistry
Helsinki University of Technology
Espoo (Otaniemi), Finland

APPLIED BIOCHEMISTRY AND BIOENGINEERING
Volume 4

ISBN 0-12-041104-0

I. INTRODUCTION

Many microorganisms have a natural attraction for surfaces (Bitton and Marshall, 1980). With few exceptions, however, this characteristic behavior has only recently been exploited for practical applications. The trickle-filter vinegar fermentation process developed in the early 1800s was based on natural microbial films, and may be regarded as the first immobilized-whole-cell process. Compared with the voluminous literature on conventional fermentation techniques, relatively little has been published on the immobilization and applications of whole cells. Hattori and Furusaka (1960, 1961) observed that *Escherichia coli* and *Azotobacter agile* cells immobilized by ionic binding were capable of oxidizing glucose and succinic acid. Mosbach and Mosbach (1966) entrapped *Umbilicaria pustulata* lichen cells in polyacrylamide gel to be used for the decarboxylation of orsellinic acid to orsinol, employing a technique first applied for the immobilization of enzymes by Bernfeld and Wan (1963). These investigations, together with the later important discovery by Mosbach and Larsson (1970) that the entrapped-whole-cell biocatalysts may be reactivated by nutrients for subsequent use in production of specific compounds, laid the groundwork for the development of modern heterogeneous immobilized-whole-cell biocatalysis.

With the first industrial applications of immobilized, isolated enzymes the concept of enzyme engineering was born in the beginning of the 1970s. Realizing the increasing importance of immobilized whole cells both in single and multienzyme conversions, the discipline should more appropriately be called biocatalyst engineering. The first commercial application of immobilized-whole-cell technology is believed to have been steroid dehydrogenation in 1964 by Squibb, employing heat-treated microbial cells (Lilly, 1978). Since 1968 Hokkaido Sugar Co., Ltd. has been using *Mortierella vinacea* mycelial pellet α-galactosidase to hydrolyze raffinose in beet sugar processing (Obara and Hashimoto, 1976). Tanabe Seiyaku Co., Ltd., however, pioneered large-scale industrial applications by introducing in 1973 a process for producing L-aspartic acid from fumarate by polyacrylamide gel-entrapped *E. coli* whole-cell aspartase (Chibata, 1980a). Published commercial-scale applications are listed in Table I. The major industrial application today is the use of glucose isomerase for high-fructose syrup (HFS) production, which was started in Japan by Sanmatsu Kogyo Co. in 1966 and in the United States by Standard Brands, Inc. in 1967 as small-scale operations, and today several companies worldwide are processing glucose employing both immobi-

TABLE I
COMMERCIAL APPLICATIONS OF IMMOBILIZED MICROBIAL CELLS

Application	Organism and enzyme	Method of immobilization	Company	Year
Steroid conversion	Whole-cell steroid dehydrogenase	Heat treatment	Squibb	1964
Raffinose hydrolysis	*Mortierella vinaceae* α-galactosidase	Mycelial pellets	Hokkaido Sugar Co.,	1968
			Great Western Sugar Co.	1974
Fructose and glucose (HFS)	*Streptomyces* sp.[a]	Heat treatment[a]	Standard Brands, Inc.[a,b]	1972
L-Aspartic acid from fumarate	*Escherichia coli* aspartase	Polyacrylamide	Tanabe Seiyaku Co., Ltd.	1973
		Carrageenan		1977
L-Malic acid from fumarate	*Brevibacterium ammoniagenes* fumarase	Polyacrylamide	Tanabe Seiyaku Co., Ltd.	1974
		Carrageenan		
Wastewater treatment	Mixed culture	Adsorption on anthracite	City of Oak Ridge and Norton Co.	1976
Prednisolone from hydrocortisone	*Curvularia lunata* 11β-hydroxylase	Polyacrylamide	(Sweden)	1978
	Corynebacterium simplex Δ′-dehydrogenase			
6-APA from penicillin G	*E. coli* penicillin acylase	Polyacrylamide	(USSR)	—

[a] For others, see Table III.
[b] Known today as Nabisco Brands, Inc.

lized-enzyme and whole-cell technology (Bucke, 1977; Anonymous, 1980a,b).

Immobilized-whole-cell applications may include the use of disrupted or lysed cells, cell fragments, and organelles, as well as both nonviable and living cells (Cheetham, 1980; Linko, 1980, 1981a, b). As an example, for single-enzyme conversions with no cofactor-regeneration requirements, nonviable immobilized cells are typically used. Thus, to enable the inversion of sucrose (P. Linko *et al.*, 1980; Y.-Y. Linko *et al.*, 1980a,b) or lactose (Weckström *et al.*, 1980) by yeast, ethanol fermentation has to be eliminated. In contrast, for the production of primary metabolites such as ethanol or lactic acid, living immobilized cells are needed (Linko, 1980, 1981a,b).

A number of reviews on immobilized whole cells have been published. These include the recent general works on immobilized biocatalysts and their applications by Chibata (1978a) and Brodelius (1978), and outstanding reviews on immobilized whole cells by Durand and Navarro (1978), Chibata *et al.* (1979c), and Cheetham (1980). Kolot (1980) has written a comprehensive survey on immobilized yeast cells. Other excellent reports include a symposium volume edited by Venkatasubramanian (1979) and reviews by Vandamme (1976), Jack and Zajick (1977b), Abbott (1977, 1978), Klein and Wagner (1979), and Venkatasubramanian and Vieth (1979). Living immobilized cells have been covered recently by Linko (1980, 1981a,b). Medical applications of immobilized biocatalysis have been comprehensively covered in a recent two-volume work edited by Chang (1977), and with a few exceptions are not included in this chapter. Wastewater treatment with natural microbial films has been considered to be beyond the scope of this work. A comprehensive survey on the developments in immobilized whole microbial cell-catalyzed biochemical conversions for applications in food and pharmaceutical industries, and for energy and fuel production will be given.

II. FOOD-RELATED CARBOHYDRATE TRANSFORMATIONS

Although the major future potential of immobilized microbial cells may well be in the area of the utilization of complex multienzyme systems of living immobilized cells with cofactor regeneration *in situ*, and of the multibiocatalyst reactors involving coimmobilized microorganisms on a heterogeneous catalysis basis in continuous biotechnical processes, the majority of current industrial-scale applications are based on single-enzyme-catalyzed transformations. Several current and potential applications involve carbohydrate conversions, of which the biggest single success story of immobilized-biocatalyst technology is the development of high-fructose syrup (HFS) production into a major industry during the decade of the 1970s. Table II summarizes most of the immobilized microbial cell-catalyzed carbohydrate transformations published in the literature.

A. Glucose Isomerization in the Starch and Sugar Industry

The history of high-fructose syrup (HFS) development has been reviewed by Barker (1975), Bucke (1977), Casey (1977), Linko *et al.* (1977a), Antrim *et al.* (1979), Hemmingsen (1979), Chen (1980a,b), and others. Since the discovery in the early 1950s of xylose isomerase capable of converting nonphosphorylated aldoses to ketoses (Hochster and Watson, 1954), and the publication by Marshall and Kooi (1957) of a glucose isomerization process, which was later patented (Marshall, 1960), glucose isomerase was developed from an expensive academic curiosity to a bulk commodity in just a decade. This breakthrough owes much to the concurrent developments in immobilized-biocatalyst technology. In a few early articles (Yoshimura *et al.*, 1966; Tsumura *et al.*, 1967; Takasaki *et al.*, 1969), the use of whole microbial cells for the isomerization of glucose to fructose was already described. The discovery by Takasaki *et al.* (1969) that glucose isomerase-active *Streptomyces albus* could be grown on crude inexpensive xylans such as cereal bran or straw allowed mass production of the biocatalyst, and the observation that the enzyme activity could be retained within cells during repeated or prolonged processing by preventing cell lysis at operating temperatures with a simple heat treatment at above 60°C for a few minutes made possible the reuse of the whole-cell biocatalyst, as well as continuous processing in a column reactor (Takasaki and Tanabe, 1971; Takasaki and Kanbayashi, 1973).

In 1966, commercial production of HFS was first started in Japan by Sanmatsu Kogyo Co., employing their own technology, with 2000 tonnes annual capacity. The Takasaki technology was applied in 1967 by Clinton Corn Processing Co. (a division of Standard Brands, Inc.) in the United States for small-scale production of 15% fructose syrup. This pioneering achievement was followed in 1972 by continuous computer-controlled processing with about 0.56 million (M) tonnes* annual capacity for 42% HFS, for which the company was granted the

* tonne ≡ metric ton.

TABLE II
FOOD-RELATED CARBOHYDRATE TRANSFORMATIONS BY IMMOBILIZED CELLS

Enzyme and organism	Method of immobilization	References
Glucose isomerase		
Actinoplanes missouriensis	Entrapment in gelatin and cross-linking with glutaraldehyde	Velzen (1974); Hupkes and van Tilburg (1976); Roels and van Tilburg (1979)
A. missouriensis	Entrapment in α-cellulose and cross-linking with glutaraldehyde	Y.-Y. Linko *et al.* (1976, 1977a,b, 1978a,b, 1979; P. Linko *et al.* 1980, 1981a,b)
A. missouriensis	Entrapment in agar	Ehrenthal and Miner (1980)
A. missouriensis	Entrapment in irradiation-modified gelatin and cross-linking with glutaraldehyde	Bachman *et al.* (1981)
Arthrobacter sp.	Flocculation by polyelectrolytes, optional extrusion, and drying	Lee and Long (1974); Long (1976, 1977); Nyström (1975)
Arthrobacter sp.	Heat treatment	Lloyd *et al.* (1972, 1974)
Bacillus coagulans	Heat treatment	Yoshimura *et al.* (1966)
B. coagulans	Glutaraldehyde treatment and drying	Amotz *et al.* (1976)
B. coagulans	Treating homogenized, extruded cell mass with glutaraldehyde	Poulsen and Zittan (1976)
Lactobacillus brevis	Flocculation with chitosan	Tsumura and Kasumi (1976)
Lactobacillus sp.	Heat treatment	Bhatia and Prabha (1980)
Streptomyces albus	Flocculation with cationic electrolyte	Masaki and Ishige (1976)
S. albus	Entrapment in drying oil	Takahashi (1978)
Streptomyces bambergensis	Glutaraldehyde treatment or entrapment in Ca alginate gel	Cizmek and Drazik (1981)
Streptomyces glaucescens	Heat treatment	Weber (1976)
Streptomyces griseus	Adsorption on anion-exchange resin	Ishimatsu *et al.* (1976)
S. griseus	Entrapment in polyacrylamide gel	Chibata *et al.* (1974d, 1976a, 1978a)
Streptomyces oblivaceus	Cross-linking with glutaraldehyde	Lantero (1978); Snell (1976)
S. olivaceus	Entrapment in polyacrylamide gel	Masaki and Ishige (1976)
S. olivaceus	Glutaraldehyde and polyamine treatment, extrusion, and drying	Chen and Jao (1980)
Streptomyces phaechromogenes	Cross-linking with aromatic primary amines	Moskovitz (1974)
S. phaechromogenes	Entrapment in polyacrylamide	Ohwaki and Minami (1975)

S. phaechromogenes	Heat treatment	Ryu *et al.* (1979)
S. phaechromogenes	Adsorption on active carbon and entrapment in polyacrylamide	Ohwaki *et al.* (1976)
S. phaechromogenes	Binding on collagen membrane	Vieth *et al.* (1976b)
S. phaechromogenes	Mixing with highly porous polymer	Kubo *et al.* (1977)
S. phaechromogenes	Flocculation with chitosan and cross-linking with glutaraldehyde; acetylated chitosan	Nishimura *et al.* (1977); Maekawa *et al.* (1977, 1978); Inoue *et al.* (1978)
S. phaechromogenes	Entrapment in methacrylate	Fukui *et al.* (1977); Kumakura *et al.* (1978, 1979a,b)
S. phaechromogenes	Adsorption in inert carrier and entrapment in polyacrylamide or methacrylate	Kumakura *et al.* (1977, 1978, 1979a,b)
S. phaechromogenes	Entrapment in carrageenan and hardening	Tosa *et al.* (1979)
S. phaechromogenes	Dry cells with Mg^{2+} and Co^{2+} entrapped in cellulose acetate	Lartigan *et al.* (1980)
Streptomyces venezuelae	Binding in collagen membrane	Vieth *et al.* (1973); Vieth and Venkatasubramanian (1976)
Streptomyces vilaceoniger	Mycelial pellets	Leroy *et al.* (1974)
Streptomyces sp.	Heat treatment	Cotter *et al.* (1971); Dworschack *et al.* (1973); Littlejohn and Dworschack (1975)
Streptomyces sp.	Entrapment in polyacrylamide gel	Taguchi *et al.* (1975)
Streptomyces sp.	Cross-linking with diazotized benzidine	Lartique and Weetall (1976)
Streptomyces sp.	Adsorption on DEAE-Sephadex	Shigesada *et al.* (1975)
Streptomyces sp.	Adsorption on Ca and Mg salts and active	Ito and Ozawa (1976); Ito *et al.* (1977)
Streptomyces sp.	Entrapment in polyester sacks	Ghose and Chand (1978)
Streptomyces sp.	Flocculation with chitosan and $CaCO_3$	Maekawa *et al.* (1979c)
Streptomyces sp.	Extrusion of a mixture of cells, chitosan, and gelatin; and drying	Kasumi *et al.* (1979)
Streptomyces sp.	Flocculation with chitosan and silicates	Maekawa *et al.* (1979a)
Streptomyces sp.	Flocculation with chitosan and $MgHPO_4$	Maekawa *et al.* (1979b)
Streptomyces sp.	Entrapment in cellulose triacetate membrane	Kolarik *et al.* (1974)
Streptomyces sp.	Extrusion with thickening agents of heat-treated dried cells, and cross-linking with glutaraldehyde	Ahn and Byun (1979a,b)

(*continued*)

TABLE II (*Continued*)

Enzyme and organism	Method of immobilization	References
Invertase		
Aspergillus oryzae	Adsorption on ECTEOLA-cellulose	Johnson and Ciegler (1969)
Saccharomyces cerevisiae	Flocculation with polyelectrolytes	Lee and Long (1974)
S. cerevisiae	Entrapment in α-cellulose or in cellulose diacetate beads	P. Linko *et al.* (1980)
S. cerevisiae	Entrapment in Ca alginate beads and cross-linking with glutaraldehyde	Y.-Y. Linko *et al.* (1980b)
S. cerevisiae	Entrapment in Ca alginate beads and cross-linking with glutaraldehyde or in cellulose diacetate beads	Y.-Y. Linko *et al.* (1980a)
S. cerevisiae	Immobilization on silicic acid hydrogel, extrusion, and drying	Meisner *et al.* (1980)
S. cerevisiae	Entrapment in gelatin, lyophilization, and treating with formaldehyde	Gianfreda *et al.* (1980)
S. cerevisiae	Entrapment in polyacrylamide with γ rays	D'Souza and Nadkarni (1980a,c)
Saccharomyces pastorianus	Entrapment in agar gel	Toda (1975); Toda and Shoda (1975)
α-Galactosidase		
Absidia lignierii	Pelletization and freezing	Sato and Terashima (1974)
Mortierella vinacea	Mycelial pellets	Yamane (1971); McGinnis (1975); Kobayashi and Suzuki (1972, 1976, 1980)
M. vinacea	Glutaraldehyde treatment	Saimaru *et al.* (1975); Chibata *et al.* (1974a)

β-Galactosidase		
Bacillus stearothermophilus	Entrapment in polyacrylamide gel	Griffiths and Muir (1978)
B. stearothermophilus	Covalent linking to DEAE-cellulose and treatment with glutaraldehyde	Griffiths and Muir (1980)
Caldariella acidophila	Entrapment in cellulose acetate membrane	DeRosa *et al.* (1980a)
C. acidophila	Entrapment in polyacrylamide gel	DeRosa *et al.* (1980b)
C. acidophila	Immobilization in albumin, and treatment with glutaraldehyde	DeRosa *et al.* (1981); Buonocore *et al.* (1981)
Escherichia coli	Entrapment in agar gel	Toda (1975)
E. coli	Entrapment in polyacrylamide gel	Guillaume and Plichon (1975); Ohmiya *et al.* (1977)
E. coli	Immobilization in bovine albumin, and treatment with glutaraldehyde	Petre *et al.* (1978)
Kluyveromyces fragilis	Entrapment in α-cellulose, cellulose diacetate, and cellulose triacetate beads, using different solvent systems	P. Linko *et al.* (1980); Weckström *et al.* (1980)
K. fragilis	Binding on collagen	Thomas *et al.* (1980)
Kluyveromyces lactis	Entrapment in cellulose triacetate fiber	Dinelli (1972)
K. lactis	Entrapment in polyacrylamide gel	Ohmiya *et al.* (1977)
K. lactis	Immobilization in polyphenylene oxide and glutaraldehyde	Jirku *et al.* (1980)
Lactobacillus bulgaricus	Entrapment in agar or after granulating in acetylcellulose	Miyata and Kikuchi (1976a,b)
L. bulgaricus	Entrapment in polyacrylamide gel	Ohmiya *et al.* (1977)

1975 IFT Industrial Achievement Award (Schnyder, 1974; Mermelstein, 1975). The same year also A. E. Staley Manufacturing Co. began the production of HFS. Since then several companies in many countries have started HFS production (Barker, 1975; Rosenzweig, 1976; Anonymous, 1976b). Hämeen Peruna Oy in Finland began the production of 42% HFS in 1974, and in 1975 became one of the first in Europe to apply continuous automated immobilized-biocatalyst technology on a commercial scale (Nevalainen, 1977).

Several possibilities for increasing the fructose content of HFS are known (Crocco, 1976, 1977; Linko *et al.*, 1977a; Karonen *et al.*, 1980), and the second-generation 60–90% fructose HFSs became commercially available in 1976 (Crocco, 1977). The introduction in 1978 of 55%-fructose HFS on the market was soon followed by the decision of major U.S. soft drink manufacturers to replace half or more of their sugar with the new product, a move that resulted in an increase in demand of about 50% within a year (Spaeth, 1980; Anonymous, 1980a). The total annual HFS capacity in 1980 of 11 U.S. manufacturers has been estimated at about 4.3 M tonnes, with a planned increase to about 6 M tonnes by 1982 (Anonymous, 1980b). Of the estimated 1980 U.S. HFS consumption of about 3.7 M tonnes, nearly 45% of the total industrial sugar market, approximately 0.5 M tonnes is used by the baking industry, 2 M tonnes by soft drink manufacturers, and 1.2 M tonnes for processed foods (Spaeth, 1980). In Japan, 16 companies currently produce approximately 0.35 M tonnes of HFS annually, accounting for more than 10% of the total sugar demand (Anonymous, 1980c). Most of the future increase in demand is believed to be in the second-generation 55% HFS by the soft drink industry.

Both immobilized-enzyme and immobilized-cell biocatalysts are used in commercial HFS manufacture. Early patents assigned to Standard Brands, Inc. described stabilizing of the enzyme within *Streptomyces* sp. (Lloyd *et al.*, 1972) and *Arthrobacter* sp. (Lloyd *et al.*, 1974) by heating at 60–80°C for a few minutes. The heat-treated mycelia or cells were mixed with filter aid for use in the isomerization reactor. The half-life of the biocatalyst was reported to be of the order of several hundred hours. A similar approach was described in a patent assigned to R. J. Reynolds Tobacco Co. (Lee and Long, 1974). *Arthrobacter* sp. cells were flocculated in the presence of polyelectrolytes, and the aggregates were used for isomerization both with and without filter aid. In a later modification (Long, 1976), cells were flocculated in the presence of metal compounds such as magnesium oxide for improved hardness of the biocatalyst pellets. Heady and Jacaway (1974) added either $Mg(OH)_2$ or diatomaceous earth filter aid to *Strep-*

tomyces sp. fermentation broth to obtain a cake suitable for the isomerization process on filtering. In a patent assigned to Miles Laboratories (Snell, 1976), *Streptomyces olivaceus* cells were dried at 60–70°C to low moisture content for use as biocatalyst. Tsumura and Kasumi (1976, 1977) found that chitosan could be employed for the immobilization of glucose isomerase and easy flocculation of *Streptomyces* sp. mycelium. The aggregate could be stabilized by combined treatment with citrate before drying at 30°C, and by bifunctional reagents such as glutaraldehyde (Tsumura *et al.*, 1978; Kasumi *et al.*, 1979). Littejohn and Dworschack (1975) improved biocatalyst bed-flow characteristics by mild proteolysis of *Streptomyces* sp. mycelium.

In a patent assigned to Baxter Laboratories, Inc., Moskowitz (1974) described covalent cross-linking of glucose isomerase to the cellular material of *Streptomyces* sp. by diazotized primary diamino compounds in order to improve stability and to minimize side reactions during isomerization. Corn Products Co. has obtained a license to a similar process patented by Lartique and Weetall (1976) for Corning Glass. In another process employed by Novo Industri A/S for commercial production of immobilized glucose isomerase (Zittan *et al.*, 1975; Poulsen and Zittan, 1976; Amotz *et al.*, 1976), dried *Bacillus coagulans* cell mass is reconditioned and pelleted in an axial extruder using a process similar to that described by Nielsen and Markussen (1974a,b). The shaped cell mass is finally cross-linked with glutaraldehyde. Oestergaard and Knudsen (1976) and Norsker *et al.* (1979) have discussed various technical aspects to be considered in large-scale applications of this biocatalyst. Vieth *et al.* (1973, 1976b) and Saini and Vieth (1975) immobilized *Streptomyces* sp. cells in hide collagen membranes that could be subsequently tanned with glutaraldehyde.

Entrapment of whole microbial cells in polymer matrix is a simple and inexpensive technique for large-scale biocatalyst preparation. Chibata *et al.* (1974d, 1975c, 1978a) entrapped *Streptomyces aureus, Streptomyces griseus, Streptomyces olivaceus,* and *B. coagulans* cells in polyacrylamide gel granules. Some activity loss is likely during immobilization as a result of the cytotoxicity of the polymerizing catalyst and the denaturing of enzyme caused by the monomer (Bagnall, 1978). More recently Chibata and co-workers (Chibata *et al.*, 1978c; Tosa *et al.*, 1979) entrapped *Streptomyces phaechromogenes* cells in κ-carrageenan gel beads. This nontoxic polysaccharide isolated from seaweed can be gelatinized under mild conditions to form rigid hard beads suitable for large-scale operations, and since 1977 κ-carrageenan-entrapped microbial cells have been employed by

Tanabe Seiyaku Co., Ltd. in commercial-scale production of L-malic acid, and since 1978 of L-aspartic acid (Chibata, 1980a). Glucose isomerase-active microbial cells have been entrapped in agar, gelatin, and konjac gels (Kanno *et al.*, 1978; Park *et al.*, 1980; Bachman *et al.*, 1981). In a patent assigned to Gist-Brokades N/V, van Velzen (1974) described a method to prepare gelatin bead-immobilized biocatalysts, and the technique has since been applied to large-scale production of *Actinoplanes missouriensis* whole-cell glucose isomerase (Hupkes and van Tilburg, 1976). The enzyme is first stabilized within the mycelium, which should not contain any residual gelatin hydrolyzing

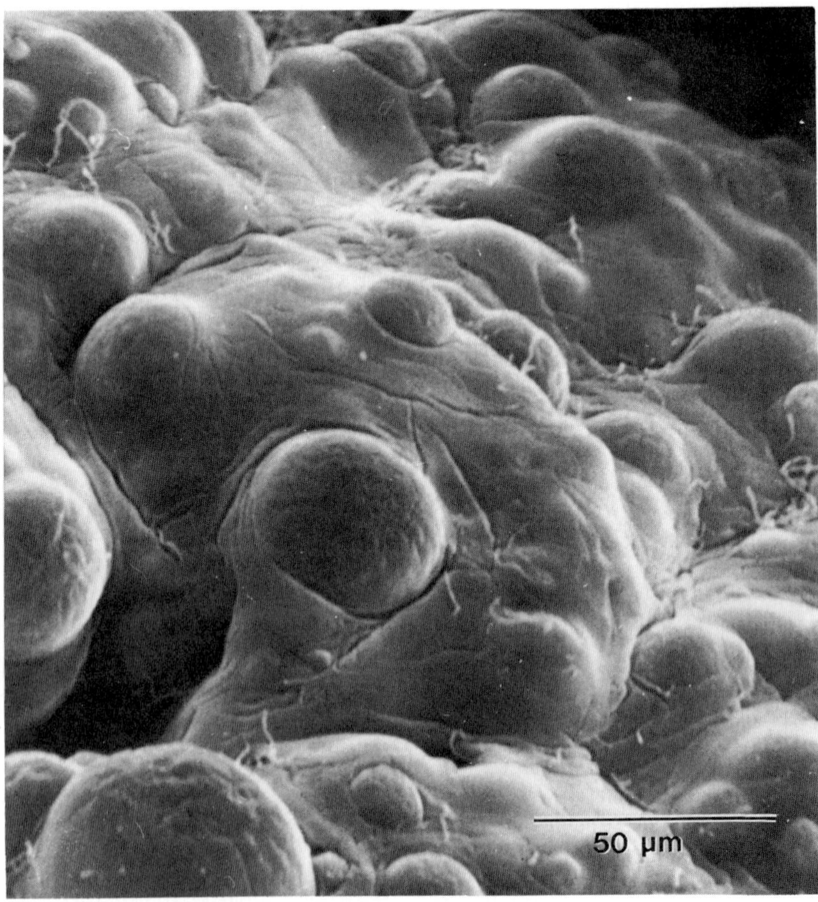

Fig. 1. Scanning electron micrograph of *Actinoplanes missouriensis* whole-cell preparation entrapped in α-cellulose beads.

proteolytic activity. A half-life of 20–40 days was reported, depending on process conditions.

In our laboratory, a novel method to solubilize commercially available α-cellulose allowed the entrapment of *A. missouriensis* cells within cellulose fibers and beads (Linko *et al.*, 1976, 1977a,b, 1978a,b). α-Cellulose is dissolved in a melt of *N*-ethylpyridinium chloride and dimethylformamide under nitrogen, dry mycelium suspended in the solution, and the suspension extruded into water to obtain biocatalyst in desired form. The method was later further improved by replacing dimethylformamide with dimethyl sulfoxide (Y.-Y. Linko *et al.*, 1980a). The biocatalyst was finally stabilized by cross-linking with glutaraldehyde. Figure 1 shows a typical scanning electron micrograph of the partially dried immobilized whole-cell biocatalyst. An example of continuous isomerization of 45% (w/v) glucose is shown in Fig. 2. In a detailed reactor performance and kinetic study (P. Linko *et al.*, 1980, 1981a,b), pressure drop in biocatalyst bed, and external mass-transport limitation were found negligible. The effectiveness factor was $\eta_{min} = 0.85$ (8% biocatalyst cell mass, $D_{eff} = 2.6 \times 10^{-6}$ cm^2, Thiele modulus $\phi = 1.6$, $V_{mf\,max} = 1 \times 10^{-3}$ mol dm^{-3} s^{-1}, $K_{mf} = 0.8$, $V_{mf} = 5 \times 10^{-4}$ mol dm^{-3} s^{-1}), and at typical process conditions at high initial glucose levels $\eta \rightarrow 1$. We could further show that about 48% conversion (X) was reached at initial glucose concentration $G_i = 2.75\ M$ (reaction times, $\tau < 3$ h), and that the rela-

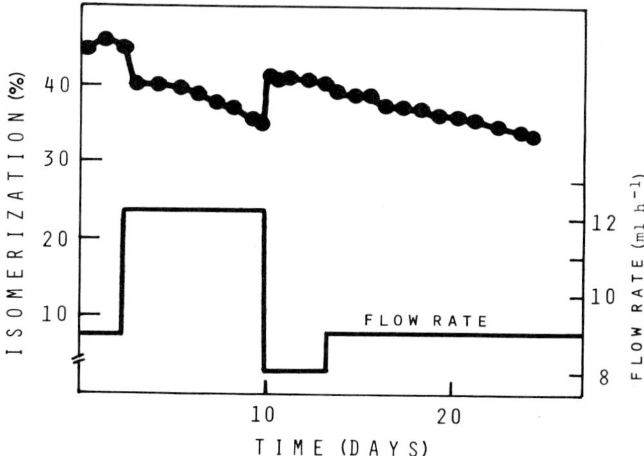

Fig. 2. Continuous isomerization of glucose (45 g liter^{-1}) with α-cellulose bead-entrapped *Actinoplanes missouriensis* cells (3 liters bed volume, 2.4 kg beads, total activity 0.71 m kat, pH 7.5, 3 mM Mg^{2+}, no Co^{2+}, estimated $t_{1/2} \sim 40$ days.

tionship between reaction time (τ) and conversion was best described by the simple equation (Pansolli *et al.*, 1976):

$$\tau = \ln(1 - 2X)\, G_i/2V_m \tag{1}$$

At more dilute concentrations (G_i = 0.275 . . . 1.1 *M*) under similar process conditions, nearly 60% conversion was reached, and the required residence time was best described by the equation:

$$\tau = \left\{\frac{K_{mf}K_{mr}}{C} + \left[\frac{K_{mf}}{C} + \frac{V_{mr}K_{mf}(K_{mr} - K_{mf})}{C^2}\right] G_i\right\}$$

$$\ln\frac{V_{mf}K_{mr}G_i}{CG - V_{mr}K_{mf}G_i} + \frac{K_{mf}K_{mr}}{C}(G_i - C) \tag{2}$$

where $C = V_{mf}K_{mr} + V_{mr}K_{mf}$, and $G = (1 - X)\, G_i$. Typical activity half-lives of $T_{1/2}$ = 150 days (50°C, 40 days (60°C), 20 days (65°C), and 9 days (70°C). Total productivity is a function of the feed rate Q:

$$P_t = (Q_i/k_d)/[1 - \exp(-k_d t)] \tag{3}$$

where the inactivation factor is $k_d = \ln 2/T_{1/2}$, was about 1 ton isomerized glucose per kilogram biocatalyst during 60 days at biocatalyst activity of 1.3 μ kat g^{-1}, 60°C. Several other outstanding process technical and kinetic investigations on other types of im-

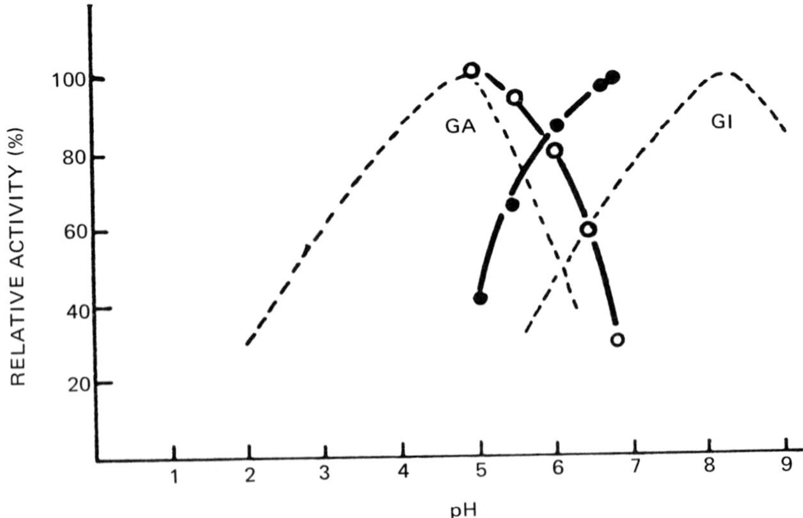

Fig. 3. Activity–pH profiles of glucoamylase (GA, ----, 50°C, 30 min), glucose isomerase (GI, ----, 65°C, 40% w/v glucose), and the immobilized GA–GI two enzyme system (——; ○, GA; ●, GI; 50°C). (From Lindroos *et al.*, 1980, with permission.)

mobilized biocatalysts have been published (Hamilton *et al.*, 1974; Lee *et al.*, 1975; Engasser and Horwath, 1976; Pitcher, 1977; Ghose and Ghand, 1978; Roels and van Tilburg, 1979; Chen *et al.*, 1981).

In addition to HFS production, immobilized whole-cell glucose isomerase can be applied to glucose conversion for recirculation to fructose-separation process, either for the second-generation HFS or for pure fructose manufacture. It has also been used to increase sweetness of β-galactosidase hydrolyzed whey lactose syrups (Poutanen *et al.*, 1978a,b). For optimum results, substrate solution should be pretreated for calcium removal. Sweetness near to that of sucrose could be obtained by isomerizing whey lactose hydrolysate after increasing its glucose level to 46%. Immobilized-whole-cell *Actinoplanes missouriensis* glucose isomerase has also been utilized in a two-enzyme reactor together with phenol formaldehyde resin-bound glucoamylase for continuous simultaneous saccharification of liquefied starch (Lindroos *et al.*, 1980). If Termamyl 60L α-amylase (Novo Industri A/S) was employed for starch liquefaction, no calcium was needed during the process, and the immobilized-whole-cell glucose isomerase was very stable. It is of interest to note that although both the pH (Fig. 3) and temperature (Fig. 4) optima of glucoamylase and glucose isomerase were quite different, DE values of up to >90 with an isomerization percentage of about 47 could be obtained in continu-

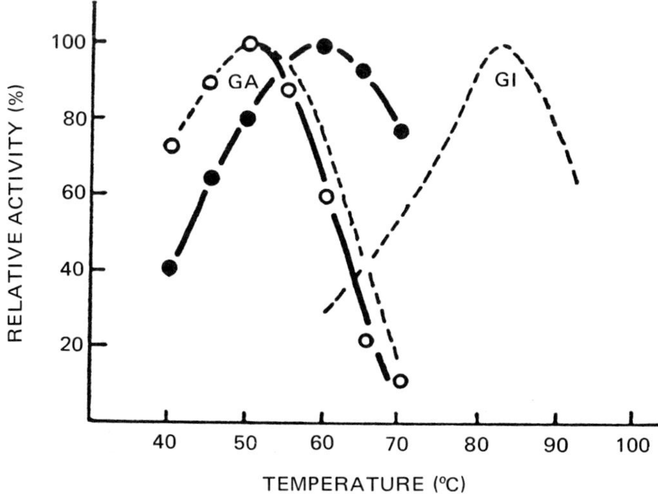

Fig. 4. Activity–temperature profiles of glucoamylase (GA, ----, pH 4.5, 30 min), glucose isomerase (GI, ----, pH 6.5), and the immobilized GA–GI two-enzyme system (——; ○, GA; ●, GI; pH 6.5). (From Lindroos *et al.*, 1980, with permission.)

ous operation. Typical results with various flow rates are further illustrated by Fig. 5.

A number of commercially available immobilized-whole-cell glucose isomerases are listed in Table III. In addition, some companies use their own biocatalysts, which are not available to others. The increase in HFS demand, particularly of the 55%-fructose type in the United States, necessitates further biocatalyst and reactor design development. New microbial strains with increased glucose isomerase activity, and with higher operational enzyme stability with less dependence on magnesium and cobalt should be developed by using both conventional and genetic engineering techniques. Increase in reactor size necessitates biocatalysts of excellent flow characteristics, with little pressure drop and diffusion resistance.

B. Inversion of Sucrose in the Sugar Industry

Much of the pioneering work on enzyme kinetics by Michaelis in the beginning of this century was carried out with yeast invertase, and as

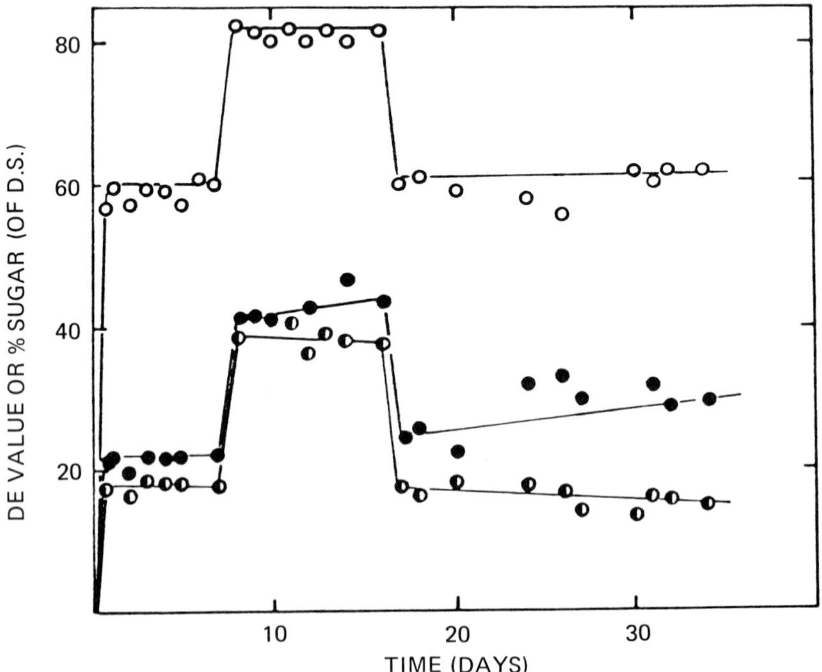

Fig. 5. Continuous simultaneous saccharification and isomerization of 33.3% (d.s.) DE 28 barley starch slurry in a packed-bed-column reactor with various flow rates (50°C; pH 6.5; ○, DE value; ●, glucose; ◐, fructose). (From Lindroos *et al.*, 1980, with permission.)

early as in 1908 Michaelis and Ehrenreich reported that invertase could be adsorbed on charcoal and aluminum hydroxide gel. Wiseman (1979) has published an excellent report on later developments on invertases and their applications, although very little attention was given to whole-cell biocatalysts. Most of the early investigations on immobilized invertases have involved isolated, partially purified enzymes, often under conditions far from those of potential industrial applications (Y.-Y. Linko *et al.*, 1980b). Johnson and Ciegler (1969) immobilized invertase-active *Aspergillus ventii* and *Aspergillus oryzae* spores by ionic binding on ECTEOLA-cellulose for continuous sucrose hydrolysis in a column reactor, but the highest substrate concentration used was only 3% (w/w). A few years later Toda and Shoda (1975) employed a similar approach to entrap *Saccharomyces pastorianus* yeast cells within agar beads as van Velzen (1974) had used for gelatin as a carrier. Invertase-active yeast cells were suspended

TABLE III
COMMERCIAL IMMOBILIZED-CELL GLUCOSE ISOMERASES

Microorganism	Method of immobilization	Company
Actinoplanes missouriensis	Entrapment in gelatin and glutaraldehyde cross-linked	Gist-Brokades N.V. [U.S. Patent No. 3,838,007 (1972)]
Arthrobacter sp.	Flocculation by polyelectrolytes	Reynolds Tobacco Co., Inc. and ICI [U.S. Patent No. 4,060,456 (1977)]
Bacillus coagulans	Cell mass homogenate extruded and cross-linked with glutaraldehyde	Novo Industri A/S [U.S. Patent No. 3,980,521 (1976)]
Streptomyces albus	Heat fixed on cells	Agency of Industrial Science and Technology [U.S. Patent No. 3,950,222 (1976)]
Streptomyces olivaceus	Cross-linking with glutaraldehyde	Miles Laboratories, Inc. and Car-Mi [U.S. Patent No. 3,974,036 (1976)]
Streptomyces sp.	Bound to anion-exchange resin	Denki Kagagu Kogyo and Nagase Sangyo [Jpn Kokai 76-44,688]
Streptomyces sp.	Heat-fixed on cells	Standard Brands, Inc. [U.S. Patent Nos. 3,817,852 (1972); 3,909,353 (1975)]

into 2.5% (w/v) agar solution at 50°C, and the mixture was injected into cold organic solvent such as toluene to form biocatalyst beads during settling of the dispersed droplets. A detailed kinetic study was published later (Toda, 1975).

Lee and Long (1974) used flocculating polyelectrolytes to obtain invertase-active *Saccharomyces cerevisiae* yeast biocatalyst, and Kierstan and Bucke (1977) entrapped *S. cerevisiae* cells in calcium alginate beads. However, unless ethanol fermentation by the immobilized yeast is prevented without affecting invertase activity, such biocatalysts cannot be used in practice for sucrose hydrolysis. D'Souza and Nadkarni (1980a,b) observed that the yeast cell itself presented no permeability barrier for the action of invertase. They entrapped *S. cerevisiae* cells in polyacrylamide gel with about 85% activity retention when γ rays (200 kR) were used for polymerization. Other methods of immobilization included entrapment in agarose gel, and covalent binding to Sepharose 4B. The immobilized biocatalyst was used for continuous inversion of 20% sucrose solution at room temperature (25–27°C) for about a month with no observable loss in activity. They also bound *Aspergillus niger* glucose oxidase with concanavalin A to *S. cerevisiae* cells previously induced for maximal invertase and catalase activities (D'Souza and Nadkarni, 1980a,b,c). Yeast cells, glucose oxidase, and concanavalin A were incubated in 0.1 *M* phosphate buffer to pH 7 containing 0.9% saline and 10^{-4} *M* Mn^{2+}, Mg^{2+}, and Ca^{2+} ions at 37°C for 1 h. The cells were subsequently entrapped in polyacrylamide beads and used for continuous conversion of 10% sucrose solution to fructose and gluconic acid in a continuous stirred-tank reactor (CSTR). Takasaki (1976) has employed bifunctional reagents to bind invertase and a number of other enzymes covalently on microbial cells to obtain multienzyme biocatalysts.

Almost 80% activity yield was reported by Meisner *et al.* (1980) on immobilization of *S. cerevisiae* cells in silicic acid hydrogel. Yeast and calcium carbonate were mixed to a paste with the hydrogel, and the hardened gel was subsequently extruded and dried at 55°C for 24 h. According to Goldstein *et al.* (1977), whole yeast cells can be immobilized on collagen, but they gave no technical details on the applications. Scardi and co-workers (Gianfreda *et al.*, 1980) entrapped *S. cerevisiae* cells in gelatin that was then rapidly cooled for gelation, lyophilized, and treated with cold 20% formaldehyde in 50% ethanol. The hardening treatment prevented ethanol fermentation, and the biocatalyst was used for continuous inversion of up to 0.1 *M* sucrose solution in a column reactor. About 13% of activity was lost during storage of the biocatalyst at 5°C for 3 months.

Once immobilized, whole-cell invertase appears to be remarkably stable. Yeast provides an inexpensive source for the enzyme, making immobilized whole-cell invertase an attractive alternative to invert sugar production by cation-exchange resin technology. Continuous biocatalysis would eliminate the need to regenerate the resin, and the formation of by-products typical of hydrogen ion catalysis could be avoided. Furthermore, processing of such substrates as beet or cane molasses would be possible. In our laboratory, *S. cerevisiae* of high invertase activity has been entrapped in cellulose (P. Linko *et al.*, 1980), cellulose diacetate (P. Linko *et al.*, 1980; Y.-Y. Linko *et al.*, 1980a), and calcium alginate (Y.-Y. Linko *et al.*, 1980a,b) beads. Cellulose and calcium alginate beads were finally treated with glutaraldehyde, and in all cases ethanol fermentation was completely inhibited. The operational stability of the biocatalysts was excellent, with little loss in activity during continuous inversion of up to 50% (w/w) sucrose at 40°C, pH 4.8. Assuming no technical problems such as contamination, the estimated biocatalyst activity half-life was of the order of 4–5 years. No special difficulties were encountered either in continuous processing of 50% (w/w) sucrose solutions or cane molasses (~33% d.s., 16.3% sucrose, 8.9% invert sugar). Figure 6 illustrates the relationships among sucrose concentration, flow rate, percentage conversion, and glucose formation (Y.-Y. Linko *et al.*, 1980a). Complete conversion of 50% (w/w) sucrose required a residence time of $\tau = 6$ h, whereas with 20% (w/w) sucrose $\tau = 0.5$ h was sufficient. Typical cane molasses could be processed without difficulty at $\tau \sim 2$ h.

Wiseman (1979) lists several potential applications for immobilized invertase, but according to Brodelius (1978), no scale-up experiments have been reported. Kolot (1980), however, refers to the industrial use of strong cation-exchange resin-bound invertase, but no details are given.

C. Hydrolysis of Raffinose in the Beet Sugar Industry

Depending on climate, beet sugar contains varying quantities of raffinose that is enriched in the mother liquor during sucrose crystallization (McGinnis, 1975). At concentrations above ~8% raffinose begins to interfere with sucrose crystallization. Consequently, the biocatalytic hydrolysis of raffinose to sucrose and galactose by an enzyme, α-galactosidase, has been considered a major breakthrough. Clearly, the organism should not contain any invertase activity. A fungus *Mortierella vinacea*, capable of producing high α-galactosidase activity without invertase, was isolated from soil (Suzuki *et al.*, 1972; Kobayashi and Suzuki, 1972), and an immobilized-cell process based

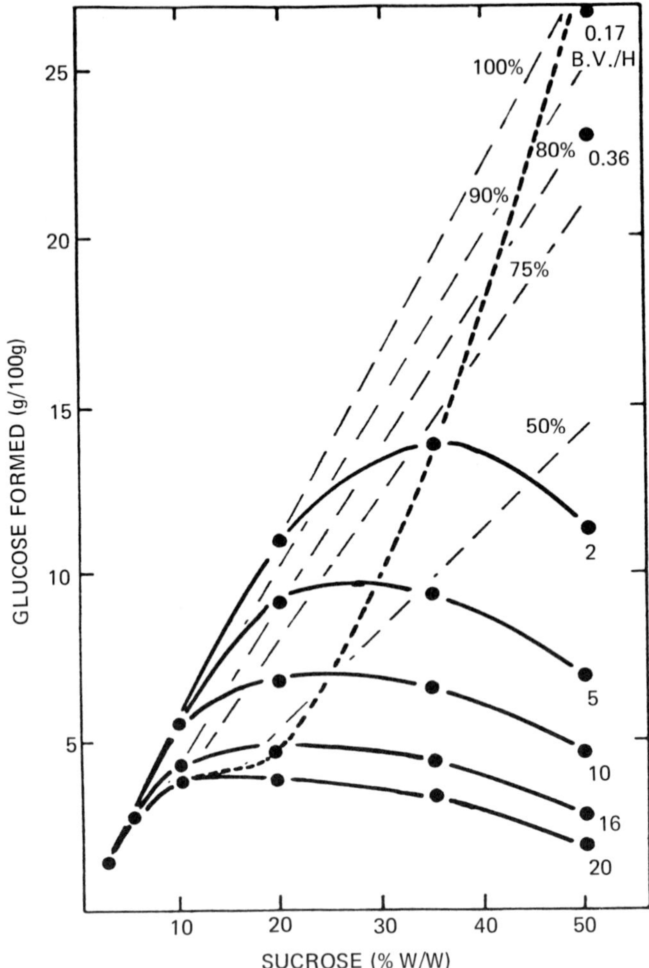

Fig. 6. The relationships among sucrose concentration, flow rate, conversion (%), and glucose formation with calcium alginate bead-entrapped *Saccharomyces cerevisiae* cells (8% alginate, 7.5% dry yeast, 40°C, pH 4.8, 120 μmol min^{-1} g^{-1}). ––, Isoconversion lines; ----, maximum glucose formed. (From Y.-Y. Linko *et al.*, 1980a, with permission.)

on the use of mycelial pellets for continuous hydrolysis of raffinose was developed (Suzuki *et al.*, 1963, 1964, 1969, 1973; Shimizu and Kaga, 1972). In the process, molasses of ~30° Brix is fed into one end of an agitated horizontal vat separated by screens into several biocatalyst pellet-containing chambers held at about 45–50°C. About 80% conversion is obtained in a typical process at τ ~ 2 h. Fresh pellets are added near the substrate inlet, and used for about 25 days.

An industrial-scale operation was started in 1968 in Japan by Hokkaido Sugar Co. (Obara and Hashimoto, 1976), and a similar process in the United States by Great Western Sugar Co. after approval in 1974 by the FDA (Yamane, 1971; Abbott, 1977).

Chibata *et al.* (1974a) and Saimaru *et al.* (1975) treated the mold pellets with glutaraldehyde, which resulted in a considerable improvement in operational stability. According to Saimaru *et al.* (1975), 72% of activity was retained after more than 250 batch operations at 50°C in a 3-week period, compared with the residual activity of 5% in untreated pellets. The enzyme has also been fixed to cellular material by heat treatment (Nishimaru *et al.*, 1975). Kobayashi and Suzuki (1976, 1980) have recently investigated the kinetics of both soluble and *M.vinacea* whole-cell α-galactosidase in detail. Among other observations they showed that product inhibition was less marked with fungal pellets than with the free enzyme.

Other α-galactosidase-active microorganisms have also been immobilized. In a patent assigned to Nippon Beet Sugar Manufacturing Co., Sato and Terashima (1974) described the preparation of *Absidia lignierii* pellets with a special machine. In continuous-column processing of diluted beet molasses containing 3.54% raffinose, about 80% of initial activity remained after 30 days at 60°C. In addition to the sugar beet industry, α-galactosidase has also been suggested for the hydrolysis of raffinose and other α-galactosides in soybean milk (Brodelius, 1978).

D. β-Galactosidase Technology in the Dairy Industry

Lactose intolerance resulting from people's inability to utilize milk sugar is widespread, and the insufficient jejunum β-galactosidase may result in serious gastrointestinal symptoms (Shukla, 1975; Linko, 1982). Low solubility and low sweetness of lactose also limits utilization of whey, a dairy industry by-product available worldwide in large quantities. Consequently, there has been considerable interest in enzymatic hydrolysis of milk and whey lactose by the enzyme β-galactosidase. Inasmuch as both of the hydrolysis products are sweeter than lactose, β-galactosidase technology can also be employed in the production of sweet syrup from whey. The early work on both soluble and immobilized β-galactosidases has been reviewed by Shukla (1975). The first 10-tonnes-per-day commercial plant for treating milk with an immobilized-enzyme reactor developed by Snamprogetti (Pastore and Morisi, 1976; Pastore *et al.*, 1976) has been operated since 1977 by Centrale del Latte in Milan, Italy (Marconi, 1978). Valio Cooperative Dairies Association has for a few years operated a plant for processing whey mainly for pig feed, using a continuous immobilized

Aspergillus niger β-galactosidase reactor developed in cooperation with our laboratory and the Technical Research Centre of Finland. Two large pilot plants based on technology developed by Corning Glass Works for whey lactose hydrolysis have recently been installed in Europe (Dohan *et al.*, 1981). Although immobilized-whole-cell biocatalysts could offer a number of advantages, no commercial-scale operation has been reported.

Employing the Snamprogetti method, Dinelli (1972) entrapped *Saccharomyces lactis* yeast in cellulose triacetate fiber. Since then, a number of techniques have been used to immobilize different β-galactosidase-active microbial cells for lactose hydrolysis. Because of its low pH optimum (~4), the fungal *A. niger* lactase is suitable for acid whey treatment, whereas for example *Escherichia coli* and *Kluyveromyces fragilis* β-galactosidases have an optimum pH of about 6.5, and thus are suitable for milk processing. Toda (1975) entrapped cells of a constitutive β-galactosidase mutant of *E. coli* in agar gel beads. In a patent assigned to Snow Brand Milk Products, Miyata and Kikuchi (1976a,b) described the use of agar-entrapped *Lactobacillus bulgaricus* in a column reactor for lactose hydrolysis. Entrapment both of bacteria and of yeasts in polyacrylamide gel has also been reported (Guillaume and Plichon, 1975; Ohmiya *et al.*, 1977; Griffiths and Muir, 1978; DeRosa *et al.*, 1980b). The whole-cell biocatalyst prepared by entrapment of the extremely acidothermophilic archaebacterium *Caldariella acidophila* in cellulose acetate (DeRosa *et al.*, 1980a) and in chicken egg white, followed by cross-linking with glutaraldehyde (Buonocore *et al.*, 1981), is of special interest because of the high optimum temperature of 80°C of *C. acidophila* β-galactosidase, thus allowing continuous processing of dairy products at temperatures sufficiently high to minimize microbial contamination. A half-life of about 30 days was reported in the presence of lactose at 70°C in continuous-column operation. Although the pH optimum was 5, lactose in commercial milk could be satisfactorily hydrolyzed.

Aspergillus niger β-galactosidase immobilized in phenol formaldehyde resin has good operational stability (Hyrkäs *et al.*, 1976), but *Kluyveromyces fragilis* lactase, which at present is typically used in soluble form to hydrolyze lactose in milk, is difficult to immobilize to obtain good activity yield and operational stability because of its sensitivity to glutaraldehyde treatment, and other techniques employed in immobilization to eliminate ethanol fermentation. Consequently, several whole-cell-entrapment techniques were investigated in our laboratory, employing cellulose and cellulose di- and triacetates in various solvent systems as carriers. Highest activity yield of 10%, with least cell leaking, and satisfactory operational stability was obtained employing cellulose di- and triacetate dissolved in dimethyl sulfoxide and precipitated as beads from water (P. Linko *et al.*, 1980; Weckström *et al.*, 1980). Acetone in the solvent system appeared to increase permeability, but at the same time resulted in considerable leakage of cells from the biocatalyst reactor bed. Glutaraldehyde treatment could not be used because of the adverse effect on yeast lactase activity. Commercial UHT skim milk could be continuously processed at 5°C, $\tau \sim 2$ h, to a 48% degree of conversion in a tetracycline hydrochloride-sterilized column reactor for at least a week without microbial contamination.

Another approach has been the immobilization of whole cells in membranes or thin films. Petre *et al.* (1978) either dried a mixture of lyophilized *E. coli* cells, bovine serum albumin, and glutaraldehyde on glass to obtain membranes, or froze the suspension at −30°C for 4 h, followed by raising the temperature to 4°C to obtain solid biocatalyst of porous spongelike structure, suitable for column reactors. Similar techniques were recently employed by DeRosa *et al.* (1980b) and Drioli *et al.* (1981) to entrap *C. acidophila* cells in membranes, which could also be used for simultaneous hydrolysis and ultrafiltration. *K. fragilis* yeast cells have been immobilized on fibrous collagen and used to hydrolyze 34% of lactose in milk used for yogurt (Thomas *et al.*, 1980). In a patent assigned to Snow Brand Milk Products (Miyata and Kikuchi, 1976b), a method is described for immobilizing granular lyophilized *L. bulgaricus* cell paste in acetylcellulose film.

Finally, some attempts have been made to bind thermostable β-galactosidase-active *Bacillus stearothermophilus* cells covalently on DEAE-cellulose (Griffiths and Muir, 1980), and *Kluyveromyces lactis* yeast cells on polyphenyleneoxide (Jirku *et al.*, 1981).

Clearly, a vast potential exists in the dairy industry for immobilized β-galactosidase biocatalysts. Immobilized-enzyme technology has already been perfected for large-scale treatment of acid whey, but more active and stable biocatalysts are still needed to treat neutral milk products.

III. AMINO ACID PRODUCTION

The increasing importance during the last 25 years of amino acids in animal feed, food, medical, and pharmaceutical applications, as well as in chemical synthesis has resulted in intensive research for better and more economic production techniques. Some amino acids are produced by classical fermentation methods. Chemical synthesis results in racemic mixtures of L- and D-amino acids, necessitating the separation of the optical isomers for many purposes. L-Amino acids can be obtained by enzymatic hydrolysis of proteins, but sophisticated purification and separation techniques are needed. The application of immobilized-enzyme technology for the production of pure L-amino acids, resulting in 1969 in the introduction by Tanabe Seiyaku Co., Ltd. of the first industrial-scale immobilized-biocatalyst process, has been recognized as a major breakthrough in applied biochemistry, for which Chibata was awarded the A. I. Virtanen medal in 1979. Recently, immobilized whole microbial cells have been shown to have marked potential in amino acid production (Table IV).

A. Optical Resolution

L-Amino acids can be obtained from racemic mixtures by selective hydrolysis of acyl-DL-amino acids with specific L-amino acid acylase. The development since the early 1960s of a continuous immobilized aminoacylase process based on a patent assigned to Tanabe Seiyaku Co., Ltd. (Chibata *et al.*, 1974a) required a considerable pioneering effort both in enzyme-immobilization technology and in chemical engineering as applied to continuous biocatalyst reactors, and has been described in detail (Chibata, 1978a, 1980a). *Aspergillus oryzae* aminoacylase was immobilized by ionic binding on DEAE-cellulose for packed-bed column operation. In commercial operation the L-amino acid is crystallized and acyl-D-amino acid remaining in the mother liquor recycled to the process through a racemization reactor. An overall reduction in production costs by about 40% in comparison to that of conventional batch processes employing soluble enzymes was reported in production of amino acids such as L-alanine, L-methionine, L-phenylalanine, L-tryptophane, and L-valine. In order to eliminate the need for costly and laborious enzyme isolation and purification, L-methionine aminoacylase-containing *Aspergillus ochraceus* mycelium pellets have been immobilized using albumin and glutaraldehyde (Hirano *et al.*, 1977a,b; Suzuki *et al.*, 1977), and *A. oryzae* by flocculating with polyelectrolytes (Lee and Long, 1974). L-Alanine aminoacylase-active *Candida* yeast has been entrapped in polyacrylamide gel (Makarova *et al.*, 1979). In both cases biocatalyst stability was markedly increased.

It is also possible to obtain L-amino acids from racemic mixtures by selective oxidation of D-amino acids to the corresponding α-keto acids by D-amino acid oxidases. *Trigonopsis variabilis* has been shown to oxidize D-histidine, D-isoleucine, D-leucine, D-methionine, D-phenylalanine, D-tryptophane, D-tyrosine, and D-valine at a relatively high rate (Brodelius *et al.*, 1981). *T. variabilis* cells were coentrapped with magnesium oxide in calcium alginate gel beads. Magnesium oxide was used to degrade hydrogen peroxide formed in the reaction. The limiting factor in the process, which allows simultaneous production of both α-keto and L-amino acids, is sufficient oxygen transport.

B. Biosynthetic Production of L-Aspartic Acid

Unlike the immobilized *Aspergillus* sp. aminoacylase, immobilized isolated aspartase for the production of L-aspartic acid from fumaric acid and ammonia was not sufficiently stable for continuous processing (Tosa *et al.*, 1973). This prompted the development of immobilized *Escherichia coli* whole-cell aspartase (Chibata *et al.*, 1974e,f), which in 1973 led to the first industrial-scale application of an immobilized-whole-cell biocatalyst reactor since the introduction of the *Acetobacter* sp. trickle-bed recycle reactor to convert ethanol to vinegar (Peppler and Perlman, 1979). *E. coli* cells were entrapped in a polyacrylamide gel lattice, and incubated for activation at weakly alkaline pH in 1 *M* ammonium fumarate containing 1 m*M* Mg^{2+}. The considerable increase in aspartase activity could be explained by cell lysis. The estimated half-life of the biocatalyst was about 120 days, a marked improvement over the immobilized enzyme (Fig. 7). A detailed engineering analysis of the process has been published (Tosa *et al.*, 1974; Sato *et al.*, 1975). Pure L-aspartic acid could be produced from ammonium fumarate at over 95% yield without recrystallization by simple adjustment of the pH to the isoelectric point. With a 1000-liter column

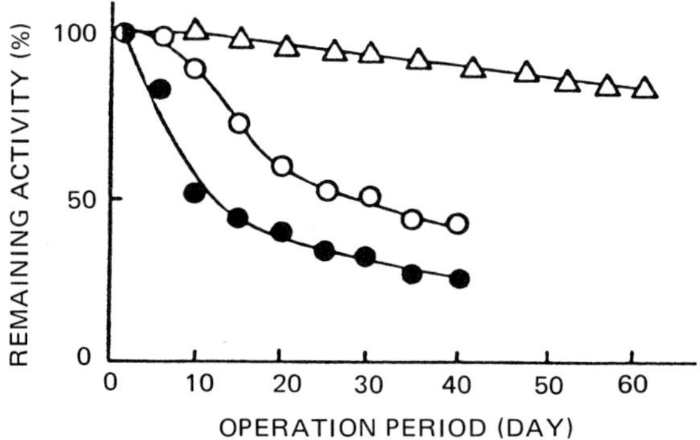

Fig. 7. Operational stability of various aspartase preparations (37°C). Δ, Immobilized cells (120 days); ○, Immobilized enzyme (30 days); ●, intact cells (10 days). (From Chibata, 1980a, with permission.)

TABLE IV
AMINO ACID PRODUCTION BY IMMOBILIZED CELLS

Organism (enzyme)	Method of immobilization	References
L-Amino acids		
— (L-amino acid acylase)	Entrapment in gelatin and treatment with glutaraldehyde	Suzuki (1977)
L-Aspartic acid		
Bacterium cadaveris (aspartase)	Binding of glutaraldehyde-treated cells on glycidyl methacrylate polymer	Nelson (1976)
Escherichia coli (aspartase)	Entrapment in polyacrylamide	Chibata *et al.* (1974e,f, 1976b, 1978c, 1979a); Tosa *et al.* (1973, 1974, 1977); Sato *et al.* (1975); Nishida *et al.* (1979); Zueva *et al.* (1980a); Takamatsu *et al.* (1980).
E. coli (aspartase)	Binding on collagen	Vieth and Venkatasubramanian (1976)
E. coli (aspartase)	Entrapment in κ-carrageenan and treatment with hexamethylenediamine and glutaraldehyde	Sato *et al.* (1979); Chibata (1980b)
E. coli (aspartase)	Coentrapment with inert carrier (zeolite, ceramic, silicate, etc.) in polyacrylamide	Zueva *et al.* (1980b)
E. coli	Entrapment in agar gel	Meng *et al.* (1978)
L-Glutamic acid		
Corynebacterium glutamicum (fermentation)	Entrapment in polyacrylamide gel	Slowinski and Charm (1973)
Corynebacterium lilium (fermentation)	Binding on collagen membrane	Vieth *et al.* (1976a); Venkatasubramanian *et al.* (1978)

L-Citrulline		
Pseudomonas putida (L-arginine deiminase)	Entrapment in polyacrylamide gel	Chibata *et al.* (1974e,g, 1975d); Yamamoto *et al.* (1974a)
Streptococcus faecalis (catabolism)	Entrapment in polyacrylamide gel	Franks (1971, 1972)
L-Alanine		
Candida sp. (L-alanine acylase)	Entrapment in polyacrylamide gel	Makarova *et al.* (1979)
Pseudomonas dacunhae, P. tagnei (L-aspartate-β-decarboxylase)	Entrapment in polyacrylamide gel	Chibata *et al.* (1974h, 1975a,f)
P. dacunhae (L-aspartate-β-decarboxylase)	Entrapment in κ-carrageenan gel and treatment with glutaraldehyde	Yamamoto *et al.* (1980)
L-Isoleucine		
Serratia marcescens (fermentation)	Entrapment in polyacrylamide gel	Chibata (1980a); Wada *et al.* (1980b,c)
Trigonopsis variabilis (D-amino acid oxidase)	Coentrapment with magnesium oxide in Ca alginate gel	Brodelius *et al.* (1980, 1981)
Leucine		
T. variabilis (D-amino acid oxidase)	Coentrapment with magnesium oxide in Ca alginate gel	Brodelius *et al.* (1980, 1981)
L-Tryptophane		
E. coli (tryptophanase)	Entrapment in polyacrylamide gel	Chibata *et al.* (1974k); Decottigneis-Le Marechal *et al.* (1979)
T. variabilis (D-amino acid oxidase)	Coentrapment with magnesium oxide in Ca alginate gel	Brodelius *et al.* (1980, 1981)
L-5-Hydroxytryptophane		
E. coli	Entrapment in polyacrylamide gel	Chibata *et al.* (1974j)

(continued)

TABLE IV (*Continued*)

Organism (enzyme)	Method of immobilization	References
L-Lysine		
Microbacterium ammoniaphilum (2,6-diaminopimelate decarboxylase)	Entrapment in polyacrylamide gel	Kanamitsu (1975)
L-Phenylalanine		
Rhodotorula gracilis (phenylalanine ammonia lyase)	Adsorption on glycidyl methacrylate polymer	Nelson (1976)
Trigonopsis variabilis (D-amino acid oxidase)	Coentrapment with magnesium oxide in Ca alginate gel	Brodelius *et al.* (1980)
L-Tyrosine		
Citrobacter freundii (β-tyrosinase)	Intact cells	Yakovleva (1980)
Erwinia herbicola (β-tyrosinase)	Binding on collagen, and entrapment in cellulose acetate or treatment with *N*-polyacrylate or with dialdehyde starch	Yamada *et al.* (1976, 1978); Hino *et al.* (1976)
T. variabilis (D-amino acid oxidase)	Coentrapment with magnesium oxide in Ca alginate gel	Brodelius *et al.* (1980)
L-Threonine		
E. coli (fermentation)	Entrapment in dialysis cell	Abbott and Gerhardt (1970a)
L-Histidine		
T. variabilis (D-amino acid oxidase)	Coentrapment with magnesium oxide in Ca alginate gel	Brodelius *et al.* (1980, 1981)

L-Valine		
T. variabilis (D-amino acid oxidase)	Coentrapment with magnesium oxide in Ca alginate gel	Brodelius *et al.* (1980, 1981)
L-Methionine		
Aspergillus ochraceus (L-methionine acylase)	Cross-linking a mixture of organism and albumin with glutaraldehyde	Suzuki *et al.* (1977); Hirano *et al.* (1977a,b)
Aspergillus oryzae (L-methionine acylase)	Flocculation with polyelectrolytes	Lee and Long (1974)
T. variabilis (D-amino acid oxidase)	Coentrapment with magnesium oxide in Ca alginate gel	Brodelius *et al.* (1980, 1981)
Glutathione		
E. coli (γ-glutamyl cysteine synthetase, glutathione synthetase, and acetyl kinase or glycolytic pathway for ATP regeneration)	Coentrapment with dextran-bound ATP in polyacrylamide, or with *Saccharomyces cerevisiae* in polyacrylamide or carrageenan, which may be optionally treated with hexamethylenediamine and glutaraldehyde	Chibata *et al.* (1979b); Murata *et al.* (1978a,b, 1979a, 1980a,b, 1981)
D-Arginine		
P. putida (L-arginine deiminase)	Entrapment in polyacrylamide gel	Chibata *et al.* (1974h)
D-Amino acids		
(L-amino acid acylase)	Entrapment in gelatin and treatment with glutaraldehyde	Suzuki (1977)
N-Carbamyl-D-amino acids (dihydropyrimidinase)	Entrapment in polyacrylamide gel	Yamada *et al.* (1980b)

reactor, the theoretical yield was 1915 kg of L-aspartic acid/day, illustrating the high productivity of the biocatalyst. Takamatsu *et al.* (1980) and Zueva *et al.* (1980a) studied the kinetics of L-aspartic acid production by entrapped-whole-cell aspartase. The inclusion of solid carriers such as zeolites, ceramics, and silicates during the entrapment in polyacrylamide gel was found to increase operational stability (Zueva *et al.*, 1980b).

Recently an improved process based on the entrapment of *E. coli* and other microbial cells in κ-carrageenan has been developed (Chibata *et al.*, 1978c, 1979a; Sato *et al.*, 1979; Nishida *et al.*, 1979). Stability was improved by treatment with glutaraldehyde. The estimated half-life of almost 2 years was about 5 times higher than that of the polyacrylamide-based biocatalyst, and the productivity was increased by about 15 times. Since 1978, Tanabe Seiyaku Co., Ltd. has used this new biocatalyst for industrial production of L-aspartic acid (Chibata, 1980a). Meng *et al.* (1978) reported the entrapment of *E. coli* AS 1.881 cells of high aspartase activity in 6% agar gel with improved yield over polyacrylamide-entrapped cells.

Venkatasubramanian *et al.* (1978) immobilized L-aspartase-active *E. coli* cells on hide collagen membrane, which was subsequently tanned with 0.1% glutaraldehyde. Using 10% fumaric acid adjusted to pH 8.5 with ammonium hydroxide, up to 98% conversion to L-aspartic acid was obtained in 20 h. A method to immobilize whole microbial cells by covalent bonding on water-insoluble glycidyl ester copolymers for the production of L-aspartic acid has been described in a patent assigned to Pfizer, Inc. (Nelson, 1976).

C. Immobilized-Whole-Cell Biocatalysis in the Production of Other Amino Acids

1. L-Glutamic Acid

L-glutamic acid is industrially manufactured by traditional fermentation techniques employing *Corynebacterium glutamicum*, which is able to accumulate glutamic acid in high quantities. Wheat endosperm proteins contain more than 30% of glutamic acid and glutamine, and consequently wheat is a good source for hydrolytic glutamic acid production. Furthermore, DL-glutamic acid can be produced synthetically. Nevertheless, immobilized-whole-cell biocatalysis has been investigated for the development of continuous-processing techniques.

Slowinski and Charm (1973) showed that L-glutamic acid can be produced from glucose in successive batch fermentations by polyacrylamide gel-entrapped *C. glutamicum* cells, but probably because of insufficient oxygen-transport productivity was not competitive with that of traditional fermentation. The biocatalyst lost little activity during 2-month storage at 4°C. An economic analysis based on hide collagen-immobilized *C. glutamicum* cells (Venkatasubramanian *et al.*, 1978) resulted, however, in higher return on investment with the immobilized-biocatalyst system than with conventional fermentation processes. Problems to be solved were related to dissolved-oxygen level, mass transport, cofactor regeneration, elimination of possible side reactions, and biocatalyst activity.

2. L-Citrulline

Chibata and co-workers (Kakimoto *et al.*, 1971) observed that *Pseudomonas putida* has high L-arginine deiminase activity, but no ornithine transcarbamylase, which would degrade citrulline to ornithine, and developed a method to entrap *Ps. putida* cells in polyacrylamide gel for continuous production of L-citrulline from L-arginine (Chibata *et al.*, 1974e,g; Yamamoto *et al.*, 1974a). Immobilization increased both optimum temperature and temperature stability. Unlike the situation with the intact cells, no surface-active agents were needed for maximum production by the immobilized-cell biocatalyst. The half-life in continuous-column operation at 37°C with 0.5 *M* L-arginine hydrochloride as substrate was estimated to be about 140 days. After concentration and separation with strong cation-exchange resin, about 96% yield of pure crystalline L-citrulline was obtained.

3. L-Alanine

Leuschner (1966) patented a method to prepare water-insoluble biocatalysts by entrapping either enzymes or enzyme-active whole cells in various polymer membranes, and described the production of L-alanine from L-glutamic acid and pyruvic acid with immobilized glutamate transaminase. *Pseudomonas dacunhae* and *P. tagnei* cells containing L-aspartate-β-decarboxylase have been entrapped in polyacrylamide (Chibata *et al.*, 1974i, 1975a,f) and in carrageenan (Yamamoto *et al.*, 1980) gel beads for continuous production of L-alanine from L-aspartic acid. Biocatalyst enzyme activity was enhanced by incubation in substrate solution at 37°C for over 20 h. The addition of 0.1 m*M* of pyridoxal 5′-phosphate to substrate solution was sufficient to maintain enzyme activity, and the treatment of the biocatalyst with glutaraldehyde significantly improved operational stability and prevented cell leakage. The increase in pH during the reaction was claimed to be of no problem because of the relatively broad pH optimum of the immobilized-cell biocatalyst, but some technical difficulties were encountered with the evolution of carbon dioxide.

4. L-Isoleucine

The attempt by Slowinski and Charm (1973) to produce L-glutamic acid from glucose by heterogeneous biocatalysis is the first to utilize immobilized whole cells for aerobic multistep production of amino acids. Wada *et al.* (1980c) entrapped living *Serratia marcescens* cells in carrageenan gel in an attempt to improve the production of L-isoleucine from D-threonine and glucose by conventional fermentation techniques.

To improve oxygen transport, reaction was carried out in a fluidized-bed reactor (Fig. 8, Chibata, 1980a). In a two-bed reactor system, steady state could be maintained in both beds for at least a month (Fig. 9). The immobilized cells grew only in the presence of oxygen and nutrients. Optimum production was obtained at 10 + 10 h residence time, with pH automatically controlled at 7.5.

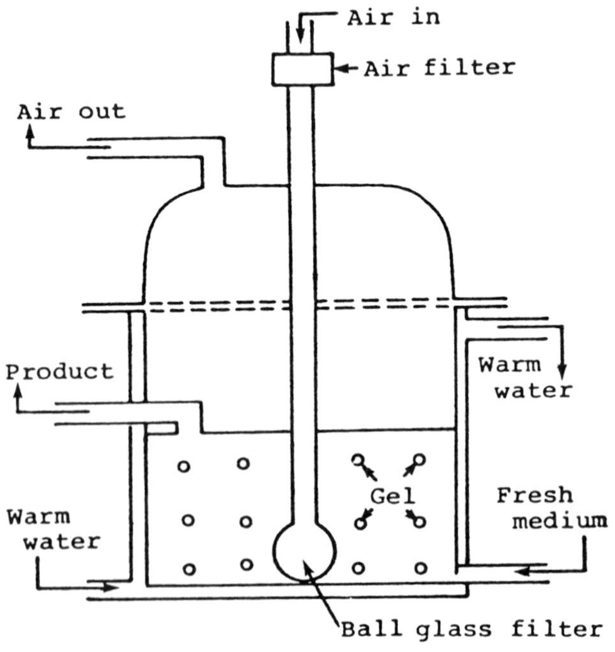

Fig. 8. Fluidized-bed reactor for continuous production of L-isoleucine with κ-carrageenan gel-immobilized *Serratia marcescens* cells. Gel, 50 ml; working volume, 150 ml; aeration rate, 200 ml/min; retention time: 10 h. (From Chibata, 1980a, with permission.)

According to Wada *et al.* (1980a), living cells were continuously released from the biocatalyst gel at a constant rate, and ATP and coenzymes were regenerated *in situ*. The productivity was judged competitive with the conventional batch process, but a disadvantage was the relatively low product concentration obtained.

5. L-Tryptophane

Chibata and co-workers (1974j,k) entrapped tryptophanase-active *Escherichia coli* cells in polyacrylamide gel. The biocatalyst could be used to produce L-tryptophane from indole and serine or ammonium pyruvate in 86% yield at 30°C. Similar biocatalyst was used by Decottignies-Le Maréchal *et al.* (1979) for continuous production of L-tryptophane employing a CSTR reactor equipped with automatic liquid level control and reagent-recycling system. Tryptophane produced was adsorbed in a charcoal–Celite column. Ethanol (10%) for indole solubilization in substrate solution prevented microbial contamination during processing. An excess of both sodium pyruvate and ammonium bicarbonate was necessary for L-tryptophane production, pyruvate concentration being most critical. At indole levels below 3 mM, 0.25 M sodium pyruvate was necessary for optimum production.

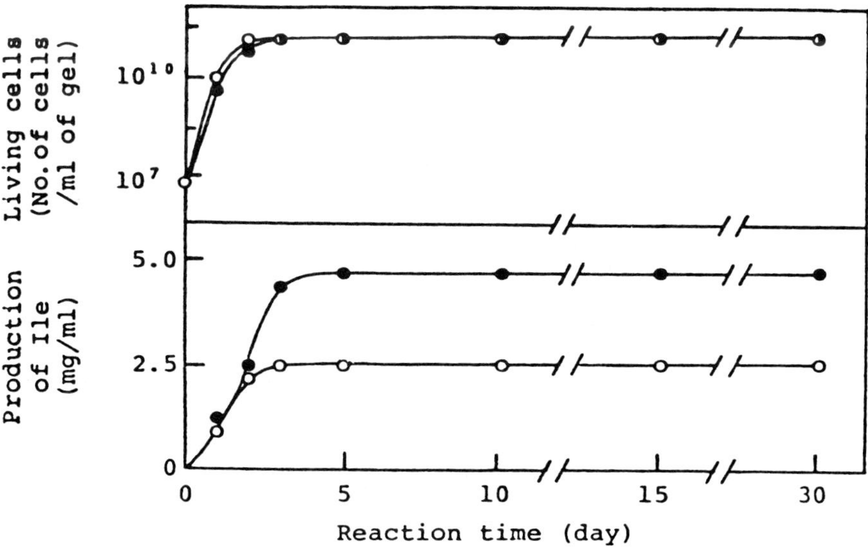

Fig. 9. Pattern of continuous production of L-isoleucine with κ-carrageenan gel-immobilized *Serratia marcescens* cells in a two-bed reactor system. ○, First bed; ●, second bed; gel, 50 ml; working volume, 150 ml; flow rate, 15 ml/h; retention time, 10 h + 10 h. (From Chibata, 1980a, with permission.)

Typical productivity in a plug-flow reactor was 0.6 mmol L-tryptophane g^{-1} h^{-1} (bacteria). Employing a similar system, Bang *et al.* (1978) obtained a half-life for polyacrylamide gel-immobilized *E. coli* whole-cell tryptophanase of more than 30 days. With 110 g of biocatalyst (5 g of wet cells) in a CSTR, and 2 g of indole, 2 g of L-serine, and 0.01 g of pyridoxal phosphate per liter of 0.1 *M* phosphate buffer (pH 8) as substrate, the maximum productivity of L-tryptophane was 0.2 g $liter^{-1}$ h^{-1}, and the yield 96%. The production of L-tryptophane has also been investigated employing chitosan bead-entrapped *E. coli* tryptophanase (Vorlop and Klein, 1981).

6. L-Lysine

The production of L-lysine by polyacrylamide-entrapped *Microbacterium ammoniaphilum* is described in a patent assigned to Asahi Chemical Co., Ltd. (Kanamitsu, 1975). *M. ammoniaphilum* contains diaminopimelate decarboxylase capable of producing L-lysine from *meso*-2,6-diaminopimelic acid.

7. L-Phenylalanine

Nelson (1976) patented a method to bind microbial cells covalently to glycidyl methacrylate polymer. Immobilized cells of *Rhodotorula gracilis* retained 49% of their phenylalanine ammonia lyase activity on immobilization, and could be used for producing L-phenylalanine from *trans*-cinnamic acid and ammonia in about 90% yield, but only 77% of initial activity was retained after one batch experiment.

8. L-Tyrosine

Erwinia herbicola enzyme tyrosine phenol lyase, β-tyrosinase, catalyzes and α,β-elimination reaction converting L-tyrosine to pyruvic acid, ammonia, and phenol. *E. herbicola* cells have been immobilized on collagen and subsequently cross-linked using dialdehyde starch (Yamada *et al.*, 1978), to be used to catalyze the reverse reaction for biosynthesis of L-tyrosine. The immobilized cells were packed into a column maintained at 30°C, and 35–70% conversion was achieved, depending on residence time. The enzyme is inhibited by relatively small levels of phenol, and the relatively low solubility of tyrosine resulted in precipitation in the reactor column. Yakovleva (1980) has reported a similar process employing intact *Citrobacter freundii* cells, but no technical details were given.

9. Glutathione

A patent assigned to Tanabe Seiyaku Co., Ltd. (Chibata *et al.*, 1979b) describes production of tripeptide glutathione from L-glutamic acid,

L-cysteine, and glycine by immobilized-whole-cell biocatalysis. The process requires two separate enzyme systems, γ-glutamylcysteine synthetase and glutathione synthetase, in addition to the necessary ATP-regeneration system. It was possible to produce glutathione in a column reactor packed with *Escherichia coli* cells coimmobilized with dextran-bound ATP in polyacrylamide gel, but the yields were relatively low (Murata *et al.*, 1979a, 1980a). Other ATP-regeneration systems were also investigated, including the glycolytic pathway of immobilized *Saccharomyces cerevisiae* cells (Murata *et al.*, 1978a,b, 1980a, 1981), and the acetylkinase system of immobilized *E. coli* cells with acetyl phosphate added in the substrate (Murata *et al.*, 1980a,b). The glycolytic pathway appeared most promising, but the productivity of the system was not yet competitive either with the conventional fermentation technique or with chemical synthesis. Nevertheless, these examples clearly illustrate the potential of immobilized-whole-cell biocatalysis in the production of relatively complicated organic chemicals.

IV. PRODUCTION OF ORGANIC ACIDS

Organic acids are widely used in food and medical applications, and many are produced by conventional fermentation (Miall, 1978; Lockwood, 1979; Nickol, 1979). In the early 1800s a vinegar-manufacturing process based on bacteria growing on wood chips was introduced, and in 1974 Tanabe Seiyaku Co., Ltd. began producing L-malic acid by employing polyacrylamide-entrapped *Brevibacterium ammoniagenes* cells (Chibata, 1980a). Immobilized whole cells have also been investigated for the production of several other organic acids (Table V).

A. Acetic Acid

Vinegar fermentation has been reviewed by Greenshield (1978). The discovery by Boerhaave and Kastner in the beginning of the last century of the trickle-filter bacterial film fermentation to produce vinegar is believed to be the first industrial application of immobilized microorganisms. In 1823 Schützenbach added aeration holes near the reactor bottom, and in 1929 forced-aeration and temperature-control systems were first included (Fetzer, 1930; Nickol, 1979). According to Nickol (1979), the lifetime of a Frings-type vinegar reactor, if properly operated, may be more than 20 years, with occasional addition of beech shavings or other packing material used. A minimum of 0.2% residual ethanol is necessary in order to keep *Acetobacter* sp. alive.

TABLE V
ORGANIC ACID PRODUCTION WITH IMMOBILIZED CELLS

Organism (enzyme)	Method of immobilization	References
Acetic acid (fermentation)		
Acetobacter sp.	Attachment to hydrous titanium (IV) oxide	Kennedy *et al.* (1976, 1980); Barker *et al.* (1979)
Acetobacter sp.	Adsorption on porous ceramic	Ghommidh *et al.* (1981)
Acetobacter aceti	Entrapment in carrageenan gel	Mori *et al.* (1980)
Bacterium schuetzenbachii	Adsorption on beech shavings	Fetzer (1930)
L-Malic acid (fumarase)		
Brevibacterium ammoniagenes	Entrapment in polyacrylamide gel	Chibata *et al.* (1975g); Yamamoto *et al.* (1976, 1977)
Corynebacterium equi		
Escherichia coli		
Microbacterium flavum		
Proteus vulgaris		
Brevibacterium flavum	Entrapment in κ-carrageenan gel	Chibata *et al.* (1978c); Chibata (1980b); Takata *et al.* (1980)
B. ammoniagenes	Entrapment in κ-carrageenan gel	Takata *et al.* (1979)
B. flavum		
Brevibacterium helvolum		

P. vulgaris		
Pseudomonas fluorescens		
Sarcina aurantica		
Sarcina flava		
Sarcina ureae		
Sarcina variabilis		
Candida rugosa	Entrapment in polyacrylamide gel	Yang and Zhong (1980)
Urocanic acid (L-histidine ammonia lyase)		
Achrobacter aquatilis	Entrapment in polyacrylamide gel	Chibata *et al.* (1975e)
Achrobacter liquidum		
Agrobacterium radiobacter		
Flavobacterium flavescens		
Sarcina lutea		
A. liquidum	Entrapment in polyacrylamide gel	Yamamoto *et al.* (1974b)
Micrococcus luteus	Covalent binding on carboxymethylcellulose	Jack and Zajic (1977a)
Ps. fluorescens	Heat-treated cells in hollow-fiber dialysis unit	Kan and Shuler (1978a,b)
Lactic acid (fermentation)		
Lactobacillus bulgaricus	Containment in dialysis unit	Stieber *et al.* (1977); Coulman *et al.* (1977)
L. bulgaricus	Entrapment in polyacrylamide gel	Divies (1977)
Lactobacillus delbrueckii	Containment in dialysis unit	Friedman and Gaden (1970)
L. delbrueckii	Entrapment in polyacrylamide gel	Divies (1977)
L. delbrueckii	Entrapment in Ca alginate gel beads	Linko (1981a,b)
Streptococcus stearothermophilus	Entrapment in polyacrylamide gel	Divies (1977)
Mixed culture	Reverse-osmosis unit connected to fermenter	Setti (1975)
Mixed culture	Adsorption on gelatin coating and cross-linking with glutaraldehyde	Griffith and Compere (1975); Compere and Griffith (1975)

(continued)

TABLE V (*Continued*)

Organism (enzyme)	Method of immobilization	References
Citric acid (fermentation)		
Aspergillus niger	Binding in collagen membrane with glutaraldehyde	Vieth and Venkatasubramanian (1978)
A. niger	Entrapment in Ca alginate (6–8%) gel	Linko (1981a,b)
Candida lipolytica	Entrapment in polyacrylamide gel	Stottmeister (1979); Berger and Langhammer (1980)
Saccharomycopsis lipolytica	Adsorption on wood chips	Briffaud and Engasser (1978)
Gluconic acid (glucose oxidase)		
A. niger	Flocculation by polyelectrolytes	Lee and Long (1974)
A. niger	Entrapment in polyacrylamide gel	Nelson (1975)
A. niger	Binding glutaraldehyde-treated cells to glycidyl methacrylate	Nelson (1976)
A. niger	Entrapment in Ca alginate (6–8%) gel beads	Linko (1981a,b)
A. niger	Binding isolated glucose oxidases to mycelial pellets	Karube *et al.* (1977a)
A. niger and *Saccharomyces cerevisiae*	Binding *A. niger* and other glucose oxidases to *S. cerevisiae* cells with concanavalin A and optional entrapment in polyacrylamide gel	D'Souza and Nadkarni (1980a,b)
Gluconobacter suboxydans	Entrapment in Ca alginate gel	Tramper *et al.* (1981)

2-Ketogluconic acid		
(fermentation)		
Serratia marcescens	Binding in collagen with glutaraldehyde	Venkatasubramanian *et al.* (1978)
Erythorbic acid		
(fermentation)		
Penicillium cyaneofulvum	Entrapment in polyacrylamide gel	Kato (1974)
2-Keto-L-gulonic acid		
(fermentation)		
Gluconobacter melanogenus	Entrapment in polyacrylamide gel	Martin and Perlman (1976a)
G. melanogenus and *Pseudomonas syringue*	Coimmobilization in polyacrylamide gel	Martin and Perlman (1976b)
α-Keto acids		
(D-amino acid oxidase)		
Trigonopsis variabilis	Entrapment in Ca alginate gel	Brodelius and Mosbach (1980); Brodelius *et al.* (1980, 1981)
Salicylic acid		
(fermentation)		
Ps. fluorescens	Containment in dialysis unit	Abbott and Gerhardt (1970b)

In a patent assigned to Gist-Brokades N/V, Barker *et al.* (1975) described a process to immobilize enzymes or cells by chelation to hydrous-transition metal oxides. The technique was applied to immobilize *Acetobacter* sp. on hydrous titanium(IV) oxide, which was employed in a tower fermenter to produce acetic acid (Kennedy *et al.*, 1976, 1980; Kennedy, 1978). Production rates higher than those obtained with free cells were reported. Acetic acid production could be further improved by mixing equal weights of chromatographic cellulose powder and titanium(IV) chloride solution (15% w/v, in 15% HCl) and stirring at room temperature for 2 h. Oxygen transport did not appear to be a limiting factor.

Ghommidh *et al.* (1981) adsorbed *Acetobacter aceti* cells on porous ceramic from Corning Glass Works. With pure oxygen, maximum production rate obtained was 10.4 g $\text{liter}^{-1}\ \text{h}^{-1}$ at a dilution rate of D = 0.513 h^{-1}, about five times higher than the values reported for conventional processes.

B. L-Malic Acid

According to Chibata (1980a), industrial production of L-malic acid by polyacrylamide gel-immobilized *Brevibacterium ammoniagenes* cells was started by Tanabe Seiyaku Co , Ltd. in 1974 (Chibata, 1980a). The fumarase activity of the immobilized biocatalyst can be markedly enhanced by incubating in a fumaric acid or fumarate solution containing surfactant (Chibata *et al.*, 1975g). The use of immobilized whole cells may be associated with unwanted side reactions, and in addition to high fumarase activity, *B. ammoniagenes* cells also formed substantial quantities of succinic acid as by-product, which is very difficult to separate from L-malic acid (Yamamoto *et al.*, 1976; Chibata *et al.*, 1978b). Succinic acid formation could not be prevented by cell lysis, heat treatment, or freeze-thawing. The most effective treatment to suppress succinic acid formation and at the same time increase fumarase activity was to incubate the immobilized cells in 1 *M* sodium fumarate (pH 7.5) containing 0.3% bile extract for 20 h at 37°C. An increase in substrate flow rate in a continuous-column reactor increased the rate of fumarase inactivation (Yamamoto *et al.*, 1977).

In 1977 Tanabe Seiyaku Co., Ltd. replaced polyacrylamide gel-immobilized *B. ammoniagenes* with carrageenan-entrapped *Brevibacterium flavum* cells treated after immobilization with bile extract (Takata *et al.*, 1979, 1980; Chibata *et al.*, 1980b). The productivity

$$P_t = \int_0^t E_0 \exp(-k_d t)\, dt \tag{4}$$

where E_0 = initial enzyme activity, k_d = activity decay constant, and t = operation period) was increased about five times in comparison to polyacrylamide-immobilized *B. ammoniagenes*. The operational stability with *B. ammoniagenes* at 37°C was $t_{1/2}$ = 53 days (polyacrylamide) and 75 days (carrageenan), and with *B. flavum* 94 days (polyacrylamide) and 160 days (carrageenan), respectively, further illustrating the superiority of κ-carrageenan as a carrier. Yang and Zhong (1980) reported a half-life of 95 days with polyacrylamide-immobilized fumarase-active *C. rugosa* at 30°C. According to Takata *et al.* (1980), theoretically 24.1 tonnes of L-malic acid is obtained from 1 *M* sodium fumarate with a 1000-liter column reactor in one month at a residence time $\tau \sim 3.3$ h. The yield of L-malic acid in the industrial process is about 70%.

C. Urocanic Acid

Urocanic acid is produced from L-histidine by L-histidine ammonia lyase-active *Achromobacter liquidum* by classical fermentation techniques to be used as a sunscreening agent by the pharmaceutical industry (Chibata, 1978a). In a patent assigned to Tanabe Seiyaku Co., Ltd., a method is described to produce urocanic acid with polyacrylamide gel-immobilized *A. liquidum* cells (Chibata *et al.*, 1975e). As with β-galactosidase, the stability of the enzyme from different microorganisms during immobilization varies considerably. Thus, although the initial activity of intact *Micrococcus ureae* cells was among the highest of several organisms tested, enzyme activity was almost totally lost during immobilization. Jack and Zajic (1977a), however, showed that *Micrococcus luteus* cells covalently bound to carboxymethylcellulose retained about 75% of L-histidine ammonia lyase activity, although cell viability was lost. The biocatalyst did not lose activity during 16 days of continuous processing of 0.25 *M* L-histidine. In this case the high costs of 1-ethyl-3-(3′-dimethylaminopropyl)carbodiimide hydrochloride (EDAC) was a limiting factor.

One of the problems encountered with the immobilized *A. liquidum* cells was that the cells also exhibited urocanase activity, converting urocanic acid further to imidazolone propionic acid, but urocanase could be quantitatively inactivated by a simple heat treatment for 30 min at 70°C, without interfering with urocanic acid production (Shibatani *et al.*, 1974). Consequently, it was possible to convert L-histidine containing Mg^{2+} ions at pH 9 continuously quantitatively to urocanic acid, using a column reactor packed with immobilized *A. liquidum* cells. Urocanic acid of high purity could be obtained from the effluent without recrystallization by adjusting the pH to about 4.7. Magnesium ions stabilized the biocatalyst, which had a half-life of about 6 months with no decrease in activity during 40 days of operation. According to Jack and Zajic (1977a), no magnesium was required with covalently bound *M. luteus* cells. In both cases, however, nearly 100% conversion of 0.25 *M* L-histidine was obtained at the relatively high residence time of $\tau \sim 17$ h.

Kan and Shuler (1978a,b) used *Pseudomonas fluorescens* cells entrapped in a hollow-fiber dialyzer reactor for the production of urocanic acid. Less than 10% of enzyme activity was lost during 1 week of operation.

D. Other Organic Acids Produced by Immobilized Microbial Cells

1. Lactic Acid

Lactic acid has been produced by fermentation since 1881 (Lockwood, 1979). Considerable quantities of lactic acid are also obtained by hydrolysis of lactonitrile. The first continuous-fermentation process for lactic acid was reported by Whittier and Rogers (1931). Friedman and Gaden (1970) retained *Lactobacillus delbrueckii* cells in a dialysate reservoir to study the growth and kinetics of lactic acid formation. Setti (1975) employed a reverse-osmosis unit connected to a fermenter for the selective permeation of the product with nearly quantitative retention of substrate, and presented a mathematical model for the production of lactic acid from cheese whey. Keller and Gerhart (1975) obtained more than 98% yield from whey lactose in a two-stage continuous-fermentation system. Griffith and Compere (1975) sprayed a mixture of 10% (w/v) gelatin and 1% (w/v) Nalco 8172 polyelectrolyte on tower fermenter packing such as berl saddles, followed by cross-linking with 5% glutaraldehyde, for better attachment of microorganisms. With *Lactobacillus* sp. attached to such packing in a column reactor, lactic acid content of whey (5% d.m.) was increased from 1.2 to 2.1% in a single pass (Compere and Griffith, 1975) through a 5 × 183-cm column reactor operated continuously for several weeks at about 10–20 h residence time. Lactic acid produced could be removed from the effluent by cation-exchange resin column. The process was also applied to produce lactic acid from fresh hardwood molasses of about 6% (w/v) sugar content, a by-product of cellulose manufacture. When calcium lime was used as buffer, about 3% (w/v) lactic acid was obtained with a residual sugar level of only 0.25% (Griffith and Compere, 1977).

We have investigated lactic acid production from glucose by heterogeneous biocatalysis employing several different lactic acid-producing bacteria. When *Lactobacillus lactis* cells were entrapped in calcium alginate gel beads (6–8% w/v sodium-alginate, 0.5 *M* $CaCl_2$, biocatalyst bead $\phi \sim 2$ mm), cells remained living and active to produce lactic acid from glucose (Linko, 1980, 1981a). In later experiments, *L. delbrueckii* was used for improved productivity (Linko, 1981b; Stenroos *et al.*, 1982). Packed-bed column reactors were operated both as a single pass and as a recycle loop. Finely powdered calcium carbonate was added to the substrate solution as buffer. The biocatalyst activity half-life in single-pass continuous-column operation was about 50 days. The same biocatalyst bed could, however, be revived and reused several times without noticeable decrease in activity, even after several weeks of storage at 4°C. Carbon dioxide formed was allowed to escape freely, and caused no major difficulties in packed-bed column operations. In batch or loop-reactor experiments, pH was also automatically controlled by the addition of sodium hydroxide. The same reactor could be used for at least 32 recycle-batch conversions with little loss in activity, further illustrating the excellent operational stability. In continuous-column operations with 4.8% (w/v) glucose (1% yeast extract, 4.8% $CaCO_3$, 43°C, pH 5.7), up to 93% conversion to lactic acid was obtained at $\tau > 8$ h. Typically, 90–95% total lactic acid formed L-lactic acid. Figure 10 shows typical results of lactic acid production by calcium alginate-entrapped *L. delbrueckii* cells. In addition to the production of lactic acid, such immobilized-whole-cell reactors can be employed in a number of food industry applications. Thus Divies (1977) demonstrated that the production of

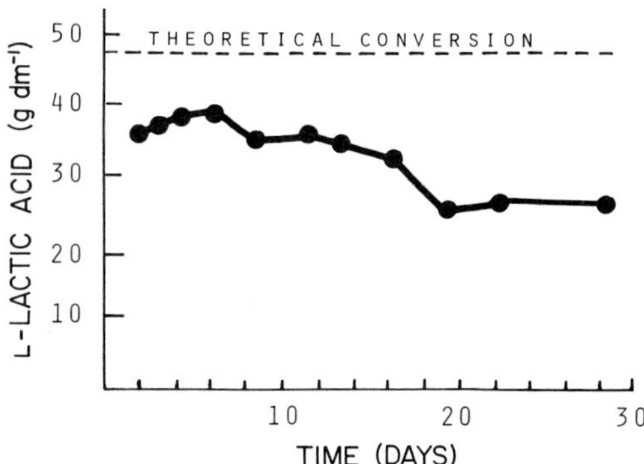

Fig. 10. Continuous production of L-lactic acid with calcium alginate bead-entrapped *Lactobacillus delbrueckii* cells (4.8% w/v glucose, pH 5.7, 43°C, $\tau \sim 10$ h).

yogurt from milk by polyacrylamide gel-entrapped *Lactobacillus bulgaricus* and *Streptococcus thermophilus* is possible.

2. Citric Acid

Citric acid is widely used in both the food and pharmaceutical industries, and it is among the few bulk chemicals produced by fermentation today. Koji, liquid surface culture, and submerged-culture techniques are all used as batch processes (Lockwood, 1979a). The rate of aeration is not critical, but continuous oxygen supply is necessary, inasmuch as a disturbance of even a few seconds in oxygen availability rapidly stops fermentation. Citric acid production is sensitive to traces of Fe^{2+}, Co^{2+}, and Ni^{2+} ions, and Cu^{2+} ion can be used as a counterion. According to Lockwood (1979), continuous-culture techniques have not been considered suitable for citric acid production, mainly because growth and metabolite production do not coincide. Nevertheless, attempts at single-stage continuous fermentation have been made with varying success (Choudhari and Pirt, 1966; Kristiansen and Sinclair, 1979, 1981; Kristiansen and Charley, 1981).

Few reports on immobilized whole cells for citric acid production have been published. Vieth and Venkatasubramanian (1978, 1979) attached *Aspergillus niger* mycelial pellets to collagen membrane with good activity retention. The membrane was used as a spirally wound multipore biocatalyst in a plug-flow-type reactor specially constructed to allow countercurrent flow of oxygen for the maintenance of 80–90% saturation level. The estimated half-life of the biocatalyst was 6–8 days, and depending on process conditions 8–40% yield of citric acid from sucrose was obtained. *A. niger* was found to be capable of reproduction both on and within the carrier matrix, preserving the cofactor-regeneration mechanism *in situ*. Limiting the nitrogen level in the substrate solution resulted in decreased cell reproduction and citric acid production. Maximum catalytic activity was obtained on immobilizing the mycelium at a proper physiological state, typically after about 72–96 h of batch fermentation. The immobilized cells exhibited about half the activity of intact free mycelium. Briffaud and Engasser (1978) grew *Saccharomycopsis lipolytica* yeast on wood chips used as trickle-flow reactor packing. In contrast to free mycelium, growth phase activated by ammonia was linear and was followed by acid production by both old and new cells. The yield of citric acid was reduced, however, by about 30% in comparison to the intact mycelium. Attempts to produce citric acid continuously by polyacrylamide-entrapped *Candida lipolytica* yeast have failed (Stottmeister, 1979; Berger and Langhammer, 1980).

In our own preliminary experiments citric acid was continuously produced by *A. niger* pellets entrapped in calcium alginate gel beads using a CSTR air-lift-type fermenter (Linko, 1981b). In typical runs with 10% (w/v) sucrose (0.03% NH_4NO_3, 0.005% $MgSO_4 \cdot 7H_2O$, 0.0001% $ZnSO_4 \cdot 7H_2O$, 0.37% $CaCl_2 \cdot 2H_2O$, 0.001% $K_4Fe(CN)_6 \cdot 2H_2O$) as substrate, about 12 g dm^{-3} yield was obtained at τ = 26–30 h, as compared with the maximum of 85–90 g dm^{-3} in corresponding batch fermentations in about 10 days. The overall fermentation efficiency in 2 weeks was 42%, with peak efficiency of 69%. The maximum rate of citric acid production obtained with the immobilized-whole-cell biocatalyst was 70 mg g^{-1} h^{-1}, about five times higher than that obtained in batch fermentations with free pellets. A small quantity of ammonium nitrate was found necessary for optimum production. Citric acid may be recovered by ion-exchange resin, and the residual substrate recycled to the process.

3. Gluconic Acid

Calcium gluconate has long been used as the source of calcium in medical and veterinary practice, and ferrous gluconate in iron therapy. Glucono-δ-lactone is widely used in baking powders and other food applications. A major use for sodium gluconate is to complex metal ions in bottle washing, and to control setting time when added to cement. Most gluconic acid is today made by submerged fermentation utilizing the ability of *Aspergillus niger* or *Gluconobacter suboxydans* to oxidize glucose to gluconic acid (Miall, 1978; Lockwood, 1979). Direct chemical oxidation is also possible.

Lee and Long (1974), in a patent assigned to R. J. Reynolds Tobacco Co., described a method for producing biocatalysts by flocculating whole microbial cells with polyelectrolytes. With *A. niger* pellets such biocatalysts could be used to convert invert sugar continuously to gluconic acid. In another patent (Nelson, 1976) assigned to Pfizer, Inc., *A. niger* pellets were covalently bound to water-insoluble glycidyl ester copolymers. In another type of approach, Karube *et al.* (1977a) employed glutaraldehyde to link glucose oxidase enzyme to *Aspergillus* sp. mycelia pellets. D'Souza and Nadkarni (1980c) used concanavalin A to bind *A. niger* glucose oxidase to *Saccharomyces cerevisiae* yeast cells, which were induced for maximal invertase and catalase activities. The multienzyme biocatalyst was used in a CSTR system to convert sucrose to fructose and gluconic acid, as an alternative to isomerizing glucose partially to fructose to be recycled to the fructose-separating process. The multienzyme whole-cell reactor exhibited no decrease in activity over a period of 20 days (10% sucrose in

0.5 *M* phosphate buffer, pH 6, 28°C, $\tau \sim 5$ h). In our own experiments (Linko, 1981b) *A. niger* pellets were entrapped in calcium alginate gel beads used in various reactor configurations to produce gluconic acid from glucose. In an example process, 15% (w/v) glucose (0.0185% corn steep liquor, 0.0017% $MgSO_4 \cdot 7H_2O$, 0.002% KH_2PO_4, 0.0042% $(NH_4)_2HPO_4$, 0.0001 % urea) was fed as a mist together with a stream of air or oxygen in a trickle-bed recycle reactor. With pure oxygen, about 93% conversion was reached in 24 h (30°C, pH 6.2–6.5 with automatic pH control, recirculation rate about 5.5 b.v.h^{-1}). Recently, Tramper and van den Tweel (1981) entrapped *G. suboxydans* cells in calcium alginate gel for continuous gluconic acid production. The retention of activity on immobilization was good, but the oxidation reaction was strongly limited by oxygen transport.

4. Keto Acids

a. 2-Ketogluconic acid: 2-Ketogluconic acid is an intermediate in the production of erythorbic (isoascorbic) acid, used as an antioxidant in the food industry. It is produced by classical fermentation techniques (Miall, 1978). *Serratia marcescens* cells have been immobilized on collagen membrane, and subsequently cross-linked with 3% glutaraldehyde (Venkatasubramanian *et al.*, 1978), but the conversion of glucose to 2-ketogluconic acid was low, only 8–10% in 96 h. Erythorbic acid has also been reported to be directly produced from glucose by polyacrylamide gel-entrapped *Penicillium cyaneofulvum* (Kato, 1974).

b. 2-Keto-L-gulonic acid: 2-Keto-L-gulonic acid is an intermediate in the manufacture of vitamin C (L-ascorbic acid). Typically, the yield in 2-keto-L-gulonic acid fermentation is low, of the order of 5%. In a patent assigned to Hoffman-La Roche, Inc., a method is given to produce 2-keto-L-gulonic acid in one step by the oxidation of L-sorbosone (L-*xylo*-hexogulose) by *Pseudomonas putida* cells (Makover and Pruess, 1975). Martin and Perlman (1975, 1976a) showed, on the other hand, that immobilized cells of *Gluconobacter melanogenus* could convert L-sorbose to L-sorbosone. The rate of conversion to 2-keto-L-gulonic acid, however, was limited by L-sorbosone oxidase. Consequently, the process was later improved by coimmobilizing *G. melanogenus* and *Pseudomonas syringus* cells in polyacrylamide gel for direct oxidation of L-sorbose to 2-keto-L-gulonic acid (Martin and Perlman, 1976b). A fourfold increase in 2-keto-L-gulonic acid production was obtained.

c. α-Keto acids: α-Keto acids are used in medical practice, especially in the treatment of uremia. α-Keto acids can be produced by transamination, oxidation of corresponding amino acids with D-amino acid oxidase, and by chemical synthesis, but the yields are often low. Brodelius *et al.* (1980, 1981; Brodelius and Mosbach, 1980) have recently developed a method for obtaining α-keto acids based on the use of calcium alginate (2%) gel-entrapped *Trigonopsis variabilis* cells to oxidize amino acids. The hydrogen peroxide formed in the reaction was efficiently eliminated by coentrapping manganese oxide. Sufficient oxygen supply could be maintained by using a trickle-bed-type reactor.

5. Salicylic Acid

Salicylic acid is normally manufactured by chemical synthesis. Yields in conventional fermentation processes have been low. Kitai and Ozaki (1969) introduced a process based on continuous production in a tower fermenter equipped with several perforated plates to produce salicylic acid from naphthalene by *Pseudomonas fluorescens*. Abbott and Gerhardt (1970a,b), in contrast, employed continuous removal of the product by dialysis to improve productivity.

V. PRODUCTION OF ALCOHOLS

The interest in biotechnical ethanol production has significantly increased during the last few years, particularly in the United States and Brazil (Marion, 1979). The current state of the art in ethanol fermentation has been reviewed by Righelato (1980), and the applications of immobilized-yeast-cell technology by Kolot (1980). The development of acetone–butanol fermentation by Weizmann to the once most important industrial fermentation process has been discussed by Hastings (1978), and the technology reviewed by Walton and Martin (1979). The new developments in immobilized-whole-cell technology in alcohol fermentation have been summarized in Table VI.

A. Ethanol

In a batch ethanol process, fermentation is responsible for 80% or more of total capital costs. Consequently, a number of continuous processes have been developed and applied both in alcoholic beverage manufacture (Hough and Button, 1972) and in industrial ethanol production (Hospodka, 1966; Rosèn, 1978). Total biotechnical ethanol-production costs are largely determined by the price of fermentable carbohydrates (Kolot, 1980). Consequently, much attention is currently focused on the utilization of inexpensive waste materials and by-products such as cellulose and whey as feedstocks (Y.-Y. Linko *et al.*, 1981b), and to the improvement of the overall economics of the process (Dellweg and Misselhorn, 1978; Marion, 1979).

TABLE VI
ALCOHOL FERMENTATIONS WITH IMMOBILIZED CELLS

Organism	Method of immobilization	References
Ethanol (beer)		
Saccharomyces sp.	Inert carrier	Intermag Getränke Technik A. G. (1969); Berdelle (1975)
Saccharomyces sp.	Yeast on layered filter elements	Narziss and Hellig (1971, 1972); Berdelle (1972)
Saccharomyces amurcae and *Saccharomyces cerevisiae*	Adsorption or covalent binding by isocyanate on porous glass, silicates, etc.	Messing and Oppermann (1979)
Saccharomyces carlsbergensis	Adsorption on PVC, porous brick. Kieselguhr, etc.	Moll *et al.* (1975); Navarro *et al.* (1976); Corrieu *et al.* (1976)
S. carlsbergensis	Adsorption on diatomaceus earth	Grinbergs *et al.* (1977)
S. carlsbergensis	Covalent binding by silanization (γ-amino-propyl-trimethoxysilan) or by glutaraldehyde on porous silica	Navarro and Durand (1977)
S. carlsbergensis	Entrapment in polyethylene, cellophane, etc.	Kolpakchi *et al.* (1976)
S. cerevisiae	Adsorption on Kieselguhr	Baker and Kirshop (1973)
S. cerevisiae	Entrapment in Ca alginate gel flocks	White and Portno (1978)
S. cerevisiae	Entrapment in Ca alginate gel beads	Linko (1980, 1981a,b); Y.-Y. Linko and Linko (1981a,b)
S. cerevisiae	Adsorption on Celite or entrapment in Ca alginate	Chiou (1979)

Ethanol (wine)		
Saccharomyces sp.	"Slant-tube" fermenter (sherry wine)	Wick and Poppe (1977)
Saccharomyces sp.		Khoroshilova (1978)
S. cerevisiae	Entrapment in polyacrylamide gel	Divies (1977)
S. cerevisiae	Entrapment in Ca alginate beads	Y.-Y. Linko and Linko (1981b)
Ethanol (potable and industrial)		
Kluyveromyces fragilis	Entrapment in polyacrylamide	Villet *et al.* (1979)
K. fragilis	Entrapment in Ca alginate (6–8%) gel	Y.-Y. Linko *et al.* (1981)
Saccharomyces sp.	Adsorption on gelatin coating	Griffith and Compere (1975, 1978)
S. carlsbergensis	Retention of flocculating yeast with a settling device	Englebart and Dellweg (1976) Sitton *et al.* (1981)
S. carlsbergensis	Entrapment in κ-carrageenan	Chibata (1978a, 1980a); Wada *et al.* (1979, 1980a,b, 1981)
S. cerevisiae	Yeast retention with a settling device for vacuum fermentation	Cysewski and Wilke (1977, 1978)
S. cerevisiae	Entrapment in Ca alginate (2%) gel	Kierstan and Bucke (1977)
S. cerevisiae	Entrapment in Ca alginate (2%) gel and partial drying	Y.-Y. Linko and Linko (1981a,b)
S. cerevisiae with magnetic particles	Coentrapment in Ca alginate (2%) gel	Larsson and Mosbach (1979); Larsson *et al.* (1981)
S. cerevisiae with β-glucanase (to ferment cellobiose)	Coentrapment in Ca alginate (2%) gel	Hägerdal and Mosbach (1980); Lòpez-Leiva *et al.* (1981)
S. cerevisiae	Entrapment in Ca alginate (2%) gel and treatment with polyamine	Birnbaum *et al.* (1981)
S. cerevisiae	Entrapment in Ca alginate (2.4–3.8%) gel	Krouwel and Kossen (1981)
S. cerevisiae	Entrapment in Ca alginate (6–8%) gel	Linko (1980, 1981a,b); Y.-Y. Linko and Linko (1981a,b); Y.-Y. Linko *et al.* (1981)

(continued)

TABLE VI (*Continued*)

Organism	Method of immobilization	References
S. cerevisiae with β-galactosidase (to ferment lactose)	Entrapment in Ca alginate (6–8%) gel; β-galactosidase (*A. niger*) bound on phenol formaldehyde resin	Linko (1981a,b)
S. cerevisiae with cellulases (to ferment cellulose and cellobiose)		Hartmeier (1981)
S. cerevisiae	Entrapment in agar gel	Krouwel and Kossen (1980); Margalith and Holcberg (1981)
S. cerevisiae	Adsorption on pretreated inert support	Ghose and Bandyopadhyay (1979)
S. cerevisiae	Adsorption on wood chips	Moo-Young *et al.* (1980a,b); Gencer and Mutharasan (1981)
S. cerevisiae	Adsorption on ion-exchange resins	Daugulis *et al.* (1981)
Zymomonas mobilis	Entrapment in Ca alginate or κ-carrageenan	Grote *et al.* (1980)
Z. mobilis	Adsorption on borosilicate glass fiber pads	Arcuri *et al.* (1980, 1981)
Z. mobilis	Entrapment in cellulose nitrate membrane, κ-carrageenan, or Ca alginate	Margaritis and Rowe (1981); Margaritis *et al.* (1981)
Z. mobilis	κ-Carrageenan–locust bean gum blend	P. Linko and Linko (1981, 1982)
Isopropanol		
Clostridium butylicum	Entrapment in Ca alginate gel	Krouwel *et al.* (1980)
n-Butanol		
Clostridium acetobutylicum	Entrapment in Ca alginate gel	Häggström and Molin (1980, 1981)
C. butylicum	Entrapment in Ca alginate gel	Krouwel *et al.* (1980)
2,3-Butanediol		
Enterobacter aerogenes	Entrapment in κ-carrageenan gel	Chua *et al.* (1980)

1. Fuel and Industrial Ethanol

Engelbart and Dellweg (1976) developed a settling device that permitted the retaining of yeast flocks in the reactor during continuous removal of the liquid, resulting in increased productivity. The reactor was optimally operated at a dilution rate of 0.1–0.5 h^{-1}, with a minimum oxygen availability of 0.2–0.4 mg h^{-1} g^{-1} (yeast, d.m.) for maximum productivity. Another approach was investigated by Cysewski and Wilke (1978). In a detailed study on combining cell recycle with continuous removal of ethanol under vacuum (50 mm Hg), they were able to increase the productivity from 7 g $liter^{-1}$ h^{-1} (CSTR, 10% w/v glucose) to 82 g $liter^{-1}$ h^{-1} (CSTR with vacuum system and cell recycle, 33.4% w/v glucose) at 35°C, pH 4. A "Bio-Reactor" (Intermag, 1969; Berdelle, 1972, 1975) based on a bed of yeast admixed with an inert carrier was recently applied to continuous fermentation of glucose to ethanol (Grinbergs *et al.*, 1977). In a patent assigned to the United States Department of Energy (Griffith and Compere, 1978), a similar process is described. A column reactor is filled with conventional distillation-column packing such as berl saddles coated with gelatin in the presence of polyelectrolyte and a cross-linking agent. The packing was dried and baked at 60–80°C for 6–8 h, and mixed with yeast culture, which was allowed to grow during a 1-month start-up period. Stable yeast population could be maintained for extended periods by restricting nutrients, and adding a small quantity of membrane-disrupting detergent to the substrate feed in order to make dead cells available as nutrient for the remaining yeast population. The process was claimed to be stable at glucose levels of 5–30% (w/v).

In another patent assigned to Sanraku-Ocean Co., Ltd. (Hino *et al.*, 1979), baker's yeast (*Saccharomyces cerevisiae*) was entrapped under mild conditions in a natural or synthetic hydrophilic gel prepared from agar, albumin, alginate, cellulose and its derivatives, chitosan, dextran, gelatin, guar gum, locust bean gum, mannan, pullulan, starch and its derivatives, tragaganth, xanthan, polyethyleneimine, poly(ethylene glycol), or poly(vinyl alcohol), the preferred compound being poly(vinyl alcohol). With relatively inexpensive wood shavings or chips as support for *S. cerevisiae* cells, about 21 g $liter^{-1}$ h^{-1} productivity at $\tau \sim 3$ h has been reported (Ghose and Bandyopadhyay, 1979; Moo-Young *et al.*, 1980a,b). Yeast has also been entrapped in polyacrylamide gel (Divies, 1977b; Villet *et al.*, 1979), alginate (Kierstan and Bucke, 1977; Larsson and Mosbach, 1979; Linko, 1980, 1981a,b; Grote *et al.*, 1980; Y.-Y. Linko *et al.*, 1981; Y.-Y. Linko and Linko, 1981a,b), agar (Margalith and Holcberg, 1981), and κ-carrageenan (Wada *et al.*, 1979, 1980a,b, 1981; Chibata, 1980a; Grote *et al.*, 1980).

According to Wada *et al.* (1980b), yeast grew well on κ-carrageenan, and high cell densities of about 4–6 × 10^9 cells/ml gel could be easily maintained in continuous operation for extended periods. After increasing substrate glucose level stepwise from 10 to 25% (w/v), 114 g liter^{-1} ethanol concentration could be maintained at least for 2 months at τ = 2.6–3.6 h (Wada *et al.*, 1981), with about 32–44 g liter^{-1} h^{-1} productivity. According to Margalith and Holcberg (1981), yeast cells immobilized in agar can tolerate higher ethanol levels than free cells in suspension, and about 14% (w/w) ethanol from 30% glucose was claimed, in comparison with only 10% with free cells. Griffith and Compere (1975, 1978) also claimed a maximum of about 122 g liter^{-1} ethanol (at ~87% yield) from 30% glucose at τ = 5.5 h, corresponding to about 22 g liter^{-1} h^{-1} productivity.

In our own experiments, 10% (w/w; ~20% d.m.) of yeast cells (*S. cerevisiae, Kluyveromyces fragilis*) suspended in 6–8% sodium alginate were extruded through hollow needles (ϕ = 0.6 mm) into 0.5 *M* calcium chloride to obtain biocatalyst beads (ϕ ~2.3 mm, cell density 2–5 × 10^9 cells/g). The high alginate concentration of 6–8%, as compared to 2% used by others (Kierstan and Bucke, 1977; Larsson and Mosbach, 1979), markedly improved biocatalyst stability. Birnbaum *et al.* (1981) have recently reported the improvement of calcium alginate stability against phosphates by polyamine treatment. We obtained significant improvement in both physical characteristics and productivity by simply drying the alginate beads to about 28% solids. Reactor sizes of up to several liters have been used, and a typical laboratory-scale process is shown in Fig. 11. Using cane molasses (17.5% w/v sugar, 25°C, pH 4.8) as substrate, productivities of 19–22 g liter^{-1} h^{-1} with undried (9% solids, τ ~ 4 h) and of 39–43 g liter^{-1} h^{-1} with partially dried (28% solids, τ ~ 2 h) biocatalyst were obtained with nearly 100% conversion. When pure glucose was used as substrate, production rate decreased rapidly, but the reactor could be easily revived by intermittent addition of nutrients under aeration (Linko, 1980). Lactose in cheese whey could be completely converted to ethanol either by employing a two-biocatalyst reactor with immobilized *S. cerevisiae* cells and phenol formaldehyde resin-bound *Aspergillus niger* β-galactosidase (Linko, 1981a,b) or by using entrapped *K. fragilis* cells (Y.-Y. Linko *et al.*, 1981). Only one other report on immobilized *K. fragilis* for whey lactose fermentation has been published (Villet *et al.*, 1979), but only kinetic measurements in batch systems were performed. Figure 12 illustrates our results using demineralized whey as substrate for immobilized *K. fragilis* in continuous ethanol production.

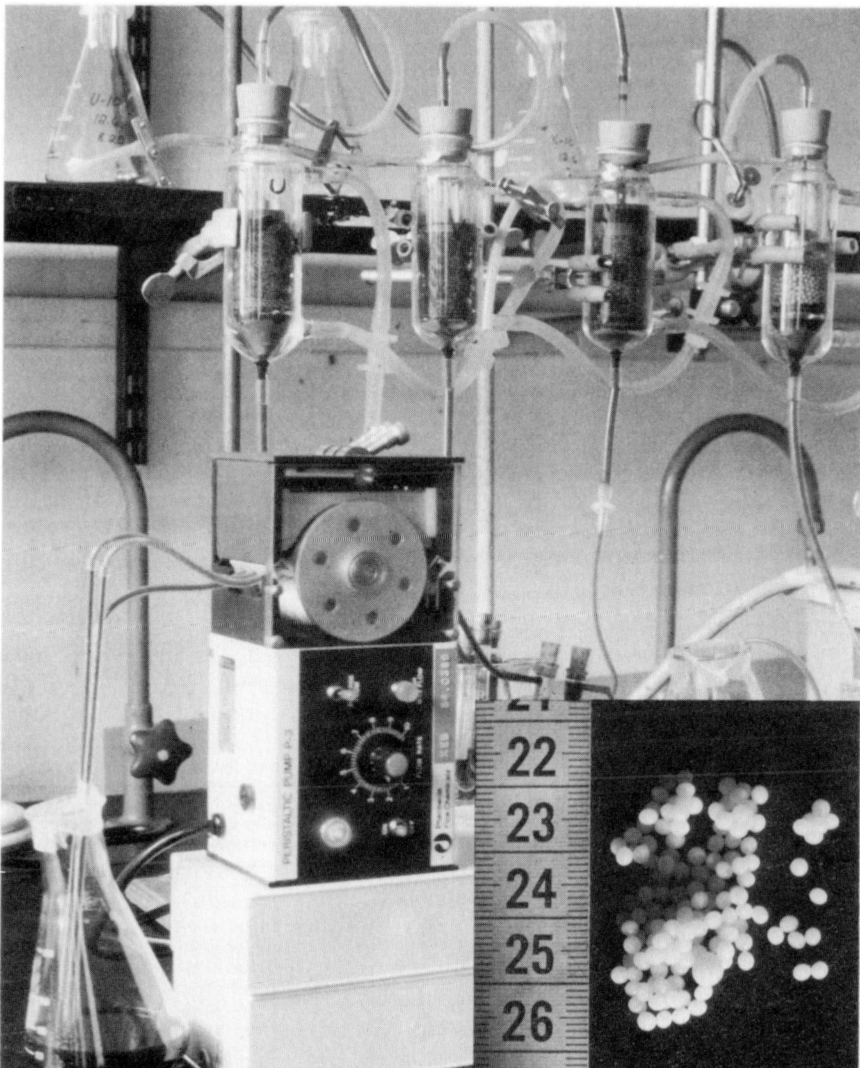

Fig. 11. Typical laboratory-scale continuous immobilized *Saccharomyces cerevisiae* cell reactors for ethanol production. Inset, enlargement of calcium alginate biocatalyst beads.

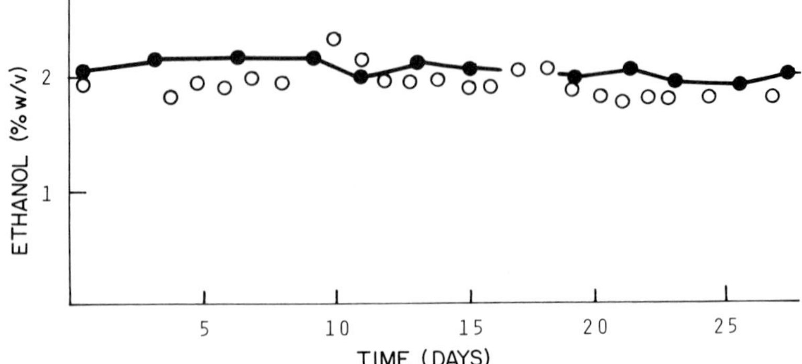

Fig. 12. Ethanol production from demineralized whey (5% w/v lactose) with calcium alginate-entrapped *Kluyveromyces fragilis* (●) yeast, and with calcium alginate-entrapped *Saccharomyces cerevisiae* yeast in a two-biocatalyst reactor with phenol formaldehyde resin-bound *Aspergillus niger β-galactosidase* (○) (25°C, pH 4.5, $\tau \sim 3.9$ h).

Economic production of ethanol from whey has been demonstrated (Hansen, 1980).

Recently also *Zymomonas mobilis* cells have been immobilized by a number of techniques for ethanol production (Arcuri *et al.*, 1980a,b); Grote *et al.* 1980; Margaritis *et al.*, 1981; P. Linko and Y.-Y. Linko, 1981, 1982). Productivities reported in laboratory-scale experiments have ranged from about 50 to 150 g liter^{-1} h^{-1}, with mean effective liquid residence times varying from about 15 min to 4 h, product ethanol levels from about 50 to 60 g liter^{-1}, and yields from about 80–90%. Although higher productivities than with yeast are obtainable under laboratory conditions, there are several problems to be solved before large-scale applications may be visualized. Grote *et al.*, (1980) reported poor biocatalyst stability, with about 30% of the activity lost within a month of continuous operation. *Z. mobilis* ferments only glucose, fructose, and sucrose, and thus cannot be utilized to convert maltose or starch to ethanol. We have been able to overcome difficulties associated with the very high rate of carbon dioxide evolution by employing κ-carrageenan–locust bean gum blend gel-entrapped cells in a specially designed tapered reactor. Nevertheless, Unisearch Ltd. of Australia has been reported to offer for licensing their process based on a special genetically engineered *Z. mobilis* strain (Anonymous, 1981).

A novel approach for continuous automatic control of ethanol fermentation has been reported by Mandenius *et al.* (1980, 1981), employing an enzyme thermistor system with immobilized glucose oxidase and catalase for glucose, and immobilized invertase for sucrose monitoring in ethanol fermentation by immobilized yeast. The principle is illustrated in Fig. 13. A two-enzyme reactor with immobilized living yeast and covalently bound β-glucosidase has been used for production of ethanol from cellobiose (Hägerdal and Mosbach, 1980; Lopéz-Leiva *et al.*, 1981).

2. Alcoholic Beverages

Slow-growing highly flocculent yeasts have been employed for continuous production of beer in a tower fermenter (Rivière, 1977), and continuous tower fermentation of cider and wine has also been reported (Anonymous, 1979). Intermag Getränke Technik A. G. (1969; Berdelle, 1972, 1975) developed a process for continuous brewing employing a fixed-bed reactor with yeast attached to an inert carrier. A somewhat different approach was presented by Narziss and Hellig (1971, 1972), who obtained stable beer in less than 2 days with yeast deposited as porous layers on vertical filter elements. Baker and Kirsop (1973) reported high beer output with a yeast–kieselguhr plug reactor, but production rate decreased to about half in a month. Others have developed similar processes for continuous brewing (Moll *et al.*, 1975, 1977; Navarro *et al.*, 1976; Corrieu *et al.*, 1976; Kolpakchi *et al.*, 1976). Chiou (1979) reported that beer indistinguishable from conventionally brewed beer could be produced with *Saccharomyces cerevisiae* immobilized either in alginate or on Celite. We observed that barley malt wort was an ideal substrate for alginate-entrapped *S. cerevisiae,* and nearly 100% conversion with a constant ethanol level of 4.5 g liter^{-1} at $\tau \sim 2$ h, and 3.8 g liter^{-1} at $\tau \sim 1$ h could be maintained for at least 3 months with productivities of up to 38 g liter^{-1} h^{-1}. A total reactor volume of 10 m^3 was calculated to suffice for annual processing of 30 million liters of beer.

As another example from our laboratory, Fig. 14 illustrates that grape juice could also be continuously processed with no observable decrease in reactor performance during 3 weeks. Production of wine by immobilized living yeast has also been reported by Divies (1977a,b), and by Wick and Popper (1977). The utilization of immobilized yeast in sherry wine production has been reported to improve quality (Khoroshilova, 1978).

An interesting approach for the production of ethanol for beverages has been published by Hough and Lyons (1972), who grafted glucoamylase on yeast cells for simultaneous saccharification of starch and ethanol fermentation.

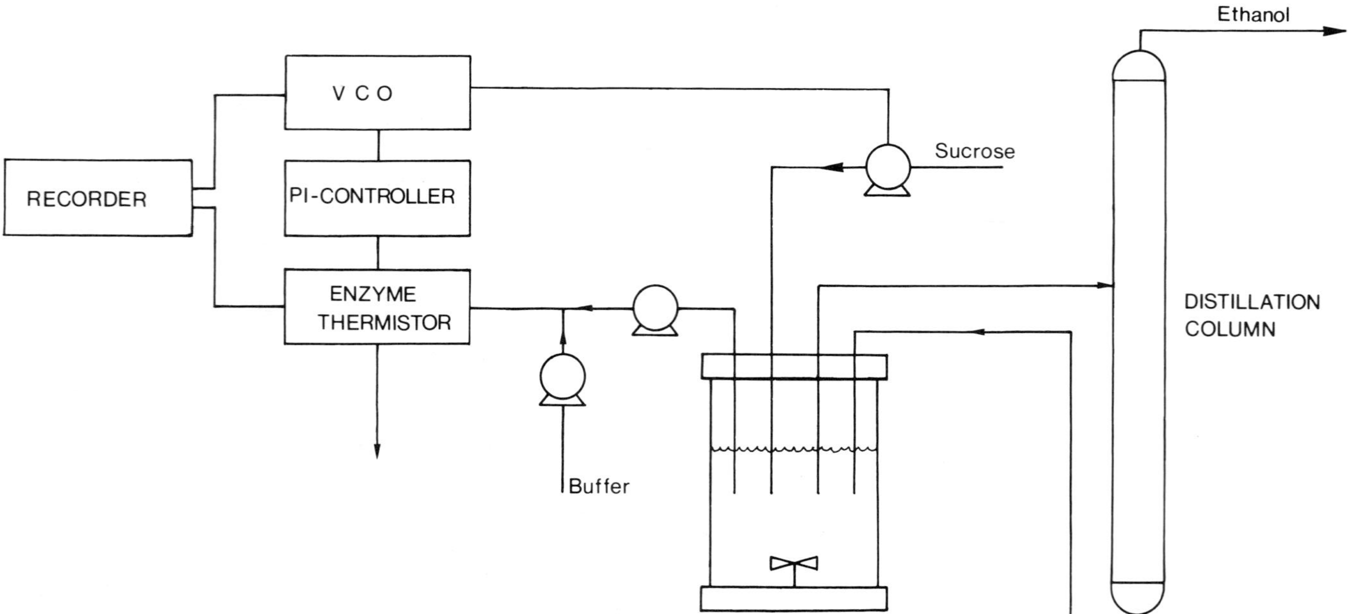

Fig. 13. A process-control system applicable to continuous ethanol production with an immobilized-yeast reactor. (Courtesy of Dr. Mandenius, University of Lund, Sweden.)

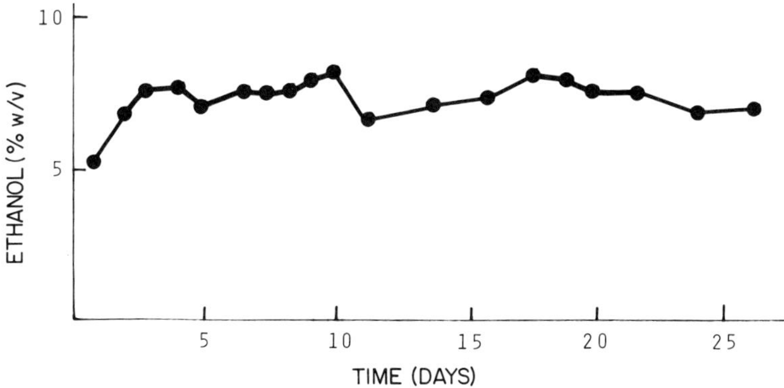

Fig. 14. Continuous production of red wine with calcium alginate bead-entrapped yeast cells (25°C, pH 4.8).

B. Other Alcohols

1. Isopropanol

Spores of a *Clostridium butylicum* strain capable of producing isopropanol and *n*-butanol with only traces of acetone have been entrapped in calcium alginate gel for solvent production (Krouwel *et al.*, 1980). The biocatalyst beads (9.4 g wet biomass/g alginate) were placed in a conical column to avoid problems associated with gas evolution. From 6% (w/v) glucose (1% yeast extract, 0.5% $CaCl_2 \cdot 2H_2O$), 2 g liter^{-1} of isopropanol could be obtained continuously for at least 9 days (37°C, $\tau \sim 3.5$ h). The mean productivity was about 0.5 g liter^{-1} h^{-1}. The total yield of isopropanol and *n*-butanol was 30–40%, with ethanol <0.1 g liter^{-1}, acetone <0.15 g liter^{-1}, and traces of acetic and butyric acids. Biomass level in products stream was about 0.85 g liter^{-1}, in comparison with about 2 g liter^{-1} in conventional batch fermentation.

2. *n*-Butanol

Krouwel *et al.* (1980) obtained mean productivity of 1 g liter^{-1} h^{-1} *n*-butanol with calcium alginate gel-immobilized *Clostridium butylicum* spores, isopropanol being the only major by-product. The *n*-butanol concentration was always <5 g liter^{-1} regardless of initial glucose level, obviously because of strong product inhibition (37°C, $\tau = 3.5$ h). Häggström and Molin (1980) and Häggström (1981) used alginate gel to entrap both *Clostridium acetobutylicum* spores and vegetative cells. If vegetative cells were immobilized in the solvent-producing phase, solvent production continued for extended periods, but the production rate was low. The yield of *n*-butanol with immobilized nongrowing cells was 17.6% (w/w) in a 48-h batch experiment, and that of acetone 3.9%. The productivity with immobilized spores was 0.48–2.8 g liter^{-1} h^{-1}, in comparison with about 0.3 g liter^{-1} h^{-1} in conventional batch fermentations (Spivey, 1978). Immobilized *Cl. butylicum* spores have also been employed in aseptic starting of a fermenter for butanol fermentation (Krouwel *et al.*, 1981).

3. 2,3-Butanediol

Enterobacter aerogenes cells capable of producing 11 g liter^{-1} of 2,3-butanediol from 5% glucose in 24 h fermentation were immobilized in κ-carrageenan and activated in the medium for growth in the matrix (Chua *et al.*, 1980). The maximum activity was reached in 1 day, after which it remained stable. It was observed that the immobilized cells were less sensitive to incubation conditions than free cells. No loss in activity was observed in 10 days with 5% glucose at 30°C, pH 7.

VI. CHEMICAL TRANSFORMATIONS FOR MEDICAL AND PHARMACEUTICAL APPLICATIONS

Immobilized-biocatalyst technology is becoming increasingly important in medical and pharmaceutical applications. Chang (1977) has described the preparation and possibilities of biologically active artificial cells by microencapsulation, including entrapment of whole microbial cells for therapeutic purposes. However, the most important current applications involve biosynthesis and transformations of fine chemicals, many of which would be difficult, uneconomical, and even impossible to obtain by current conventional methods. One of the most promising areas of biocatalysis is the microbial transformation of steroids, which has been reviewed by Kieslich and Sebek (1979) and by Kieslich (1980). Another important area is the biomodification of antibiotics for subsequent preparation of synthetic derivatives (Shibata and Uyeda, 1978; Sebek, 1980). Recent developments in the applications of immobilized whole cells in this field have been summarized in Table VII.

A. Steroid Transformations

According to Lilly (1978), Squibb was the first to commercialize in 1964 a steroid-conversion process employing heat-treated whole-microbial-cell steroid dehydrogenase. Cheetham (1980) reported that the industrial production of prednisolone by polyacrylamide-entrapped whole cells was started in Sweden in 1978 by Δ^1 dehydrogenation of hydrocortisone. Prednisolone is up to five times as active as hydrocortisone.

Mosbach and Larsson (1970) employed polyacrylamide gel-immobilized *Curvularia lunata* lichen cell 11β-hydroxylase to convert 11-deoxycortisone (Reichstein's compound S) to cortisol, which was subsequently oxidized to prednisolone by immobilized *Corynebacterium simplex* Δ^1-dehydrogenase. This was one of the first applications of the combined use of immobilized whole cells and an immobilized enzyme as a multienzyme system. In a later modification, *C. lunata* mycelium and *Arthrobacter simplex* cells were coentrapped in polyacrylamide (Larsson and Mosbach, 1976). Larsson *et al.* (1978) showed that the immobilized-cell reactor could be reactivated by the addition of nutrients. Frozen, activated preparations of immobilized *A. simplex* showed only little loss in activity (Ohlson *et al.*, 1978). *A. simplex* cells were later also entrapped in calcium alginate gel in the living state (Ohlson *et al.*, 1979, 1980a,b).

Skryabin *et al.* (1975) and Koshcheyenko *et al.* (1976) have also investigated the possibility of utilizing polyacrylamide gel-immobilized fungal and bacterial cells for the production of prednisolone.

Venkatasubramanian *et al.* (1978) employed collagen membrane-immobilized *Corynebacterium simplex* cells to convert cortisol to prednisolone in a partially nonaqueous system. Cortisol was dissolved in an organic solvent and dispersed as microdroplets in a buffered aqueous immobilized-cell slurry to maintain maximal dissolved substrate level. Almost 80% initial conversion was obtained at $\tau = 0.5$ h. The biocatalyst half-life was estimated to be about 130 h.

Fukui and co-workers (Omata *et al.*, 1979b; Sonomoto *et al.*, 1979; Tanaka *et al.*, 1979; Fukui *et al.*, 1980a,b,c) have employed photo-cross-linkable prepolymers to entrap *A. simplex* cells to convert hydrocortisone to prednisolone, and *Nocardia rhodochrous* cells to carry out a number of Δ^1 dehydrogenations in a water-saturated benzene–*n*-heptane system. Both hydrophilic and hydrophobic photo-cross-linkable and urethane prepolymers were employed in converting 3β-hydroxy-Δ^5-steroids to corresponding 3-keto-Δ^4-steroids in water-saturated organic solvent systems. Cells entrapped in hydrophobic gels had steroid-transforming activities comparable to that of free cells, and the operational stability was improved by immobilization, particularly in conversion of dehydroepiandrosterone (Omata *et al.*, 1979a). Duarte and Lilly (1980) also investigated steroid oxidation by alginate- and polyacrylamide-entrapped *N. rhodochrous* cells in water-saturated organic solvent systems, using the oxidation of cholesterol to

TABLE VII

Chemical Transformations by Immobilized Cells for Medical and Pharmaceutical Applications

Product (enzyme or reaction)	Organism	Method of immobilization	References
Steroid transformations			
Cortisol (11β-hydroxylase)	*Curvularia lunata*	Entrapment in polyacrylamide or Ca alginate gel	Mosbach and Larsson (1970); Larsson and Mosbach (1976); Ohlson *et al.* (1980a,b); Skryabin *et al.* (1974)
	Mycobacterium globiforme	Adsorption on cellulose	Skryabin *et al.* (1976)
	Pseudomonas testosteroni	Entrapment in polyacrylamide	Yang and Studebaker (1978)
Prednisolone (Δ′-dehydrogenase)	*Arthrobacter simplex*	Entrapment in polyacrylamide or Ca alginate gel	Larsson *et al.* (1976); Ohlson *et al.* (1978, 1979); Wiersma *et al.* (1981)
	A. simplex	Photo-cross-linked prepolymer	Sonomoto *et al.* (1979); Fukui *et al.* (1980d); Omata *et al.* (1981)
	A. simplex	Meleic acid polybutadiene gel	Omata *et al.* (1979b)
	Arthrobacter globiformis	Binding in silica gel activated with $CrCl_3$ or $TiCl_3$	Arinbasarova and Koshcheyenko (1980)
	Corynebacterium simplex	Binding in collegen and cross-linking with glutaraldehyde	Vieth *et al.* (1973, 1976a,b); Venkatasubramanian *et al.* (1978); Constantinides (1980)
	M. globiforme	Entrapment in polyacrylamide	Skryabin *et al.* (1976)
	Nocardia rhodochrous	Hydrophilic urethane prepolymer	Tanaka *et al.* (1979)
11-Desoxyprednisolone	*M. globiforme*	Entrapment in polyacrylamide	Skryabin *et al.* (1976)
20-Hydroxyprogesterone	*M. globiforme*	Entrapment in polyacrylamide	Skryabin *et al.* (1976)
Progesterone (from pregnenolone)	*N. rhodochrous*	Photo-cross-linked gel or polymethane	Fukui *et al.* (1980c)

Androst-1,4-dienedione (from progesterone)	*Fusarium solani*	Recycling of spores	Zedan and El-Tayeb (1977)
$\Delta^{1,4}$-3-Ketosteroids (from Δ^5-3β-acetoxy-Δ^5-3β-hydroxy-, and Δ^4-3-ketosteroids)	*Arthrobacter* sp., *Mycobacterium* sp., and *Nocardia* sp.	Entrapment in polyacrylamide	Voishvillo *et al.* (1976); Omata *et al.* (1979b); Fukui *et al.* (1980c)
4-Androstene-3,17-dione, etc. (from testosterone)	*N. rhodochrous*	Hydrophilic and lipophilic gels, and photo-cross-linkable and urethane prepolymers	Fukui *et al.* (1980b,c,d)
Cholesterone (from cholesterol)	*Nocardia erythropolis*, *N. rhodochrous*	Adsorption on DEAE-cellulose, entrapment in polyacrylamide or Ca alginate gel, or polyethylene maleic anhydride	Atrat *et al.* (1980a,b); Duarte and Lilly (1980)
Secoketol (from secodione)	*Saccharomyces paradoxus*	Binding on β-alanyl derivative of aminoseparon	Gulaya *et al.* (1979)
Biosynthesis of antibiotics			
Ampicillin (from 6-APA and D-phenylglycine methyl ester)	*Escherichia coli*, *Erwininia aroidae*, *Fusarium* sp. (spores)	Entrapment in polyacrylamide	Chibata *et al.* (1974c); Dinelli (1972); Marconi *et al.* (1975)
	Kluyvera citrophila	Entrapment in polyacrylamide	Morikawa *et al.* (1980a)
Cephalexin (from 7-ADCA and D-phenylglycine methyl ester)	*Achromobacter* sp.	Adsorption on DEAE-cellulose	Fujii *et al.* (1973); Marconi *et al.* (1975)
Penicillin G (from glucose)	*Penicillium chrysogenum*	Entrapment in polyacrylamide or Ca alginate gel, or binding in collagen membrane	Morikawa *et al.* (1979)
Bacitracin (from peptone)	*Bacillus* sp.	Entrapment in polyacrylamide	Morikawa *et al.* (1980b)
Colistin	*Bacillus polymyxa*	Entrapment in polyacrylamide	Suzuki *et al.* (1979)
Nisin	*Streptococcus lactis*		Kozlova *et al.* (1980)

(*continued*)

TABLE VII (*Continued*)

Product (enzyme or reaction)	Organism	Method of immobilization	References
Deacylation of β-lactam antibiotics			
6-APA (from penicillin G)	*E. coli, Streptomyces griseus, Nocardia gardneri*	Entrapment in polyacrylamide	Chibata *et al.* (1974c); Sato *et al.* (1976)
	E. coli	Entrapment in Ca alginate, polymethacrylamide, polyurethane, and epoxide	Klein and Wagner (1980, 1981)
	E. coli	Cells mixed with polyethylene-imine and glutaraldehyde, and entrapped in cellulose diacetate	Giovenco and Maimone (1981)
	E. coli	Entrapped in polyurethane foam or gel	Klein and Kluge (1981)
	E. coli	Entrapped in 4% agar gel and cross-linked with 0.5–1.0% glutaraldehyde	Sun *et al.* (1980)
	E. coli	Entrapped in gelatin gel and cross-linked with glutaraldehyde	Wang *et al.* (1980)
	Proteus rettgeri	Glutaraldehyde-treated cells bound to glycidyl methacrylate polymer	Fukushima *et al.* (1976)
	Streptomyces sp.	Adsorption on cellulose, etc.	Amin *et al.* (1980)
6-APA (from penicillin V)	*E. coli*, etc.	Entrapment in polyacrylamide	Chibata *et al.* (1976c)
	E. coli	Entrapped in epoxy matrix	Klein and Wagner (1981)
	Fusarium sp. (spores)	Several methods	Charles (1980)
	Aerobic gram-negative coccus	Cells mixed with polyethyleneimine and glutaraldehyde, mixed, extruded, and dried	Gestrelius (1979, 1980)

7-ACA (from 3-acetoxy-methyl-7-(4-carboxybutaneamide)-3-cephem-4-carboxylic acid)	*Comamonas* sp., *Pseudomonas* sp.	Entrapment in cellulose acetate	Fukushima *et al.* (1976)
7-ACA (from cephalosporin C)	*Bacillus subtilis*	Binding on brick powder treated with polyethyleneimine, etc.	Konecny and Sieber (1980)
7-ACA (deamination by D-amino acid oxidase)	*Trigonopsis variabilis*	Treatment with glutaraldehyde	Yamanouchi Pharmaceutical Co., Ltd. (1979)
Production of fine chemicals			
Pantothenic acid (from β-alanine, potassium pantoate, and ATP by pantothenic acid synthetase)	*E. coli*	Entrapment in agar gel	Kawabata and Demain (1979)
Coenzyme A (from pantothenic acid and ATP)	*Brevibacterium ammoniagenes*	Entrapment in polyacrylamide	Shimizu *et al.* (1975)
	B. ammoniagenes	Coentrapped with *Saccharomyces cerevisiae* in polyacrylamide	Samejima *et al.* (1978)
	B. ammoniagenes	Entrapment in polyacrylamide (two-reactor system)	Yamada *et al.* (1980a)
Pyridoxal 5′-phosphate (from pyridoxine)	*Kloeckera* sp. *Pseudomonas polycolor*	Entrapment in polyacrylamide	Chu *et al.* (1977)
	Pseudomonas fluorescens	Entrapment in polyacrylamide	Yamada *et al.* (1980a)
ATP (and ATP regeneration)	*S. cerevisiae*	Encapsulation in ethylcellulose	Samejima *et al.* (1978); Ado *et al.* 1980
	S. cerevisiae	Entrapment in polyacrylamide or carrageenan	Chibata *et al.* (1979b); Murata *et al.* (1980a,b, 1981)
	S. cerevisiae	Binding in hydroxyethylacrylate	Asada *et al.* (1979)
	S. cerevisiae	Entrapment in polyurethane or photo-cross-linked gel	Fukui *et al.* (1980d)

(continued)

TABLE VII (*Continued*)

Product (enzyme or reaction)	Organism	Method of immobilization	References
NADP	*Achromobacter aceris*	Entrapment in polyacrylamide	Chibata *et al.* (1975c); Uchida *et al.* (1978)
NADH	*B. ammoniagenes*	Entrapment in polyacrylamide	Murata *et al.* (1979b); Godbole *et al.* (1980)
	B. ammoniagenes	Microencapsulation in cellulose acetate	Ado *et al.* (1980)
NAD^+, FMN, etc. (hydrogenase)	*Alcaligenes eutrophus*	Entrapment in polyacrylamide, Ca alginate, or carrageenan	Klibanov and Puglisi (1980)
CDP-choline	*Hansenula jardinii*	Entrapment in polyurethane, etc.	Fukui *et al.* (1979, 1980d)
	H. jardinii	In polyethylene glycol or hydroxyethylacrylate	Kimura *et al.* (1978)
	S. cerevisiae	In polyethylene glycol or hydroxyethylacrylate	Kimura *et al.* (1981)
	S. cerevisiae	Entrapment in polyacrylamide	Samejima *et al.* (1978)
Glucose 1-phosphate and glucose 6-phosphate (acid phosphatase)	*Escherichia freundii*	Entrapment in polyacrylamide	Seif *et al.* (1975)
(polyphosphate glucokinase)	*Achromobacter butyri*	Entrapment in polyacrylamide	Murata *et al.* (1979c)
Amylases and proteases	*B. polymyxa* *Clostridium acetobutylicum* *Serratia marcescens* *Streptococcus thermophilus*	Entrapment in dialysis tubing	Fogarty and Griffin (1973)

α-Amylase	*Bacillus subtilis*	Entrapment in polyacrylamide, agar, Ca alginate, or collagen	Kokubu *et al.* (1978); Suzuki and Karube (1979a,b)
Asparaginase	*Alcaligenes faecalis*	Entrapment in gelatin with glutaraldehyde, etc.	Nakajima and Suzuki (1979)
Catalase	*Kloeckera* sp. (microbodies)	Entrapment in albumin with glutaraldehyde, etc.	Tanaka *et al.* (1978)
Lytic enzymes	*Streptomyces* sp.	Cross-linking with glutaraldehyde	Antczak *et al.* (1981)
Protease	*Clostridium histolyticum*	Entrapment in UF unit	Wang *et al.* (1970)
	Streptomyces fradiae	Entrapment in polyacrylamide	Kokubu and Suzuki (1981)
Dihydroxyacetone	*Acetobacter xylinum* *Acetobacter suboxydans* *Gluconobacter melanogenus*	Entrapment in polyacrylamide or carrageenan	Nabe *et al.* (1979)
	Gluconobacter suboxydans	Entrapment in polyacrylamide	Sonaer and Çağlar (1981)
	G. suboxydans	Entrapment in Ca alginate gel	Holst *et al.* (1981)
D-Fructose, D-ribulose, L-sorbose (from alditols)	*A. suboxydans*	Entrapment in polyacrylamide	Schnarr *et al.* (1977)
Porphobilinogen (from δ-aminolevulinic acid)	*Chromatium vinosum*	Dehydratase fixed within cells	Vogelmann *et al.* (1975)
Polyribosinic (from inosine) and polyribocitidylic (from cytidine diphosphate) acids	*Micrococcus luteus*	Covalent binding to CNBr-activated cellulose	Hoffman *et al.* (1970)
Gibberellic acid (from *n*-alkanes)	*Fusarium moniliforme*	Attachment on glass rings	Heinrich and Rehm (1981)
Cycloheximide (from glucose)	*Streptomyces griseus*	Entrapment in dialysis tubing	Kominek (1975a,b)
Formaldehyde (from CH_4)	*Hansenula polymorpha*	Entrapment in polyacrylamide	Couderc and Baratti (1980)

Δ^4-cholestenone by cholesterol oxidase as a model system. The hydrogen peroxide formed in the reaction was decomposed by *N. rhodochrous* catalase *in situ*. The immobilized preparation had better stability than free cells. According to Bhasin *et al.* (1976), the rate of enzyme catalysis in similar mixed-solvent systems appeared to be related to the mass transport of steroid from the organic to the water phase.

By employing immobilized *Mycobacterium globiforme* or *Corynebacterium simplex*, Voishvillo *et al.* (1976) obtained 70–85% yield in converting Δ^5-3β-hydroxy-, Δ^5-3β-actoxy-, and Δ^4-3-ketosteroids to $\Delta^{1,4}$-3-ketosteroids. The rate of steroid degradation was decreased by immobilization, resulting in increased yield.

Kondo and Matsuo (1960) and Kondo (1962) had observed that *C. simplex* Δ^1-dehydrogenase does not appear to be inhibited by high steroid concentrations, and applying this discovery to immobilized-whole-cell biocatalysis, Constantinides (1980) was able to obtain 85% yield of prednisolone in batch experiments, suggesting the feasibility of large-scale processing. It has been calculated that a single 1- to 2-kg immobilized *C. simplex* reactor would suffice to supply the whole annual Swedish demand of prednisolone, about 250 kg (Larsson *et al.*, 1976).

Germain *et al.* (1981) isolated mutants of *Nocardia* sp. unable to grow on steroids, but accumulating in high yields of various catabolic intermediates, and immobilized whole cells in various carriers to investigate production of intermediates. In the presence of hexane, methyl perhydroindanone propionic acid accumulating mutant produced androst-1,4-dienedione instead. Some results from these experiments are illustrated in Fig. 15.

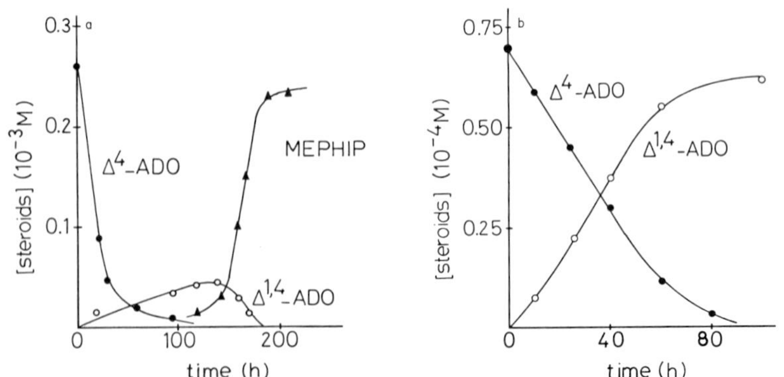

Fig. 15. Conversion of androst-4-ene-3,17-dione to methyl-perhydroindanone in aqueous solution (a) and to androst-1,4-diene-3,17-dione ($\Delta^{1,4}$-ADO) in hexane (b) by polyacrylamide gel-entrapped *Nocardia restricta*.

B. Antibiotic Production and Modifications

1. Biosynthesis of Antibiotics

The biosynthesis of ampicillin from 6-aminopenicillanic acid (6-APA) and cephalexin from deacetyl-7-aminocephalosporanic acid (7-ADCA) by DEAE-cellulose-adsorbed *Bacillus megaterium, Achromobacter* sp., and *Escherichia coli* has been reported (Dinelli, 1972; Marconi *et al.,* 1975). Suzuki and co-workers (Suzuki and Karube, 1979a,b; Morikawa *et al.,* 1979) entrapped *Penicillium chrysogenum* in polyacrylamide gel, alginate gel, and collagen membrane for the production of penicillin G and bacitracin. The antibiotic-producing activity was lowest with collagen-immobilized mycelium, and alginate beads were too fragile for repeated operations. The latter may, however, have been due to the low alginate concentration (2%) employed. With polyacrylamide-entrapped *P. chrysogenum,* penicillin G could be produced in nine consecutive 1 day batch experiments from 1% glucose containing 0.5% ammonium sulfate and 0.01% phenylacetate in 0.05 *M* phosphate buffer (pH 7). No details on productivity were given. Freeman and Aharonowitz (1980) employed linear, water-soluble polyacrylamide chains, partially substituted with acylhydrazide groups for entrapment of streptomycetes under milder conditions. The preparation was subsequently cross-linked with multifunctional carbonyl compounds to obtain gel of properties similar to polyacrylamide. β-Lactam antibiotics were continuously produced for 96 h with yields similar to those of resting cells.

Suzuki and Karube (1979a,b) reported a slightly higher rate of bacitracin production by polyacrylamide gel-immobilized *Bacillus* sp. than that of conventional fermentation, using peptone as substrate. The half-life of the immobilized-whole-cell system was about 10 days, and high bacitracin productivity (~28 U ml^{-1} h^{-1}) from 0.5% peptone was retained for at least 8 days. Recently, Suzuki and co-workers have employed polyacrylamide gel-entrapped *Kluyvera citrophila* for ampicillin (Morikawa *et al.,* 1980a), and *Bacillus* sp. for bacitracin (Morikawa *et al.,* 1980b) production.

In a patent assigned to Kyowa Hakko Kogyo Co., Ltd. (Suzuki *et al.,* 1979), a method is described to produce colistin from meat extract as substrate with polyacrylamide gel-entrapped *Bacillus polymyxa.* The production was slow, however, in the beginning yielding only 100 U ml^{-1} after 4 h at 30°C; but after 12 repetitive batches as much as 520 U ml^{-1} was obtained.

Kozlova *et al.* (1980) have reported the production of nisin by immobilized *Streptococcus lactis,* but structural changes in cell walls resulted in lysis after 3 months.

2. Deacylation of β-Lactam Antibiotics

6-Aminopenicillanic acid (6-APA) and 7-aminocephalosporanic acid (7-ACA) are important intermediates for semisynthetic penicillins and cephalosporins, and they can be produced by both enzymatic and chemical hydrolysis from the parent compounds. An excellent general review on penicillin acylases (amidases) has been written by Vandamme and Voets (1974), but most work on immobilized enzymes and whole cells has been done more recently. In a recent article on immobilized-enzyme applications, Brodelius (1978) covers most early developments in immobilized-enzyme technology involving penicillins and cephalosporins. Several immobilized-enzyme-based processes have been commercialized (Chibata, 1978a), and immobilized biocatalyst technology has been industrially applied for 6-APA production since 1973 (Chibata, 1980a).

Sato *et al.* (1976) investigated in detail the production of 6-APA from penicillin G by polyacrylamide gel-immobilized *Escherichia coli* cells (Chibata *et al.*, 1974c). One of the limitations with whole-cell preparations is the presence of β-lactamase (penicillinase) activity, which decomposes both penicillin and 6-APA. By optimizing the process, 6-APA could be continuously produced by immobilized-whole-cell column reactor in about 80% yield. The half-life of the column was about 17 days at 40°C, and 42 days at 30°C. Morkawa *et al.* (1980a,b) entrapped whole cells of *Kluyvera citrophila* in polyacrylamide gel, obtaining about 60–70% penicillin acylase activity retention. The biocatalyst also exhibited excellent ampicillin productivity. Klein and Wagner (1980) obtained good biocatalyst characteristics by immobilizing penicillin acylase-active *E. coli* cells by polycondensation of epoxides. High cell loading, excellent mechanical properties, and good specific reaction rate (U ml^{-1}) were reported. The half-life was 40 days in comparison to 1 day for free suspended cells. By cloning the penicillin G acylase gene of *E. coli* ATCC 11105 on multicopy plasmids (Mayer *et al.*, 1980), a new hybrid strain *E. coli* 5K (pHM:12) of high activity was obtained, and the work of Klein and Wagner (1980) is believed to be the first application of immobilized-whole-cell technology to a microorganism modified by genetic engineering to increase productivity. In a later work, Klein and Wagner (1981) immobilized *E. coli* cells in porous polyurethane foams and gel.

In a patent application (Gestrelius, 1979), microbial cells containing penicillin V acylase activity have been reacted in aqueous medium with glutaraldehyde in the presence of a polyamine with >5% of amino groups. The dried multicell preparation had an activity of 14.5 U g^{-1} with about 30% activity yield. In another patent (Vojtiset *et al.*,

1980), a similar process is described. Glutaraldehyde-treated cells were permeated by butanol and surfactant, followed by treatment with polyethyleneimine and glutaraldehyde, freezing and cutting to suitable particle size, and thawing. After filtering and treating with acetone–water (1 : 1), the biocatalyst is rehydrated in pH 7.6. Giovenco and Maimone (1981) first aggregated *E. coli* cells in the fermentation broth with a mixture of branched polyethyleneimine and glutaraldehyde. The aggregates were suspended in cellulose diacetate in acetone, and biocatalyst was obtained as small beads from water. In our own earlier experience with entrapment of *Saccharomyces cerevisiae* cells, better results without preaggregation were obtained, when a mixture of acetone and dimethyl sulfoxide (2 : 3) of dimethyl sulfoxide alone was used as a solvent for cellulose acetates (P. Linko *et al.*, 1980).

Charles (1981) obtained an increase in operational stability of *Fusarium* sp. penicillin V acylase on immobilization.

Until recently, 7-ACA could be manufactured only via chemical deacylation of the α-aminoadipic acid side chain of cephalosporin C. In a patent assigned to Toyo Jozo Co., Ltd. (1974; Fukushima *et al.*, 1976), an alternate method based on immobilized *Pseudomonas* sp. or *Comanomonas* sp. cells was described. 3-Acetoxy-methyl-7-(4-carboxybutaneamide)-3-cephem 4-carboxylic acid is used as substrate for the immobilized-whole-cell biocatalyst to obtain 7-ACA in good yield. According to another patent (Yamanouchi Pharmaceutical Co., Ltd., 1979), *Trigonopsis variabilis* D-amino acid oxidase was activated and the cells immobilized by glutaraldehyde treatment for deamination of cephalosporin C, 7-(*O*-5-amino-5-carboxyvaleramido)-3-(1-methyl-1*H*-tetrazol-5-yl)thiomethyl-7-methoxy-3-cephem 4-carboxylic acid.

C. Fine Chemicals

1. Pantothenic Acid

Pantothenic acid is an important vitamin, and a part of coenzyme A structure. It is currently produced by chemical synthesis, which results in a racemic mixture. Most microorganisms can produce pantothenic acid from pantoic acid and β-alanine by pantothenic acid synthetase, which requires ATP. Kawabata and Demain (1979) entrapped frozen and thawed *Escherichia coli* cells in agar gel (3%). Pantothenic acid was produced from a mixture of β-alanine, potassium pantoate, and ATP. The half-life of the biocatalyst was about 4 days, and about 40 days were required for complete inactivation. Exogenous ATP was essential for the production.

2. Coenzyme A

In a patent assigned to Tanabe Seiyaku Co., Ltd. (Chibata *et al.*, 1975b), a method is described to produce coenzyme A (CoA) from pantothenic acid and ATP by polyacrylamide-entrapped *Sarcina lutea* cells. Shimizu *et al.* (1975) immobilized *Brevibacterium ammoniagenes* cells in polyacrylamide gel to produce CoA in a packed-bed column reactor from pantothenic acid, cysteine, ATP, and magnesium sulfate in phosphate buffer (pH 6.5), accumulating under optimum conditions 500 μg ml^{-1} of CoA. The half-life of the biocatalyst system was about 1 week. Samejima *et al.* (1978) employed a similar system equipped with an immobilized-lysozyme column before and after the main reactor. At 37°C molar conversion to CoA (10 m*M* = 2.4 g liter^{-1} sodium pantothenate, 15 m*M* disodium ATP, 10 m*M* cysteine, 10 m*M* $MgSO_4$ in 0.15 *M* phosphate buffer, pH 6.8) was 17% (0.9 g liter^{-1}), decreasing to 12.5% after 9 days of operation. When ATP was replaced by coimmobilizing *B. ammoniagenes* and *Saccharomyces cerevisiae*, a significant improvement in CoA production was obtained. When, instead of ATP, glucose (155 m*M*), inorganic phosphate (400 m*M*), and AMP (15 m*M*) were used in the substrate, CoA production approached that by free cells. According to Yamada *et al.* (1980a), continuous production in a column reactor minimizes the feedback inhibition of panthonate kinase by CoA, thus improving CoA productivity. The system was further improved by employing two columns in a series. In the first reactor, pantothenic acid was nearly quantitatively phosphorylated to *P*-pantothenic acid, after which cysteine and ATP were added for CoA synthesis in the second reactor. The total productivity in a 7-day period was almost double as compared with the single-reactor system.

3. Flavin Adenine Dinucleotide

Yamada *et al.* (1980a) immobilized *Arthrobacter oxydans* cells of high flavin adenine dinucleotide (FAD) pyrophosphorylase activity in a film of poly(vinyl alcohol) cross-linked with tetraethylsilicate. Polyethyleneimine was added to prevent cell leakage. More than 1 mg ml^{-1} of FAD could be continuously produced from 5 μmol ml^{-1} of FMN (15 μmol ATP, 10 μmol $MgSO_4$, 200 μmol imidazole-HCl buffer pH 7, 1 mg sodium lauryl sulfate per milliliter) at 34°C in the dark, $\tau \sim$ 10 h.

4. Pyridoxal 5′-Phosphate

Pseudomonas polycolor and *Kloeckera* sp. cells have been immobilized in polyacrylamide gel for the production of pyridoxal 5′-

phosphate from pyridoxine 5′-phosphate (Chu *et al.*, 1977). Yamada *et al.* (1980b) employed poly(vinyl alcohol) cross-linked with tetraethylsilicate to immobilize *Pseudomonas fluorescens* cells for the same reaction in a batch system (0.6 μmol ml^{-1} pyridoxine 5′-phosphate, 37°C). The treatment of the biocatalyst with polyethyleneimine significantly improved operational stability in comparison with both free and immobilized nontreated cells.

5. Nucleotides

Coenzymes and their regeneration systems are the limiting factor in many immobilized-enzyme and whole-cell applications. Consequently, the production of coenzymes *in situ* in particular is of major importance. A few reports for the production of pure nucleotides and related compounds have also been published.

Several articles have been published on the regeneration of ATP. Samejima *et al.* (1978) employed ethylcellulose-encapsulated *Saccharomyces cerevisiae* cells to produce ATP from AMP in a reactor system consisting of two columns in a series, each equipped with a gas-purge tube to eliminate carbon dioxide. If necessary, the entire process line could be sterilized by including an immobilized-lysozyme column in the system. The initial conversion from AMP (43 m*M* = 15 mg ml^{-1} AMP, 166 m*M* glucose, 4 m*M* $MgSO_4$, 1 *M* phosphate buffer pH 8, space velocity 0.2 h^{-1}, 35°C) to ATP was over 70%, decreasing after 10 days of continuous operation to about 50%. In a later publication (Ado *et al.*, 1980), the optimization of the ATP-regeneration system by cellulose acetate-butylate-microencapsulated *S. cerevisiae* cells was investigated. Chibata and co-workers (1979b; Murata *et al.*, 1980a,b, 1981) compared several different ATP-regeneration systems in the production of glutathione from L-glutamic acid, L-cysteine, and glycine, including the glycolytic pathway of immobilized *S. cerevisiae* cells and the acetyl kinase system of immobilized *E. coli*. The results indicated that the ATP produced by the glycolytic pathway of *S. cerevisiae* cells could be supplied to the glutathione synthetase in air-dried *Escherichia coli* cells (Murata *et al.*, 1980b), but a constant NAD supply was necessary (Murata *et al.*, 1981). An immobilized *S. cerevisiae* ATP-regenerating system has also been utilized to synthesize CDP-choline from CTP and phosphorylcholine (Ado *et al.*, 1979). CDP-choline, which is an important intermediate in the biosynthesis of phospholipids, and in the treatment of brain injuries, has also been produced in a similar system by Kimura *et al.* (1978, 1981) utilizing poly(ethylene glycol) hydroxyethylacrylate-immobilized dried *S. cerevisiae*. Also in this system, the addition of NAD was necessary.

Uchida *et al.* (1978) observed that the polyacrylamide gel-immobilized *Achromobacter aceris* NAD kinase was more stable than that of free cells, with an operational half-life of about 20 days in a column system at 37°C in comparison to 6 days with free suspended cells. NAD-kinase of the immobilized cells could be activated by treatment with acetone, heat, or acid, and the ATP-degrading activity was completely inhibited by acid treatment. Consequently, acid-treated immobilized *A. aceris* cells could be employed in the production of NADP from NAD. Godbole *et al.* (1980) observed that, unlike with purified alcohol dehydrogenase, whole *S. cerevisiae* cell NADH could be entrapped in polyacrylamide gel to obtain a stable biocatalyst. It was also found that the NADH formed was not destroyed in the system. Ado *et al.* (1980) have recently described a system based on coimmobilizing *S. cerevisiae* cells for high ATP-regeneration activity and *Brevibacterium ammoniagenes* for high NAD-kinase activity by microencapsulation with cellulose acetate-butylate to produce NADP by the immobilized-cell system at rates similar to those of *B. ammoniagenes* alone in the presence of ATP (6 g liter^{-1}).

6. Glucose Phosphates

Escherichia freundii cells have been shown to retain their acid phosphatase activity for transphosphorylation on immobilization in polyacrylamide (Saif *et al.*, 1975). This could be utilized for the production of glucose 1- and glucose 6-phosphates from glucose and *p*-nitrophenyl phosphate. Murata *et al.* (1979c), in contrast, employed polyacrylamide-entrapped *Achromobacter butyri* cells activated with organic solvents such as acetone, or continuous production of glucose 6-phosphate from glucose and metaphosphate. Immobilization considerably increased stability of polyphosphate glucokinase. The half-life in continuous-column operation was about 20 days.

7. Enzymes

Some enzymes such as *Aspergillus* proteases and pectinases are still produced on semisolid media, but the bulk is manufactured by batch aerobic submerged-culture techniques (Aunstrup, 1977). *Bacillus coagulans* glucose isomerase is apparently the only enzyme produced by continuous culture today (Diers, 1976). It has been shown recently that some enzymes may also be produced by immobilized-whole-cell systems. Fogarty and Griffin (1973) investigated the possibility of producing extracellular proteases and amylases by cultivating microorganisms such as *Bacillus polymyxa, Clostridium acetobutylicum, Serratia marcescens,* and *Streptococcus thermophilus* in a dialysis tube placed in the fermenter containing nutrients. High enzyme concentration was obtained inside the tube, but total productivity on the basis of the fermenter volume was small.

Similarly, cultivation of *Clostridium histolyticum* on peptone in an ultrafiltration unit to obtain a fourfold increase in protease production has been reported (Wang *et al.*, 1970).

The first attempt to produce extracellular enzymes by carrier-immobilized whole cells was that of Suzuki and others (Kokubu and Suzuki, 1981; Suzuki and Karube, 1979b), who entrapped *Bacillus subtilis* cells in polyacrylamide for α-amylase production. Under optimum conditions, about 16,000 U ml^{-1} of α-amylase in each batch filtrate (30°C, 5 h, 1% meat extract, 0.5% yeast extract) was obtained, a value about three times higher than that obtained with free cells. Similarly immobilized *Streptomyces fradiae* cells were used for protease production, employing a medium of 0.5% starch, 0.5% meat extract, and 0.05% yeast extract. The half-life of the biocatalyst was about 30 days, and the protease production of 12,000 U ml^{-1} h^{-1} was about four times higher than that obtained by submerged culture (Kokubu and Suzuki, 1981).

Recently, Antczak *et al.* (1981) reported the immobilization *in situ* of *Streptomyces* sp. lytic enzymes by treating the cells with 1% glutaraldehyde. The immobilized-whole-cell preparation was capable of lysing baker's yeast cells. Such biocatalysts could also have potential applications in continuous sterilization of liquids.

L-Asparaginase is important in tumor therapy, and a number of immobilized L-asparaginase preparations have been developed and evaluated (Chibata *et al.*, 1976a, 1977). A method to produce L-asparaginase-active immobilized cells has been described in a patent assigned to Mitsui Sugar Co., Ltd. (Nakajima and Suzuki, 1979). Lyophilized cells of *Alcaligenes faecalis* were defrosted and kneaded to a paste with 10% gelatin solution, and extruded through a nozzle (ϕ = 0.6 mm) to acetone containing 5% glutaraldehyde, dried, and cut to suitably sized particles. About 80% of activity was retained after 96 h continuous-column operation.

8. Oxidation of Polyhydric Alcohols

Dihydroxyacetone (DHA) is used as a pharmaceutical intermediate and as a suntanning agent. Nabe *et al.* (1979) investigated the possibility of producing DHA by glycerol-oxidizing bacteria *Acetobacter xylinum* immobilized in polyacrylamide gel. Complete oxidation of glycerol to DHA by the immobilized-whole-cell biocatalyst was realized in 40 h. In this case, κ-carrageenan was not a suitable carrier because of the detrimental effect of metal ions necessary to gelatinize the carrier on enzyme activity and stability. In two recent articles, immobilized *Gluconobacter suboxydans* cells have been employed for DHA production. Holst *et al.* (1981) used calcium alginate gel as the carrier. Oxygen-transport problems were eliminated by adding hydrogen peroxide to substrate (0.1 *M* glycerol, 0.05 *M* H_2O_2). Hydrogen peroxide was rapidly decomposed by *G. suboxydans* catalase to water and oxygen inside the biocatalyst bead. To minimize detrimental effects of hydrogen peroxide to the cells, hydrogen peroxide was continuously fed to a stirred-tank reactor. The oxidation was complete in <10 h, corresponding to a productivity of 0.9 g liter^{-1} h^{-1}, in comparison to less than 1% conversion in oxygen-saturated glycerol (Fig. 16).

Sonaer and Çağlar (1981), in contrast, used polyacrylamide gel-entrapped *G. suboxydans* in an air-lift-type reactor. The operational stability was not satisfactory, but the biocatalyst could be repeatedly reactivated before complete inactivation.

An interesting approach has been presented by Schnarr *et al.* (1977), who employed polyacrylamide gel-entrapped *Acetobacter suboxydans* to convert a number of polhydric alcohols to the corresponding ketoses. Thus, L-sorbose was produced from D-sorbitol, D-fructose from D-mannitol, and L-ribulose from ribitol. The oxidation of D-mannitol to D-fructose by growing cells was complete in 12 h, with no by-products, but with immobilized cells 34 h was required with some further oxidation of D-fructose.

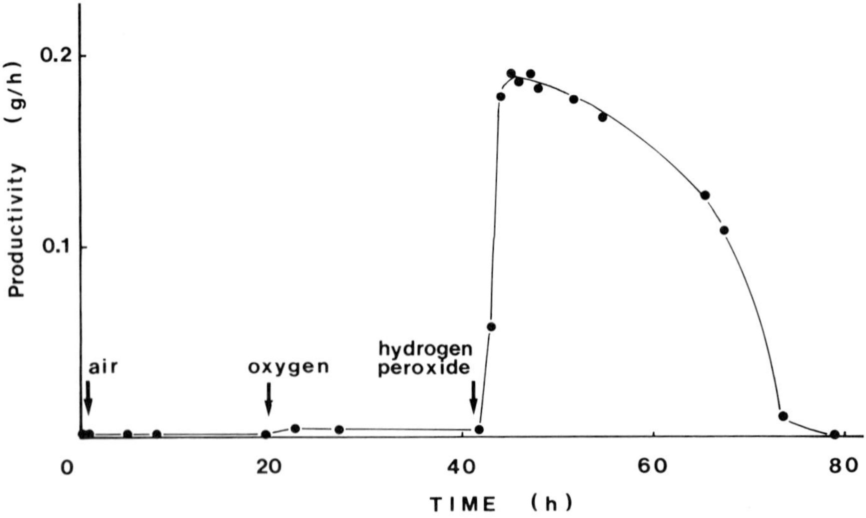

Fig. 16. Continuous production of dihydroxyacetone from glycerol by calcium alginate-entrapped *Gluconobacter oxydans* (ATCC 621) cells.

9. Miscellaneous Synthetic Applications

Immobilized whole cells have also been investigated for the formation of porphobilinogen from δ-aminolevulinic acid by *Chromatium vinosum* cells (Vogelmann *et al.*, 1975), gibberellic acid from *n*-alkanes by glass ring-attached *Fusarium moniliformas* (Heinrich and Rehm, 1981), polyribosinic and polyribocitidylic acids from inosine diphosphate and cytidine diphosphate, respectively, by *Micrococcus luteus* covalently coupled to cyanogen bromide-activated cellulose (Hoffman *et al.*, 1970), cycloheximide from glucose by *Streptomyces griseus* entrapped in dialysis tubing (Kominek, 1975a,b), and formaldehyde from methanol by polyacrylamide gel-entrapped *Hansenula polymorpha* (Couderc and Baratti, 1980). A novel approach, applying polyacrylamide gel-entrapped *Staphylococcus aureus* cells for affinity chromatographic purification of human serum albumins, has also been presented (Mattiasson *et al.*, 1980).

VII. OTHER POTENTIAL APPLICATIONS OF IMMOBILIZED MICROBIAL CELLS

A number of other potential applications of immobilized microbial cells are summarized in Table VIII.

A. Hydrocarbon Oxidation

Cetus Corporation has recently developed a process for biotechnical oxidation of ethylene and propylene (Cape *et al.*, 1980). Pure fructose has been mentioned as a possible by-product. Until today, few data have been published on the application of microbial-cell biocatalysts in such conversions.

In a patent assigned to Bio Research Center, Ltd. (Furuhashi *et al.*, 1980), a method is described for utilizing polyacrylamide gel-entrapped *Nocardia corallina* cells to produce propylene oxide from propylene (1.74 g liter^{-1} K_2HPO_4, 1.50 g liter^{-1} $MgSO_4$, and 50 mg liter^{-1} $FeSO_4$ in the medium) at 30°C, Propylene gas was added to a partial pressure of 150 mm Hg to the partially evacuated reactor system (550 mg Hg). A maximum yield of 0.41 g liter^{-1} of propylene oxide was obtained during the second 24-h cycle.

De Bont *et al.* (1981) employed *Mycobacterium* sp. immobilized in calcium alginate gel, and on sand to oxidize ethylene and propylene. Propionaldehyde as a cosubstrate considerably improved ethylene oxide production.

TABLE VIII
MISCELLANEOUS POTENTIAL APPLICATIONS OF IMMOBILIZED MICROBIAL CELLS

Product or application (enzyme or reaction)	Organism	Method of immobilization	References
Hydrocarbon oxidation			
Propylene oxide (and other alkene oxides)	*Nocardia corallina*	Entrapment in polyacrylamide	Furuhashi *et al.* (1978)
	Mycobacterium sp.	Entrapment in Ca alginate or attachment on sand, etc.	de Bont *et al.* (1981)
Production of hydrogen			
H_2	*Anabaena cylindrica*	Adsorption on glass beads	Lambert *et al.* (1979)
	Anacystis nidulans and *Rhodospirillum rubrum*	Entrapment in agar (a two-reactor system)	Weetall and Krampitz (1980)
	Clostridium butyricum	Entrapment in agar or polyacrylamide	Karube *et al.* (1976, 1977b); Suzuki *et al.* (1980)
	R. rubrum	Entrapment in agar	Weetall and Bennett (1976)
Nitrogen fixation			
N_2 fixation	*Azotobacter vinelandii*	Binding on anion-exchange resin	Seyhan and Kirwan (1979)
	A. vinelandii	Covalent binding on cellulose	Gainer *et al.* (1980)
	Rhizobium japonicum	Entrapment in polyacrylamide	Dommergues *et al.* (1979)
NH_3 (from N_2)	*Klebsiella pneumoniae*	Binding on collagen membrane	Venkatasubramanian and Toda (1980)
Degradation reactions			
Wastewater treatment, denitrification	Mixed cultures	Attachment on cellophane	Goto *et al.* (1971)
	Activated sludge	Retention in UF unit	Hardt *et al.* (1971)
	Mixed cultures	Adsorption on sand	Anonymous (1976a, 1977)
	Mixed cultures	Adsorption on anthracite	Scott and Hancher (1976); Genung *et al.* (1979, 1980)

	Mixed cultures	Attachment on aluminum or polypropylene pall rings treated with nitric acid	Compere and Griffith (1977)
	Hyphomicrobium sp.	Entrapment in polyacrylamide	Divies (1977)
	Micrococcus denitrificans	Entrapment in liquid membrane	Mohan and Li (1975)
	Pseudomonas aeruginosa	Entrapment in polyacrylamide	Divies (1977)
	Pseudomonas denitrificans	Entrapment in Ca alginate	Nilson *et al.* (1980, 1981); Nilson and Molin (1981)
Methane production	Mixed culture	Adsorption on column packing, etc.	Anonymous (1976b); van den Berg *et al.* (1981); Genung *et al.* (1979)
Phenol degradation	*Candida tropicalis*	Entrapment on polyacrylamide, Ca alginate, etc.	Hackel *et al.* (1975); Klein *et al.* (1976, 1978, 1979, 1980)
	Pseudomonas sp.	Adsorption on anthracite	Anonymous (1976a); Scott and Hancher (1976); Arcuri *et al.* (1981); Scott *et al.* (1981)
Orsellinic acid decarboxylation	*Umbillicaria pustulata*	Entrapment in polyacrylamide	Mosbach and Mosbach (1966)
L-Malic acid catabolism	*Lactobacillus casei*	Entrapment in polyacrylamide	Divies and Siess (1976); Divies (1977)
L-Arginine catabolism	*Streptococcus faecalis*	Entrapment in polyacrylamide	Franks (1971, 1972)
ε-Aminocaproic acid dimer hydrolysis	*Achromobacter guttatus*	Entrapment in polyacrylamide	Kinoshita *et al.* (1975)
Octanoic acid degradation	*Penicillium roquefortii* (spores)		Johnson and Ciegler (1969)
Inulin hydrolysis	*Kluyveromyces marxianus*	Entrapment in Ca alginate gel	Kierstan and Bucke (1977)

(continued)

TABLE VIII (*continued*)

Product or application (enzyme or reaction)	Organism	Method of immobilization	References
Cellulose and cellobiose breakdown	*Thermonospora* sp.	Entrapment in κ-carrageenan, and binding on anion exchangers	Phillips and Humphrey (1980)
β-Glucan breakdown	*Alcaligenes faecalis*	Entrapment in polyacrylamide	Wheatley and Phillips (1981a,b)
Benzene degradation	*Pseudomonas putida*	Entrapment in polyacrylamide	Somerville *et al.* (1977); Mason *et al.* (1978)
Diacetyl degradation	*Saccharomyces carlsbergensis*	Adsorption on diatomaceus earth	Tolls *et al.* (1970)
Urea hydrolysis	*Sarcina ureae*	Entrapped in cellulose triacetate fiber	Ghose and Kannan (1979)
Tributyrin and tributyrin and triacetin	*Pseudomonas mephitica*	Heat treatment	Kosugi and Suzuki (1973)
Other applications			
Affinity chromatography (purification of proteins, etc.)	*Hansenula polymorpha*	Entrapment in polyacrylamide	Mattiasson *et al.* (1980)
Electricity (fuel cells)	*Clostridium butyricum*	Entrapment in polyacrylamide	Karube *et al.* (1977b)
Bioluminescence (for fishing applications)	*Photobacterium phosphoreum*	Entrapment in Ca alginate gel or in polyurethane	Makiguchi *et al.* (1980a,b)
Bread baking	*Saccharomyces cerevisiae*	Entrapment in Ca alginate	Chibata *et al.* (1974h)
Ethylene from acetylene	*Azotobacter vinelandii*	Adsorption on anion-exchange resin	Kirwan and Gainer (1976)
Optical resolution; (other than amino acids)	*Alginomonas* sp.	Entrapment in polyacrylamide	Nonomura *et al.* (1976)
L-menthol production	*Rhodotorula minuta*	Entrapment in polymethane or photo-cross-linkable gel	Omata *et al.* (1981); Fukui *et al.* (1981)
Ferrous iron-oxidation investigations	*Thiobacillus ferrooxidans*	Entrapment in polyacrylamide	Kutsal and Çağlar (1981); Kuyucak *et al.* (1981)

B. Production of Hydrogen

Weetall and Bennett (1976) and Bennett and Weetall (1976) agar-entrapped *Rhodospirillum rubrum* bacterial cells used in order to produce hydrogen from 0.01 *M* malate in a continuous process. A reactor containing 4 g of bacteria produced 10^5 μl of hydrogen over 150 h, with a half-life of 115 h. In a later study, Weetall and Krampitz (1980) described a system in which the agar gel-immobilized photosynthetic bacterium *R. rubrum* produced molecular hydrogen and at the same time was able to reoxidize NADPH, and an agar gel-immobilized cyanobacterium (blue-green algae) *Anacystis nidulans* produced molecular oxygen from water, and at the same time reduced exogenous NADP. The net result was the production of hydrogen and oxygen from water. The system was continuously flushed with argon, and NADP-containing substrate was pumped to the *A. nidulans* reactor and recycled back via the *R. rubrum* reactor.

Suzuki *et al.* (1980) employed *Clostridium butyricum* cells entrapped in agar gel for continuous production of hydrogen from alcohol factory wastewater. The immobilized-cell reactor was able to produce hydrogen at a rate of ~6 ml min^{-1} kg^{-1} (wet gel) for at least 20 days with little decrease in activity. When the pure hydrogen was transferred to a fuel cell, 550–500 mA current was continuously obtained over a 7-day period with about 1 kg of immobilized bacteria.

C. Nitrogen Fixation

There has been an increasing interest in biological nitrogen fixation during the last few years. Dommergues *et al.* (1979) was able to show that polyacrylamide gel-entrapped *Rhizobium japonicum* cells could be used as an inoculant for soybeans instead of commonly used peat as a carrier. Venkatasubramanian and Toda (1980) used a mutant strain of *Klebsiella pneumoniae* immobilized on collagen membrane to convert nitrogen to ammonia. In a batch system, ammonia production rates of 25–33 μmol g^{-1} (dry cell) h^{-1} were obtained, and for continuous operation the technique could be applied to a spiral-wound reactor system. Gainer *et al.* (1980) have studied the ion-exchange resin-immobilized *Aspergillus vinelandii* for the fixation of nitrogen to ammonia. Another immobilization technique employed was covalent bonding to cellulose with cyanuric chloride (Seyhan and Kirwan, 1979; Gainer *et al.*, 1980). In both cases, cell concentrations were higher than when free cells were used in suspension, but the nitrogenase activity of ion-exchange resin-bound cells decreased rapidly after about 60–100 h.

The activity of covalently immobilized biocatalyst could be maintained for over a month.

D. Miscellaneous Degradation Reactions

1. Special Processes in Wastewater Treatment

Wastewater treatment is the major application of surface-bound microbial cell systems. Although a detailed survey of conventional wastewater treatment technology is beyond the scope of this chapter, a number of recent developments based on immobilized whole cells will be described. One of the most interesting achievements in this field is the ANFLOW plant developed at Oak Ridge National Laboratory (Scott and Hancher, 1976; Genung *et al.*, 1979, 1980). A large number of waste-treatment microorganisms were adsorbed on anthracite coal particles (Scott and Hancher, 1976) and applied to a tapered up-flow fixed-film anaerobic bioreactor, which had been investigated in detail on the laboratory scale by Young and McCarty (1976). They used high concentrations of synthetic feeds with low levels of suspended solids at typical residence times of $\tau \sim 36$ h for an ~27-liter reactor. A large-scale unit was installed in Oak Ridge by Norton Co. in 1976, and the plant proved relatively insensitive to ambient temperature changes during the first 16 months of operation (Genung *et al.*, 1979). The methane content of gas produced has been as high as 70%. A 13,000 liter d^{-1} demonstration plant is currently being planned (Genung *et al.*, 1980).

Rittermann and McCarthy (1980a,b) developed a model of steady-state biofilm kinetics for a column reactor inoculated with methylene chloride-utilizing bacteria.

Most work on wastewater denitrification has been done with relatively dilute wastewater streams. Griffith and Compere (1977) studied the behavior of a continuous fixed-film packed-bed-type reactor for denitrification of high-nitrate wastes. The system was able to treat nitrate levels of 15.5–24 g $liter^{-1}$. Nilsson *et al.* (1980a,b) and Nilsson and Molin (1981) employed *Pseudomonas denitrificans* entrapped in calcium alginate gel for continuous denitrification of drinking water. Incubation in nutrient solution increased the original activity of the immobilized biocatalyst about 300 times due to the growth of cells. It was shown that 100 g of the biocatalyst could denitrify 30 liters of high-nitrate drinking water (100 mg NO_3 $liter^{-1}$) per day when aspartate was used as the carbon source, and about 30% less if the carbon source was ethanol.

Van den Berg *et al.* (1981) observed that methane production rates with fixed-film reactors equaled or exceeded those obtained with completely mixed continuous fermenters. Both with simulated sewage sludge and with bean-blanching waste, the fixed-film reactor produced 3.8–5.3 m^3 (STP) CH_3/m^3/day, as compared to a maximum of 2 m^3/m^3/day for the CSTR.

2. Phenol Degradation

Phenol degradation is a typical biocatalytic process involving a multienzyme system. Klein *et al.* (1976) and Klein and Schara (1981) used several methods to immobilize *Candida tropicalis* yeast cells for phenol degradation. Scott *et al.* (1981) reported that with a 20 × 800-cm fixed-film column reactor, rates of up to 25 g $liter^{-1}$ day^{-1} at $\tau <$ 30 min were obtained. It has also been shown with mixed cultures of bacteria and filamentous fungi attached on anthracite coal (30–60 mesh) that up to 30 g of synthetic phenol per liter bed volume could be removed daily with a fluidized-bed reactor (Arcuri *et al.*, 1980).

3. Other Degradation Reactions

Decarboxylation of orsellinic acid with polyacrylamide-immobilized *Umbillicaria pustulata* lichen cells by Mosbach and Mosbach (1966) was among the first immobilized-whole-cell-catalyzed reactions reported in literature. Since then polyacrylamide gel-entrapped microorganisms have been used for a number of degradation reactions. Franks (1972) studied the catabolism of L-arginine by *Streptococcus faecalis* cells, Kinoshita *et al.* (1975) employed *Achromobacter guttatus* for hydrolysis of the cyclic dimer of ε-aminocaproic acid, and Divies and Siess (1976) investigated the catabolism of L-malic acid by *Lactobacillus casei.* Somerville *et al.* (1977) showed that the ability of *Pseudomonas putida* cells to oxidize benzene was stabilized by immobilization in polyacrylamide gel, and that decreased reactor activity could be revived by the addition of nutrients.

Octanoic acid degradation has been studied with immobilized spores of *Penicillium roquefortii* (Johnson and Ciegler, 1969), and Tolls *et al.* (1970) used *Saccharomyces carlsbergensis* yeast on a diatomaceus earth filter bed to decompose diacetyl.

Urease-active *Sarcina ureae* cells entrapped within spun-cellulose triacetate fibers have been employed for the hydrolysis of urea for various potential applications (Ghose and Kannan, 1979).

VIII. CONCLUSIONS AND FUTURE PROSPECTS

Conventional fermentation technology, developed to a major industry in the beginning of this century, was in many cases replaced soon after World War II by chemical synthesis as a result of the spectacular growth of the petrochemical industry and, consequently, the availability of inexpensive bulk chemicals. The necessity of increasing productivity, however, intensified work on continuous-fermentation processes and automation; moreover, improved processes for amino acid and ethanol fermentation, as well as baker's yeast manufacture were developed. Citric acid remained the only bulk chemical produced by batch-fermentation techniques. After World War II much research effort was directed to single-cell protein (SCP) production as a solution to world hunger problems, and petrochemicals were seen as a major carbon source for many fermentations. The recent spectacular increases in material and energy costs, together with the developments in immobilized-biocatalyst engineering to provide specific, stable catalysts for economical continuous processing has resulted in the second generation of biotechnology, and in the consideration of renewable carbohydrate resources as feedstock for chemical synthesis via heterogeneous biocatalysis. Immobilized-whole-cell technology, in particular, may be seen as a major development for the applications of biotechnology, and savings of up to 40% have been realized in actual industrial practice as compared with conventional technology (Chibata and Tosa, 1977; Chibata *et al.*, 1979c).

Immobilized-whole-cell biocatalysts offer a number of advantages over the use of soluble and immobilized enzymes, as well as over chemical synthesis, although for many applications different alternatives should always be considered, and the final choice of a process should be based on a comprehensive economic feasibility study. Cheetham (1980) has provided an expert discussion of the advantages and disadvantages of immobilized whole cells for various applications. In general, biocatalysts can be used under mild reaction conditions, and the energy requirements are frequently low. In comparison with chemical synthesis, the use of specific biocatalysts results in fewer by-products, although side reactions catalyzed by undesirable enzymes when whole cells are employed may have to be eliminated by special techniques (Shibatani *et al.*, 1974; Yamamoto *et al.*, 1977). With the use of whole-cell preparations, tedious and costly enzyme isolation and purification are not necessary, but the specific activity of the biocatalyst is generally lower in comparison with immobilized enzymes, except when enzyme activity may be increased *in situ* by immobilized living cells. Commercial applications employing nonviable cell preparations have been limited to relatively simple single-enzyme-catalyzed conversions, but the recent developments in immobilized-living-cell biocatalysts would allow applications based on the complex multienzyme systems coupled with cofactor regeneration *in situ* of the living organisms, and the application of heterogeneous catalysis principles to fermentation.

Immobilized-whole-cell biocatalysts render possible in many cases continuous processing with relatively small initial investment. Easy process control by flow rate, small reactor volumes, high dilution rates, the possibility of using very low or high substrate concentrations, high productivity, and asymmetric biosynthesis are among the advantages. Mass-transport limitations restrict the applications in most cases, however, to relatively small-molecular substrates and products, although production of extracellular enzymes by immobilized-whole-cell biocatalysts has been reported (Kokubu *et al.*, 1978; Suzuki *et al.*, 1979). Major difficulties may also be encountered with oxygen transport in aerobic processes. However, immobilized living whole cells are much less vulnerable to microbial contamination than soluble or immobilized enzymes, and the requirements for aseptic processing are thus less critical. Immobilization frequently significantly improves operational stability, which is critical for overall production costs.

In spite of the relatively few commercial applications of immobilized-whole-cell biocatalysts today (Table I), this survey clearly illustrates the vast future potential. It is likely that the main applications in the near future will continue to be in the food industry and in the production of fine chemicals, but favorable economics and recent developments in immobilized-living-cell technology offer many possibilities even for bulk chemicals, in particular for improved ethanol production from low-cost renewable carbon feedstocks and waste materials. More work is needed to develop better, more stable, more active specific biocatalysts, to investigate in detail the kinetics and process dynamics of immobilized-whole-cell systems, necessary for process control and optimization, and to develop biocatalysts and reactor designs with improved oxygen-transport characteristics. One report has already been published on the production of oxygen inside the biocatalyst particle (Holst *et al.*, 1981). The possibilities both of conventional and genetic engineering techniques for improved biocatalysts should be fully exploited; indeed, the first successful application of genetic engineering to improve immobilized-whole-cell biocatalyst activity has been reported (Klein and Wagner, 1980).

It should be finally emphasized that biocatalysis will not be the solution to all problems, but vast new possibilities may exist in areas such as hydrocarbon oxidation, hydrogen production, and nitrogen fixation. Through combined application of chemical and genetic engineering, biochemistry, biotechnology, and microbiology, immobilized-whole-cell biocatalyst engineering is likely to lead to major breakthroughs in the near future.

REFERENCES

Abbott, B. J. (1977). *Annu. Rep. Ferment. Processes* **1,** 205–233.

Abbott, B. J. (1978). *Annu. Rep. Ferment. Processes* **2,** 91–123.

Abbott, B. J., and Gerhardt, P. (1970a). *Biotechnol. Bioeng.* **12,** 501.

Abbott, B. J., and Gerhardt, P. (1970b). *Biotechnol. Bioeng.* **12,** 603.

Ado, T., Kimura, K., and Samejima, H. (1980). *Enzyme Eng.* **5,** 295–304.

Ado, Y., Suzuki, Y., Tadokoro, T., Kimura, K., and Samejima, H. (1979). *J. Solid-Phase Biochem.* **4,** 43.

Ahn, B. Y., and Byun, S. M. (1979a). *Korean J. Food Sci. Technol.* **11,** 192.

Ahn, B. Y., and Byun, S. M. (1979b). *Korean J. Food Sci. Technol.* **11,** 249.

Amin, M., Zedan, H., and El-Tayeb, O. (1980). *Proc. Int. Ferment. Symp., 6th, 1980* Abstracts, p. 121.

Amotz, S., Nielsen, T. K., and Thiessen, N. O. (1976). U.S. Patent 3,980,521.

Anonymous (1976a). *Chem. Eng. (N.Y.)* **83,** 87.

Anonymous (1976b). *Z. Zuckerind.* **26,** 470.

Anonymous (1977). *New Sci.* **74,** 401.

Anonymous (1979). *Food Eng. Int.* **4**(5), 19.

Anonymous (1980a). *F. O. Licht's Int. Sweetener Rep.* **3,** 45.

Anonymous (1980b). *Food Eng. Int.* **5**(4), 11.

Anonymous (1980c). *F. O. Licht's Int. Sweetener Rep.* **3,** 53.

Anonymous (1981). *Biomass Dig.* **2,** 1.

Antczak, T., Bielicki, S., and Galas, E. (1981). *Abstr. Commun. Eur. Congr. Biotechnol., 2nd, 1981* p. 155.

Antrim, R. L., Colilla, W., and Schnyder, B. J. (1979). *Appl. Biochem. Bioeng.* **2,** 98–156.

Arcuri, E. J., Worden, R. M., and Shumate, S. E., II (1980a). *Biotechnol. Lett.* **2,** 499.

Arcuri, E. J., Shumate, S. E., II, and Gibson, S. M. (1980b). *Proc. Int. Ferment. Symp., 6th, 1980* Abstracts, p. 59.

Arinbasarova, A. Y., and Koshcheyenko, K. A. (1980). *Prikl. Biokhim. Mikrobiol.* **16,** 854.

Asada, M., Morimoto, K., Nakanishi, K., Matsuno, R., Tanaka, A., Kimura, A., and Kamikubo, T. (1979). *Agric. Biol. Chem.* **43,** 1773.

Atrat, P., Hueller, E., and Hoerhold, C. (1980a). *Z. Allg. Mikrobiol.* **20,** 79.

Atrat, P., Hueller, E., Hoerhold, C., Buchar, M. J., Arinba, A. Y., and Koscheenko, K. A. (1980b). *Z. Allg. Mikrobiol.* **20,** 159.

Aunstrup, K. (1977). *In* "Biotechnological Applications of Proteins and Enzymes" (Z. Bohak and N. Sharon, eds.), pp. 39–62. Academic Press, New York.

Bachman, S., Gebicka, L., and Gasyna, Z. (1981). *Stärke* **33,** 63.

Bagnall, R. (1978). *Chem. Ind. (London)* **14,** 598.

Baker, D. A., and Kirsop, B. H. (1973). *J. Inst. Brew.* **79,** 487.

Bang, W. G., Lang, S., and Wagner, F. (1978). *Prepr., Eur. Congr. Biotechnol., 1st, 1978,* Part 1, pp. 186–189.

Barker, S. A. (1975). *Process Biochem.* **10**(10), 39.
Barker, S. A., Kay, I. M., and Kennedy, J. F. (1975). U.S. Patent 3,912,593.
Barker, S. A., Greenshields, R. N., Humphreys, J. D., and Kennedy, J. F. (1979). U.S. Patent 4,155,813.
Bennett, M. A., and Weetall, H. H. (1976). *J. Solid-Phase Biochem.* **1**, 137.
Berdelle, H. P. (1972). *Brauwelt* **112**, 24.
Berdelle, H. P. (1975). U.S. Patent 3,769,175.
Berger, U., and Langhammer, G. (1980). *Z. Allg. Mikrobiol.* **20**, 69.
Bernfeld, P., and Wan, J. (1963). *Science* **142**, 678.
Bhasin, D. P., Cryte, C. C., and Studebaker, J. F. (1976). *Biotechnol. Bioeng.* **18**, 1777.
Bhatia, M., and Prabha, K. A. (1980). *Biotechnol. Bioeng.* **22**, 1957.
Birnbaum, S., Larsson, P.-O., and Mosbach, K. (1981). *Abstr. Commun., Eur. Congr. Biotechnol., 2nd, 1981* p. 150.
Bitton, G., and Marshall, K. C. (1980). "Adsorption of Microorganisms to Surfaces." Wiley, New York.
Briffaud, J., and Engasser, J. M. (1978). *Prepr., Eur. Congr. Biotechnol., 1st, 1978* Part 2, pp. 133–134.
Brodelius, P. (1978). *Adv. Biochem. Eng.* **10**, 75–129.
Brodelius, P., and Mosbach, K. (1980). European Patent Appl. 18,333.
Brodelius, P., Hägerdal, B., and Mosbach, K. (1980). *Enzyme Eng.* **5**, 383–387.
Brodelius, P., Nilsson, K., and Mosbach, K. (1981). *Adv. Biotechnol.* **3**, 373–376.
Bucke, C. (1977). *Top. Enzyme Ferment. Biotechnol.* **1**, 147–171.
Buonocore, V., DeRosa, M., Gambacorta, A., Lama, L., and Nicolaus, B. (1981). *Abstr. Commun. Eur. Congr. Biotechnol., 2nd, 1981* p. 291.
Cape, R. E., Amon, W. F., and Neidelman, S. I. (1980). *Biotechnol. Lett.* **2**, 199.
Casey, J. P. (1977). *Staerke* **29**, 196.
Chang, T. M. S., ed. (1977). "Biomedical Applications of Immobilized Enzymes and Proteins," Vols. 1 and 2. Plenum, New York.
Charles, M. (1980). *Proc. Int. Ferment. Symp., 6th, 1980* Abstracts, p. 120.
Cheetham, P. S. J. (1980). *Top. Enzyme Ferment. Biotechnol.* **4**, 189–258.
Chen, A. H., and Jao, Y. C. (1980). Ger. Offen. 2,936,486.
Chen, L. F., Gong, C. S., and Tsao, G. T. (1981). *Stärke* **33**, 58.
Chen, W.-P. (1980a). *Process Biochem.* **15**(5), 30.
Chen, W.-P. (1980b). *Process Biochem.* **15**(6), 36.
Chibata, I., ed. (1978a). "Immobilized Enzymes. Research and Development." Halsted Press, New York.
Chibata, I. (1978b). U.S. Patent 4,138,292.
Chibata, I. (1980a). *In* "Food Process Engineering" (P. Linko and J. Larinkari, eds.), pps. 1–26. Applied Science Publishers, London.
Chibata, I. (1980b). *Enzyme Eng.* **5**, 393–400.
Chibata, I., and Tosa, T. (1977). *Adv. Appl. Microbiol.* **22**, 1–27.
Chibata, I., Tosa, T., and Sato, T. (1974a). *Appl. Microbiol.* **27**, 878.
Chibata, I., Tosa, T., Sato, T., and Mori, T. (1974b). U.S. Patent 3,816,254.
Chibata, I., Osaka, S., Tosa, T., Sato, T., and Tadashi, T. (1974c). German Patent 2,414,128.
Chibata, I., Tosa, T., and Sato, T. (1974d). German Patent 2,420,102.
Chibata, I., Tosa, T., Sato, T., Mori, T., and Yamamoto, K. (1974e). *Enzyme Eng.* **2**, 303–313.
Chibata, I., Tosa, T., and Sato, T. (1974f). Japan Kokai 74/25,189.
Chibata, I., Tosa, T., Sato, T., and Yamamoto, K. (1974g). Japan Kokai 74/26,486.

Chibata, I., Tosa, T., and Mori, T. (1974h). Japan Kokai 74/30,582.
Chibata, I., Tosa, T., Sato, T., and Yamamoto, K. (1974i). Japan Kokai 74/75,782.
Chibata, I., Kakimoto, K., and Nabe, K. (1974j). Japan Kokai 74/81,590.
Chibata, I., Kakimoto, T., and Nabe, K. (1974k). Japan Kokai 74/81,591.
Chibata, I., Tosa, T., Sato, T., and Yamamoto, K. (1975a). Japan Kokai 75/100,289.
Chibata, I., Kakimoto, T., Nishimura, N., and Nabe, K. (1975b). Japan Kokai 75/126,884.
Chibata, I., Kato, J., Watanabe, T., and Uchida, T. (1975c). Japan Kokai 75/135,290.
Chibata, I., Tosa, T., Sato, T., and Yamamoto, K. (1975d). U.S. Patent 3,886,040.
Chibata, I., Tosa, T., Sato, T., and Yamamoto, K. (1975e). U.S. Patent 3,898,127.
Chibata, I., Tosa, T., Sato, T., and Yamamoto, K. (1975f). U.S. Patent 3,898,128.
Chibata, I., Tosa, T., Sato, T., and Yamamoto, K. (1975g). U.S. Patent 3,922,195.
Chibata, I., Tosa, T., and Mori, T. (1976a). Japan Kokai 76/63,985.
Chibata, I., Tosa, T., and Sato, T. (1976b). *In* "Methods in Enzymology" (K. Mosbach, ed.), Vol. 44, pp. 739–746. Academic Press, New York.
Chibata, I., Tosa, T., and Sato, T. (1976c). U.S. Patent 3,953,291.
Chibata, I., Tosa, T., and Mori, T. (1977). *In* "Biomedical Applications of Immobilized Enzymes and Proteins" (T. M. S. Chang, ed.), Vol. 1, pp. 257–279. Plenum, New York.
Chibata, I., Tosa, T., and Sato, T. (1978a). U.S. Patent 4,081,327.
Chibata, I., Tosa, T., and Yamamoto, K. (1978b). *Enzyme Eng.* **3,** 463–468.
Chibata, I., Tosa, T., Sato, T., Yamamoto, K., Takata, I., and Nishida, Y. (1978c). *Enzyme Eng.* **4,** 335–337.
Chibata, I., Koto, J., Wada, M., Uchida, Y., and Murata, K. (1979a). Japan Kokai 79/135,295.
Chibata, I., Kato, J., and Murata, K. (1979b). Japan Kokai 79/138,190.
Chibata, I., Tosa, T., and Sato, T. (1979c). *In* "Microbial Technology" (H. J. Peppler and D. Perlman, eds.), 2nd ed., Vol. 2, pp. 433–461. Academic Press, New York.
Chiou, C.-J. (1979). *Chung-kuo Yeh Hua Hseuh Hai Chih* **17**(3–4), 138.
Choudhari, Q. A., and Pirt, S. J. (1966). *J. Gen. Microbiol.* **43,** 71.
Chu, Y., Tani, Y., Lee, T., and Yu, T. (1977). *Hanguk Sikp'um Kwanakhoe Chi* **9,** 183.
Chua, J. W., Erarslan, A., Kinoshita, S., and Taguchi, H. (1980). *J. Ferment. Technol.* **58,** 123.
Cizmek, S., and Drazik, M. (1981). *Abstr., Eur. Congr. Biotechnol., 2nd, 1981* p. 151.
Compere, A. L., and Griffith, W. L. (1975). *Dev. Ind. Microbiol.* **17,** 247.
Compere, A. L., and Griffith, W. L. (1977). *Dev. Ind. Microbiol.* **18,** 717.
Constantinides, A. (1980). *Biotechnol. Bioeng.* **22,** 119.
Corrieu, G., Blachere, H., Ramirez, A., Navarro, J. M., Durand, G., Duteutre, B., and Moll, M. (1976). *Abstr. Pap., Int. Ferment. Symp., 5th, 1976* p. 294.
Cotter, W. P., Lloyd, N. E., and Hinman, C. W. (1971). U.S. Patent 3,623,953.
Couderc, R., and Baratti, J. (1980). *Biotechnol. Bioeng.* **22,** 1135.
Coulman, G. A., Stieber, R. W., and Gerhardt, P. (1977). *Appl. Environ. Microbiol.* **34,** 725.
Crocco, S. (1976). *Food Eng. Int.* **1**(12), 30.
Crocco, S. (1977). *Food Eng. Int.* **2**(2), 41.
Cysewski, G. R., and Wilke, C. R. (1977). *Biotechnol. Bioeng.* **19,** 1125.
Cysewski, G. R., and Wilke, C. R. (1978). *Biotechnol. Bioeng.* **20,** 1421.
Daugulis, A. J., Brown, N. M., Cluett, W. R., and Dunlop, D. B. (1981). *Biotechnol. Lett.* **3,** 651.

de Bont, J. A. M., Tramper, J., and Luyben, K. C. A. M. (1981). *Abstr. Commun., Eur. Congr. Biotechnol., 2nd, 1981* p. 158.
Decottignies-Le Maréchal, P., Calderón, R., Vandecaateele, J. P., and Azerad, R. (1979). *Eur. J. Appl. Microbiol. Biotechnol.* **7,** 33.
Dellweg, H., and Misselhorn, K. (1978). *DECHEMA-Mongr.* **83,** 35–42.
DeRosa, M., Gambacorta, A., Nicolaus, B., and Buonocore, V. (1980a). *Biotechnol. Lett.* **2,** 29.
DeRosa, M., Gambacorta, A., Esposito, E., Drioli, E., and Gaeta, S. (1980b). *Biochimie* **62,** 517.
DeRosa, M., Gambacorta, A., and Esposito, E. (1981). *Biotechnol. Bioeng.* **23,** 221.
Diers, I. (1976). *In* "Continuous Culture" (A. C. R. Dean, D. C. E. Wood, C. G. T. Evans, and J. Melling, eds.), pp. 208–223. Ellis Horwood, Chichester.
Dinelli, D. (1972). *Process Biochem.* **7**(8), 9.
Divies, C. (1977a). French Demande 2,320,349.
Divies, C. (1977b). Belgian Patent 844,766.
Divies, C., and Siess, M. H. (1976). *Ann. Microbiol. (Paris)* **127B,** 525.
Dohan, L. A., Baret, J. L., Pain, S., and Delalande, P. *In* "Food Process Engineering" (P. Linko and J. Larinkari, eds.), pps. 137–147. Applied Science Publishers, London.
Dommergues, Y. R., Diem, H. G., and Divies, C. (1979). *Appl. Environ. Microbiol.* **37,** 779.
Drioli, E., Iorjo, G., Molinari, R., DeRosa, M., Gambacorto, A., and Esposito, E. (1981). *Biotechnol. Bioeng.* **23,** 221.
D'Souza, S. F., and Nadkarni, G. B. (1980a). *Enzyme Microb. Technol.* **2,** 217.
D'Souza, S. F., and Nadkarni, G. B. (1980b). *Biotechnol. Bioeng.* **17,** 2179.
D'Souza, S. F., and Nadkarni, G. B. (1980c). *Biotechnol. Bioeng.* **17,** 2191.
Duarte, J. M. C., and Lilly, M. D. (1980). *Enzyme Eng.* **5,** 363–367.
Durand, G., and Navarro, J. M. (1978). *Process Biochem.* **14**(9), 14.
Dworschack, R. G., Chen, J. C., Khwaja, A., and White, W. H. (1973). Ger. Offen. 2,251,855.
Ehrenthal, I., and Miner, K. E. (1980). U.S. Patent 4,208,482.
Engasser, J.-M., and Horwath, C. (1976). *Appl. Biochem. Bioeng.* **1,** 127–220.
Engelbart, W., and Dellweg, H. (1976). *Abstr. Pap., Int. Ferment. Symp., 5th, 1976,* p. 376.
Fetzer, W. R. (1930). *Food Ind.* **2,** 130.
Fogarty, W. K., and Griffin, P. J. (1973). *Biotechnol. Bioeng.* **23,** 401.
Franks, M. E. (1971). *Biochim. Biophys. Acta* **252,** 246.
Franks, N. E. (1972). *Biotechnol. Bioeng. Symp.* **3,** 327.
Freeman, A., and Aharonowitz, Y. (1980). *Int. Ferment. Symp., 6th, 1980* Abstracts, p. 121.
Friedman, M. R., and Gaden, E. L., Jr. (1970). *Biotechnol. Bioeng.* **12,** 961.
Fujii, T., Hanamitsu, K., Izumi, R., Yamaguchi, T., and Watanabe, T. (1973). Japan Kokai 73/99,393.
Fukui, S., Yamamoto, T., and Iida, I. (1977). Ger. Offen. 2,629,692.
Fukui, S., Tanaka, A., and Gellf, G. (1979). *Enzyme Eng.* **4,** 299–306.
Fukui, S., Omata, T., Yamane, T., and Tanaka, A. (1980a). *Enzyme Eng.* **5,** 347–353.
Fukui, S., Ahmed, A., Omata, T., and Tanaka, A. (1980b). *Eur. J. Appl. Microbiol. Biotechnol.* **10,** 289.
Fukui, S., Soda, E., Tanaka, A., Yamane, T., and Komata, T. (1980c). Japan Kokai 80/15,760.
Fukui, S., Sanomoto, K., Itok, N., and Tanaka, A. (1980d). *Biochimie* **62,** 381.

Fukui, S., Omata, T., Imamoto, N., Kimura, T., and Tanaka, A. (1981). *Abstr. Commun., Eur. Congr. Biotechnol., 2nd, 1981* p. 44.
Fukushima, M., Fujii, T., Matsumoto, L., and Morishita, M. (1976). Japan Kokai 76/70,884.
Furuhashi, K., Taoka, A., and Uchida, S. (1980). Ger. Offen. 2,931,148.
Gainer, J. L., Kirwan, D. J., Foster, J. A., and Seyhan, E. (1980). *Biotechnol. Bioeng. Symp.* **10**, 35.
Genser, M. A., and Mutharsan, R. (1981). *Adv. Biotechnol.* **1**, 627–633.
Genung, R. K., Pitt, W. W., Jr., Davis, M., and Koon, J. H. (1980). *Biotechnol. Bioeng. Symp.* **12**, 295.
Germain, P., Miclo, R., Glomon, C., Engasser, J. M., and Nguyen, Q. T. (1981). *Abstr. Commun., Eur. Congr. Biotechnol., 2nd, 1981* p. 245.
Gestrelius, S. M. (1979). British Patent 2,019,410.
Gestrelius, S. M. (1980). *Enzyme Eng.* **5**, 439–442.
Ghommidh, C., Navarro, J. M., and Durand, G. (1981). *Biotechnol. Lett.* **3**, 93.
Ghose, T. K., and Bandyopadhyay, K. K. (1979). *Proc. Natl. Semin. Immobilized Enzyme Eng., 1st, 1979* pp. 135–138.
Ghose, T. K., and Ghand, S. (1978). *J. Ferment. Technol.* **18**, 53.
Ghose, T. K., and Kannan, V. (1979). *Enzyme Microbiol. Technol.* **1**, 47.
Gianfreda, L., Parascandola, P., and Scardi, V. (1980). *Eur. J. Appl. Microbiol. Biotechnol.* **11**, 6.
Giovenco, B., and Maimone, A. (1981). *Abstr. Commun., Eur. Congr. Biotechnol., 2nd, 1981* p. 131.
Godbole, S. S., D'Souza, S. F., and Nadkarni, G. B. (1980). *Enzyme Microb. Technol.* **2**, 223.
Goldstein, H., Barru, P. W., Rizzuto, A. B., Venkatasubramanian, K., and Vieth, W. R. (1977). *J. Ferment. Technol.* **55**, 516.
Goto, S., Enomoto, S., Takahashi, Y., and Motomatsu, R. (1971). *Jpn. J. Microbiol.* **15**, 317.
Greenshield, R. N. (1978). *Econ. Microbiol.* **2**, 121–186.
Griffith, W. L., and Compere, A. L. (1975). *Dev. Ind. Microbiol.* **17**, 241.
Griffith, W. L., and Compere, A. L. (1977). *Dev. Ind. Microbiol.* **18**, 723.
Griffith, W. L., and Compere, A. L. (1978). U.S. Patent 4,127,447.
Griffiths, M. W., and Muir, D. D. (1978). *J. Sci. Food Agric.* **29**, 753.
Griffiths, M. W., and Muir, D. D. (1980). *J. Sci. Food Agric.* **31**, 397.
Grinbergs, M., Hildebrand, R. B., and Clarke, B. J. (1977). *J. Inst. Brew.* **83**, 25.
Grote, W., Lee, K. J., and Rogers, P. L. (1980). *Biotechnol. Lett.* **2**, 481.
Guillaume, J., and Plichon, B. (1975). French Demande 2,277,097.
Gulaya, V. E., Turková, J., Jirku, V., Frydrychová, A., Coupek, J., and Ananchenko, S. W. (1979).
Hackel, U., Klein, J., Magnet, R., and Wagner, F. (1975). *Eur. J. Appl. Microbiol.* **1**, 291.
Hägerdal, B., and Mosbach, K. (1980). *Food Process Eng. [Proc. Int. Congr.], 2nd, 1979* Vol. 2, pp. 129–131.
Häggström, L., and Molin, N. (1980). *Biotechnol. Lett.* **2**, 241.
Häggström, L. (1981). *Adv. Biotechnol.* **2**, 79–83.
Hamilton, B. K., Cotton, C. K., and Cooney, C. L. (1974). *In* "Immobilized Enzymes in Food and Microbial Processes" (A. C. Olson and C. L. Cooney, eds.), pp. 85–131. Plenum, New York.
Hansen, R. C. (1980). *North Eur. Dairy. J.* **46**(1–2), 10.

Hardt, F. R., Young, J. C., Clesceri, L. S., and Washington, D. R. (1971). *Environ. Sci. Technol.* **5**, 345.
Hartmeier, W. (1981). *Adv. Biotechnol.* **3**, 377–382.
Hastings, J. J. H. (1978). *Econ. Microbiol.* **2**, 30–45.
Hattori, T., and Furusaka, C. (1960). *J. Biochem. (Tokyo)* **48**, 831.
Hattori, T., and Furusaka, C. (1961). *J. Biochem. (Tokyo)* **50**, 312.
Heady, R. E., and Jacaway, W. A., Jr. (1974). U.S. Patent 3,847,740.
Heinrich, M., and Rehm, H. J. (1981). *Eur. J. Appl. Microbiol. Biotechnol.* **11**, 139.
Hemmingsen, S. H. (1979). *Appl. Biochem. Bioeng.* **2**, 157–184.
Hino, T., Yamada, H., Okamura, S., Kojima, H., Okamoto, Y., and Ito, Y. (1976). Japan Kokai 76/144,780.
Hino, T., Yamada, H., and Okamura, S. (1979). U.S. Patent 4,148,689.
Hirano, K., Karube, I., and Suzuki, S. (1977a). *Biotechnol. Bioeng.* **19**, 311.
Hirano, K., Karube, I., Matsunaga, T., and Suzuki, S. (1977b). *J. Ferment. Technol.* **55**, 401.
Hochster, R. M., and Watson, R. W. (1954). *Arch. Biochem. Biophys.* **48**, 120.
Hoffman, C. H., Harris, E., Chodroff, S., Micholson, S., Rothbrock, J. W., and Paterson, E. (1970). *Biochim. Biophys. Acta* **41**, 710.
Holst, O., Enfors, S. O., and Mattiasson, B. (1981). *Abstr. Commun., Eur. Congr. Biotechnol., 2nd, 1981* p. 156.
Hospodka, J. (1966). *In* "Theoretical and Methodological Basis of Continuous Culture of Microorganisms" (I. Màlek and Z. Fencl, eds.), pp. 495, 545. Academic Press, New York.
Hough, J. S., and Button, R. H. (1972). *Prog. Ind. Microbiol.* **11**, 89–132.
Hough, J. S., and Lyons, T. P. (1972). *Nature (London)* **235**, 389.
Hupkes, J. V., and van Tilburg, R. (1976). *Staerke* **28**, 356.
Hyrkäs, K., Viskari, R., Linko, Y.-Y., and Linko, M. (1976). *Milchwissenschaft* **31**, 129.
Inoue, H., Maekawa, N., Takenaka, N., Otaki, H., and Matsumoto, H. (1978). Japan Kokai 78/20,484.
Intermag Geträнke Technik A. G. (1969). Ger. Offen. 1,517,814.
Ishimatsu, Y., Shigesada, S., and Kimura, S. (1976). Japan Kokai 76/44,688.
Ito, H., Ijuku, Y., Sugihara, S., and Ozawa, S. (1977). Japan Kokai 77/125,696.
Ito, Y., and Ozawa, O. (1976). Japan Kokai 76/104,046.
Jack, T. R., and Zajic, J. E. (1977a). *Biotechnol. Bioeng.* **19**, 631.
Jack, T. R., and Zajic, J. E. (1977b). *Adv. Biochem. Eng.* **5**, 125–245.
Jirku, V., Turkova, J., Veruovič, C., and Kubánek, V. (1980). *Biotechnol. Lett.* **2**, 451.
Johnson, D. E., and Ciegler, A. (1969). *Arch. Biochem. Biophys.* **130**, 384.
Kakimoto, T., Shibatani, T., Nishimura, N., and Chibata, I. (1971). *Appl. Microbiol.* **22**, 992.
Kan, J. K., and Shuler, M. L. (1978a). *AIChE Symp. Ser.* **172**(74), 31.
Kan, J. K., and Shuler, M. L. (1978b). *Biotechnol. Bioeng.* **20**, 217.
Kanamitsu, O. (1975). Japan Kokai 75/132,181.
Kanno, T., Watanabe, H., and Sano, S. (1978). U.S. Patent 4,106,987.
Karonen, R., Poutanen, K., Linko, Y.-Y., and Linko, P. (1980). *Food Process Eng. [Proc. Int. Congr.], 2nd, 1979,* Vol. 2, pp. 123–128.
Karube, I., Matsunaga, T., Tsuru, S., and Suzuki, S. (1976). *Biochim. Biophys. Acta* **444**, 338.
Karube, I., Hirano, K., and Suzuki, S. (1977a). *Biotechnol. Bioeng.* **19**, 1233.
Karube, I., Matsunaga, T., Tsuru, S., and Suzuki, S. (1977b). *Biotechnol. Bioeng.* **19**, 1727.

Kasumi, T., Tsumura, N., and Kobayashi, T. (1979). *Staerke* **31,** 25.
Kato, E. (1974). *Meiji Daigaku Kagaku Gijutsu Kenkyusho Nempo* **16,** 26.
Kawabata, Y., and Demain, A. L. (1979). *ACS Symp. Ser.* **106,** 133–146.
Keller, A. K., and Gerhart, P. (1975). *Biotechnol. Bioeng.* **18,** 997.
Kennedy, J. F. (1978). *Enzyme Eng.* **4,** 323–328.
Kennedy, J. F., Barker, S. A., and Humphreys, J. D. (1976). *Nature (London)* **262,** 242.
Kennedy, J. F., Humphreys, J. D., Barker, S. A., and Greenshield, R. N. (1980). *Enzyme Microb. Technol.* **2,** 209.
Khoroshilova, V. G. (1978). *Pishch. Prom-st., Ser. 1* No. 12, p. 22.
Kierstan, M., and Bucke, C. (1977). *Biotechnol. Bioeng.* **19,** 387.
Kieslich, K. (1980). *Econ. Microbiol.* **5,** 370–467.
Kieslich, K., and Sebek, O. K. (1979). *Annu. Rep. Ferment. Processes* **3,** 275–304.
Kimura, K., Tatsuyomi, Y., Mizushima, N., Tanaka, A., Matsuno, R., and Fukuda, H. (1978). *Eur. J. Appl. Microbiol. Biotechnol.* **5,** 13.
Kimura, K., Tatsutomi, Y., Matsuno, R., Tanka, A., and Fukuda, H. (1981). *Eur. J. Appl. Microbiol. Biotechnol.* **11,** 78.
Kinoshita, S., Masaru, M., and Okada, H. (1975). *J. Ferment. Technol.* **53,** 223.
Kirwan, D., and Gainer, J. (1976). *Natl. Sci. Found., Res. Appl. Natl. Needs [Rep.] NSF/RA (U.S.)* **NSF/RA-760180,** 31.
Kitai, A., and Ozaki, A. (1969). *J. Ferment. Technol.* **47,** 527.
Klein, J., and Kluge, M. (1981). *Biotechol. Lett.* **3,** 65.
Klein, J., and Schara, P. (1981). *Appl. Biochem. Biotechnol.* **6,** 91.
Klein, J., and Wagner, F. (1978). *DECHEMA-Mongr.* **82,** 142–164.
Klein, J., and Wagner, F. (1980). *Enzyme Eng.* **5,** 335–345.
Klein, J., and Wagner, F. (1981). *Abstr. Commun., Eur. Congr. Biotechnol, 2nd, 1981* p. 118.
Klein, J., Hackel, U., Schara, P., Washausen, P., Wagner, F., and Martin, C. K. A. (1976). *Abstr. Pap., Int. Ferment. Symp., 5th, 1976* p. 295.
Klein, J., Hackel, U., Schara, P., Washausen, P. Wagner, F., and Martin, C. K. A. (1978). *Enzyme Eng.* **4,** 339–341.
Klein, J., Hackel, U., and Wagner, F. (1979). *ACS Symp. Ser.* **106,** 101–118.
Klibanov, A. M., and Puglisi, A. V. (1980). *Biotechnol. Lett.* **2,** 445.
Kobayashi, H., and Suzuki, H. (1972). *J. Ferment. Technol.* **50,** 625.
Kobayashi, H., and Suzuki, H. (1976). *Biotechnol. Bioeng.* **18,** 37.
Kobayashi, H., and Suzuki, H. (1980). *Biotechnol. Bioeng.* **22,** 505.
Kokubu, T., and Suzuki, S. (1981). *Biotechnol. Bioeng.* **23,** 29.
Kokubu, T., Karube, I., and Suzuki, S. (1978). *Eur. J. Appl. Microbiol. Biotechnol.* **5,** 233.
Kolarik, M. J., Chen, B. J., Emery, A. H., and Lim, H. C. (1974). *In* "Immobilized Enzymes in Food and Microbial Processes" (A. Olson and C. L. Cooney, eds.), pp. 71–83. Plenum, New York.
Kolot, F. B. (1980). *Process Biochem.* **15**(7), 2.
Kolpakchi, A. P., Isaeva, V. S., Zhvirblyanskaya, A. Y., Kazantsev, E. N., Serova, E. N., and Rettel, E. N. (1976). *Prikl. Biokhim. Mikrobiol.* **12**(6), 866.
Kominek, L. (1975a). U.S. Patent 3,915,802.
Kominek, L. (1975b). *Antimicrob. Agents Chemother.* **7,** 861.
Kondo, E. (1962). *Agric. Biol. Chem.* **27,** 69.
Kondo, E., and Matsuo, E. (1960). *J. Gen. Appl. Microbiol.* **7,** 113.
Konecny, J., and Sieber, M. (1980). *Enzyme Eng.* **5,** 415–422.

Koshcheyenko, K. A., Krasnova, L. A., Garcia-Rodriques, L. K., Surovtsev, V. I., and Skryabin, G. K. (1976). *Appl. Biochem. Microbiol.* **12,** 259.

Kosugi, Y., and Suzuki, H. (1973). *J. Ferment. Technol.* **51,** 895.

Kozlova, Y. I., Baranova, I. P., and Egorov, N. S. (1980). *Antibiotiki* **25,** 870.

Kristiansen, B., and Charley, R. (1981). *Adv. Biotechnol* **1,** 221–227.

Kristiansen, B., and Sinclair, C. G. (1979). *Biotechnol. Bioeng.* **21,** 297.

Kristiansen, B., and Sinclair, C. G. (1981). *Abstr. Commun., Eur. Congr. Biotechnol., 2nd, 1981* p. 268.

Krouwel, P. G., and Kossen, N. W. (1980). *Biotechnol. Bioeng.* **22,** 681.

Krouwel, P. G., and Kossen, N. W. (1981). *Abstr. Commun., Eur. Congr. Biotechnol., 2nd, 1981* p. 117.

Krouwel, P. G., van der Laan, W. F. M., and Kossen, N. W. F. (1980). *Biotechnol. Lett.* **2,** 253.

Krouwel, P. G., van der Laan, W. F. M., and Kossen, N. W. F. (1981). *Biotechnol. Lett.* **3,** 158.

Kubo, T., Komatsu, T., Kawai, K., Sugihara, T., and Yamada, Y. (1977). Japan Kokai 77/61,243.

Kumakura, M., Yoshida, M., and Kaetsu, I. (1977). Japan Kokai 77/15,888.

Kumakura, M., Yoshida, M., and Kaetsu, I. (1978). *J. Solid-Phase Biochem.* **3,** 175.

Kumakura, M., Yoshida, M., and Kaetsu, I. (1979a). *Appl. Environ. Microbiol.* **37,** 310.

Kumakura, M., Yoshida, M., and Kaetsu, I. (1979b). *Biotechnol. Bioeng.* **21,** 679.

Kutsal, T., and Çağlar, M. A. (1980). *Int. Ferment. Symp., 6th, 1980* Abstracts, p. 122.

Kuyucak, N., Kutsal, T., and Caglar, M. A. (1981). *Abstr. Commun. Eur. Congr. Biotechnol., 2nd, 1981* p. 93.

Lambert, G. R., Dadey, A., and Smith, G. D. (1979). *FEBS Lett.* **101,** 125.

Lantero, O. J. (1978). *Enzyme Eng.* **4,** 349–352.

Larsson, P.-O., and Mosbach, K. (1976). *In* "Methods in Enzymology" (K. Mosbach, ed.), Vol. 44, pp. 183–190. Academic Press, New York.

Larsson, P.-O., and Mosbach, K. (1979). *Biotechnol. Lett.* **1,** 501.

Larsson, P.-O., Ohlson, S., and Mosbach, K. (1976). *Nature (London)* **263,** 796.

Larsson, P.-O., Ohlson, S., and Mosbach, K. (1978). *Enzyme Eng.* **4,** 317–322.

Larsson, P.-O., Birnbaum, S., and Mosbach, K. (1981). *Adv. Biotechnol* **1,** 717–720.

Lartigan, G., Bouniot, A., and Guerineau, M. (1980). French Demande 2,430,976.

Lartique, D. J., and Weetall, H. H. (1976). U.S. Patent 3,939,041.

Lee, C. K., and Long, M. E. (1974). U.S. Patent 3,821,086.

Lee, Y.-Y., Wun, F., and Tsao, G. T. (1975). *In* "Immobilized Enzyme Technology" (H. H. Weetall and S. Suzuki, eds.), pp. 129–149. Plenum, New York.

Leroy, P., Devos, F., and Hucheite, M. (1974). Ger. Patent 2,417,642.

Leuschner, F. (1966). Ger. Patent 122,855.

Lilly, M. D. (1978). *DECHEMA Monogr.* **82,** 165–180.

Lindroos, A., Linko, Y.-Y., and Linko, P. (1980). *Food Process Eng.* [*Proc. Int. Congr.*], *2nd, 1979* Vol 2, pp. 92–102.

Linko, P. (1980). *In* "Food Process Engineering" (P. Linko and J. Larinkari, eds.), pps. 27–39. Applied Science Publishers, London.

Linko, P. (1981). *Adv. Biotechnol.* **1,** 711–716.

Linko, P. (1982). *In* "Nutritive Sweeteners" (G. Birch, ed.), pp. 109–131. Appl. Sci. Publ., London.

Linko, P., and Linko, Y.-Y. (1981). *74th Annu. AIChE Meet. 1981* Paper 69a.
Linko, P., and Linko, Y.-Y. (1982). *Enzyme Eng.* **6,** 335–342.
Linko, P., Poutanen, K., Weckström, L., and Linko, Y.-Y. (1980). *Biochimie* **62,** 387.
Linko, P., Poutanen, K., and Linko, Y.-Y. (1981a). *Abstr. Commun., Eur. Congr. Biotechnol., 2nd, 1981* p. 116.
Linko, P., Poutanen, K., and Linko, Y.-Y. (1981b). *J. Mol. Catal.* **13,** 263.
Linko, Y.-Y., and Linko, P. (1981a). *Biotechnol. Lett.* **3,** 21.
Linko, Y.-Y., and Linko, P. (1981b). *Abstr. Commun., Eur. Congr. Biotechnol., 2nd, 1981* p. 56.
Linko, Y.-Y., Pohjola, L., Viskari, R., and Linko, M. (1976). *FEBS Lett.* **62**(1), 77.
Linko, Y.-Y., Pohjola, L., and Linko, P. (1977a). *Process Biochem.* **12**(6), 14.
Linko, Y.-Y., Viskari, R., Pohjola, L., and Linko, P. (1977b). *J. Solid-Phase Biochem.* **2,** 203.
Linko, Y.-Y., Pohjola, L., and Linko, P. (1978a). *In* "Bioconversion in Food Technology" (P. Linko, ed.), pp. 78–85. Offsetpress, Technical Research Centre in Finland, Espoo.
Linko, Y.-Y., Viskari, R., Pohjola, L., and Linko, P. (1978b). *Enzyme Eng.* **4,** 345–347.
Linko, Y.-Y., Poutanen, K., Weckström, L., and Linko, P. (1979). *Enzyme Microb. Technol.* **1,** 26.
Linko, Y.-Y., Weckström, L., and Linko, P. (1980a). *In* "Food Process Engineering" (P. Linko and J. Larinkari, eds.), pps. 81–91. Applied Science Publishers, London.
Linko, Y.-Y., Weckström, L., and Linko, P. (1980b). *Enzyme Eng.* **5,** 355–358.
Linko, Y.-Y., Jalanka, H., and Linko, P. (1981). *Biotechnol. Lett.* **3,** 263.
Littlejohn, J. H., and Dworschack, R. G. (1975). U.S. Patent 3,909,355.
Lloyd, N. E., Lewis, L. T., Logan, R. M., and Patel, D. N. (1972). U.S. Patent 3,694,314.
Lloyd, N. E., Lewis, L. T., Logand, R. N., and Pate, D. N. (1974). U.S. Patent 3,817,832.
Lockwood, L. B. (1979). *In* "Microbial Technology" (H. J. Peppler and D. Perlman, eds.), 2nd ed., Vol. 1, pp. 355–387. Academic Press, New York.
Long, M. E. (1976). U.S. Patent 3,935,069.
Long, M. E. (1977). U.S. Patent 4,060,456.
Lopèz-Leiva, M., Hägerdal, B., and Mattiasson, B. (1981). *Abstr. Commun., Eur. Congr. Biotechnol., 2nd, 1981* p. 167.
McGinnis, R. (1975). *Sugar J.* **38,** 8.
Maekawa, N., Inoue, H., Otaki, M., and Matsumoto, H. (1977). Japan Kokai 77/130,979.
Maekawa, N., Inoue, H., Takanaka, N., and Matsumoto, H. (1978). Japan Kokai 78/32,190.
Maekawa, N., Takenaka, N., and Matsumoto, H. (1979a). Japan Kokai 79/05095.
Maekawa, N., Takenaka, N., and Matsumoto, H. (1979b). Japan Kokai 79/05097.
Maekawa, N., Takenaka, N., and Matsumoto, H. (1979c). Japan Kokai 79/05,098.
Makarova, E. N., Melkonya, A. B., and Markosyan, L. S. (1979). *Biol. Zh. Arm.* **32,** 860.
Makiguchi, N., Arita, M., and Asai, Y. (1980a). *J. Ferment. Technol.* **58,** 17.
Makiguchi, N., Arita, M., and Asai, Y. (1980b). *J. Ferment. Technol.* **58,** 167.
Makover, S., and Pruess, D. L. (1975). U.S. Patent 3,907,639.
Mandenius, C. F., Danielsson, B., and Mattiasson, B. (1980). *Acta Chem. Scand.* **34,** 463.
Mandenius, C. F., Danielsson, B., and Mattiasson, B. (1981). *Abstr. Commun., Eur. Congr. Biotechnol., 2nd, 1981* p. 143.
Marconi, W. (1978). *DECHEMA Monogr.* **82,** 88–141.
Marconi, W., Bartoli, F., Cecere, F., Galli, G., and Morisi, F. (1975). *Agric. Biol. Chem.* **39,** 277.

Margalith, P., and Holcberg, I. (1981). *Abstr. Commun., Eur. Congr. Biotechnol., 2nd, 1981* p. 160.

Margaritis, A., and Rowe, C. E. (1981). *Adv. Biotechnol. [Proc. Int. Ferment. Symp.], 6th, 1980* p. 130.

Margaritis, A., Bajpai, B. K., and Wallace, J. B. (1981). *Biotechnol. Lett.* **3,** 613.

Marion, L. (1979). *Chem. Eng. (N.Y.)* **86**(5), 78.

Marshall, R. O. (1960). U.S. Patent 2,950,228.

Marshall, R. O., and Kooi, E. R. (1957). *Science* **125,** 648.

Martin, C. K. A., and Perlman, D. (1976a). *Biotechnol. Bioeng.* **18,** 217.

Martin, C. K. A., and Perlman, D. (1976b). *Abstr. Pap., Int. Ferment. Symp., 5th, 1976* p. 297.

Masaki, M., and Ishige, F. (1976). Japan Kokai 76/98,381.

Mason, J. R., Pirt, S. J., and Sommerville, H. J. (1978). *Enzyme Eng.* **4,** 343–347.

Mattiasson, B., Ramstorp, M., Wideback, K., and Kronvall, G. (1980). *J. Appl. Biochem.* **2,** 321.

Mayer, H., Collins, J., and Wagner, F. (1980). *Enzyme Eng.* **5,** 61–69.

Meisner, R., Merz, G., and Klefentz, H. (1980). Ger. Offen. 2,912,827.

Meng, G.-Z., Yang, L.-W., Kou, X.-F., and Zhan, Y.-Y. (1978). *Wei Sheng Wu Hsueh Pao* **18,** 39.

Mermelstein, N. H. (1975). *Food Technol.* **29,** 20.

Messing, R., and Oppermann, R. (1979). U.S. Patent 4,149,937.

Miall, L. M. (1978). *Econ. Microbiol.* **2,** 48–120.

Michaelis, L., and Ehrenreich, M. (1908). *Biochem. Z.* **10,** 283.

Miyata, N., and Kikuchi, T. (1976a). Japan Kokai 76/128,477.

Miyata, N., and Kikuchi, T. (1976b). Japan Kokai 76/133,484.

Mohan, R. R., and Li, N. N. (1975). *Biotechnol. Bioeng.* **17,** 1137.

Moll, M., Blachere, H., and Durand, G. (1975). Ger. Offen. 2,429,574.

Moll, M., Durand, G., and Blachere, H. (1977). U.S. Patent 4,009,286.

Moo-Young, M., Lamptey, J., and Robinson, C. M. (1980a). *Biotechnol. Lett.* **2,** 541.

Moo-Young, M., Lamptey, J., and Robinson, C. (1980b). *Proc. Int. Ferment. Symp., 6th, 1980* Abstracts, p. 78.

Mori, A., Suzue, H., Osuga, J., and Wada, M. (1980). *Proc. Int. Ferment. Symp., 6th, 1980* p. 122.

Morikawa, Y., Karube, I., and Suzuki, S. (1979). *Biotechnol. Lett.* **21,** 261.

Morikawa, Y., Karube, I., and Suzuki, S. (1980a). *Eur. J. Appl. Microbiol. Biotechnol.* **10,** 23.

Morikawa, Y., Karube, I., and Suzuki, S. (1980b). *Biotechnol. Bioeng.* **22,** 1015.

Mosbach, K., and Larsson, P.-O. (1970). *Biotechnol. Bioeng.* **12,** 19.

Mosbach, K., and Mosbach, R. (1966). *Acta Chem. Scand.* **20,** 2807.

Moskowitz, G. I. (1974). U.S. Patent 3,843,442.

Murata, K., Tani, K., and Chibata, I. (1978a). *Agric. Biol. Chem.* **42,** 2221.

Murata, K., Tani, K., Kato, J., and Chibata, I. (1978b). *Eur. J. Appl. Microbiol. Biotechnol.* **6,** 23.

Murata, K., Tani, K., Kato, J., and Chibata, I. (1979a). *J. Appl. Biochem.* **1,** 283.

Murata, K., Kato, J., and Chibata, I. (1979b). *Biotechnol. Bioeng.* **21,** 887.

Murata, K., Uchida, T., Tani, K., Kato, J., and Chibata, I. (1979c). *Eur. J. Appl. Microbiol. Biotechnol.* **7,** 45.

Murata, K., Tani, K., Kato, J., and Chibata, I. (1980a). *Biochimie* **62,** 347.

Murata, K., Tani, K., Kato, J., and Chibata, I. (1980b). *Eur. J. Appl. Microbiol. Biotechnol.* **10,** 11.

Murata, K., Tani, K., Kato, J., and Chibata, I. (1981). *Eur. J. Appl. Microbiol. Biotechnol.* **11,** 72.

Nabe, K., Izuo, N., Yamada, S., and Chibata, I. (1979). *Appl. Environ. Microbiol.* **38,** 1056.

Nakajima, Y., and Suzuki, K. (1979). Japan Kokai 79/105,286.

Narziss, L., and Hellig, P. (1971). *Brauwelt* **111,** 1491.

Narziss, L., and Hellig, P. (1972). *Brew. Dig.* **47**(9), 116.

Navarro, J. M., and Durand, G. (1977). *Eur. J. Appl. Microbiol.* **4,** 243.

Navarro, J. M., Durand, G., Duteutre, B., Moll, M., and Corrieu, G. (1976). *Ind. Aliment. Agric.* **93**(6), 695.

Nelson, R. P. (1975). Ger. Offen. 2,513,529.

Nelson, R. P. (1976). U.S. Patent 3,957,580.

Nevalainen, P. (1977). Kem.—Kemi **4,** 162.

Nickol, G. B. (1979). *In* "Microbial Technology" (H. J. Peppler and D. Perlman, eds.), Vol. 2, pp. 155–172. Academic Press, New York.

Nielsen, T. K., and Markussen, E. K. (1974a). British Patent 1,361, 387.

Nielsen, T. K., and Markussen, E. K. (1974b). British Patent 1,362,365.

Nilsson, I., and Molin, K. (1981). *Abstr. Commun., Eur. Congr. Biotechnol., 2nd, 1981* p. 159.

Nilsson, I., Ohlson, S., Häggström, L., Molin, N., and Mosbach, K. (1980a). *Eur. J. Appl. Microbiol. Biotechnol.,* **10,** 261.

Nilsson, I., Häggström, I., and Molin, N. (1980b). *Proc. Int. Ferment. Symp., 6th, 1980* p. 124.

Nishida, Y., Sato, T., Tosa, T., and Chibata, I. (1979). *Enzyme Microb. Technol.* **1,** 95.

Nishimaru, H., and Shimizu, J. (1977). Japan Kokai 77/120, 182.

Nishimaru, H., Izumi, C., Narita, S., and Yamada, M. (1975). Japan Kokai 75/140,680.

Nonomura, S., Watanabe, T., and Inagaki, T. (1976). Japan Kokai 76/48, 488.

Norsker, O., Gibson, K., and Zittan, L. (1979). *Stärke* **51,** 13.

Nyström, C. W. (1975). U.S. Patent 3,935,068.

Obara, J., and Hashimoto, S. (1976). *Sugar Technol. Rev.* **4,** 209.

Oestergaard, J., and Knudsen, S. I. (1976). *Stärke* **28,** 350.

Ohlson, S., Larsson, P.-O., and Mosbach, K. (1978). *Biotechnol. Bioeng.* **20,** 1267.

Ohlson, S., Larsson, P.-O., and Mosbach, K. (1979). *Eur. J. Appl. Microbiol. Biotechnol.* **7,** 103.

Ohlson, S., Flygare, S., Larsson, P.-O., and Mosbach, K. (1980a). *Eur. J. Appl. Microbiol. Biotechnol.* **10,** 1.

Ohlson, S., Larsson, P.-O., and Mosbach, K. (1980b). *Int. Ferment. Symp., 6th, 1980* Abstracts, p. 121.

Ohmiya, U., Ohashi, H., Kobayashi, T., and Shimizu, S. (1977). *Appl. Environ. Microbiol.* **33,** 137.

Ohwaki, J., and Minami, Z. (1975). Japan Kokai 75/29786.

Ohwaki, I., Minami, Z., Nakai, T., and Kosai, K. (1976). Japan Kokai 76/133,485.

Omata, T., Iida, T., Tanaka, A., and Fukui, S. (1979a). *Eur. J. Appl. Microbiol. Biotechnol.* **6,** 207.

Omata, T., Tanaka, A., Yamane, T., and Fukui, S. (1979b). *Eur. J. Appl. Microbiol. Biotechnol.* **8,** 143.

Omata, T., Imamoto, N., Kimura, T., Tanaka, A., and Fukui, S. (1981). *Eur. J. Appl. Microbiol. Biotechnol.* **11,** 199.
Pansolli, P., Giovenco, S., Dinelli, D., and Morisi, F. (1976). *In* "Analysis Control of Immobilized Enzyme Systems" (D. Thomas and J. P. Kernevez, eds.), pp. 237–245. North-Holland Publ., Amsterdam.
Park, Y. H., Chung, T. W., and Han, W. H. (1980). *Enzyme Microb. Technol.* **2,** 227.
Pastore, M., and Morisi, F. (1976). *In* "Methods in Enzymology" (K. Mosbach, ed.), Vol. 44, pp. 822–830. Academic Press, New York.
Pastore, M., Morisi, P., and Leali, L. (1976). *Milchwissenschaft* **31,** 362.
Peppler, H. J., and Perlman, D., eds. (1979). "Microbial Technology," 2nd ed. Academic Press, New York.
Petre, D., Noel, C., and Thomas, D. (1978). *Biotechnol. Bioeng.* **20,** 127.
Phillips, J. H., and Humphrey, A. E. (1980). *Proc. Int. Ferment. Symp., 6th, 1980* Abstracts, p. 131.
Pitcher, W. R., Jr. (1977). *Adv. Biochem. Eng.* **10,** 1–26.
Poulsen, P. B., and Zittan, L. (1976). *In* "Methods in Enzymology" (K. Mosbach, ed.), Vol. 44, pp. 768–775. Academic Press, New York.
Poutanen, K., Linko, Y.-Y., and Linko, P. (1978a). *Milchwissenschaft* **33**(7), 435.
Poutanen, K., Linko, Y.-Y., and Linko, P. (1978b). *North Eur. Dairy J.* **44**(4), 90.
Righelato, R. C. (1980). *Philos. Trans. R. Soc. London, Ser. B* **290,** 303.
Rittermann, B. E., and McCarthy, P. L. (1980a). *Biotechnol. Bioeng.* **22,** 2343.
Rittermann, B. E., and McCarthy, F. L. (1980b). *Biotechnol. Bioeng.* **22,** 2359.
Rivière, J. (1977). "Industrial Applications of Microbiology." Surrey Univ. Press, London.
Roels, J. A., and van Tilburg, R. (1979). *Stäerke* **31,** 17.
Rosèn, K. (1978). *Process Biochem.* **13**(5), 25.
Rosenzweig, M. D. (1976). *Chem. Eng. (N.Y.)* **83**(20), 54.
Ryu, D. Y., Chumg, S. H., and Katoh, K. (1979). *Biotechnol. Bioeng.* **19,** 159.
Saif, S., Tani, Y., and Ogata, K. (1975). *Hakko Kogaku Zasshi* **53,** 380.
Saimaru, H., Izumi, C., Narita, S., and Yamada, M. (1975). Ger. Offen. 2,518,280.
Saini, R., and Vieth, W. R. (1975). *J. Appl. Chem. Biotechnol.* **25,** 115.
Samejima, H., Kimura, K., Ado, Y., Suzuki, Y., and Tadokoro, T. (1978). *Enzyme Eng.* **4,** 237–244.
Sanomoto, K., Iin, I., Tanaka, A., and Fukui, S. (1980). *Agriic. Biol. Chem.* **44,** 1109.
Sato, K., and Terashima, M. (1974). Japan Kokai 74/66,886.
Sato, T., Mori, T., Tosa, T., Chibata, I., Furui, M., Yamashita, K., and Sumi, A. (1975). *Biotechnol. Bioeng.* **17,** 1797.
Sato, T., Tosa, T., and Chibata, I. (1976). *Eur. J. Appl. Microbiol.* **2,** 153.
Sato, T., Nishida, Y., Tosa, T., and Chibata, I. (1979). *Biochim. Biophys. Acta* **570,** 179.
Schnarr, G. R. W., Szarel, W. A., and Jones, J. K. N. (1977). *Appl. Environ. Microbiol.* **33,** 732.
Schnyder, B. J. (1974). *Stärke* **26,** 409.
Scott, C. D., and Hancher, C. W. (1976). *Biotechnol. Bioeng.* **18,** 1393.
Scott, C. D., Hancher, C. W., and Arcuri, E. J. (1981). *Adv. Biotechnol.* **1,** 651–656.
Sebek, G. K. (1980). *Econ. Microbiol.* **5,** 576–612.
Seif, S. R., Tani, Y., and Ogata, K. (1975). *J. Ferment. Technol.* **53,** 380.
Setti, D. (1975). *Proc. Int. Congr. Food Sci. Technol., 4th, 1974* Vol. 4, pp. 289–295.
Seyhan, E., and Kirwan, D. J. (1979). *Biotechnol. Bioeng.* **21,** 271.

Shibata, M., and Uyeda, M. (1978). *Annu. Rep. Ferment. Processes* **2,** 267–304.
Shibatani, T., Nishimura, N., Nabe, K., Kakimoto, T., and Chibata, I. (1974). *Appl. Microbiol.* **27,** 688.
Shigesada, S., Ishimatsu, Y., and Kimura, S. (1975). Japan Kokai 75/160,475.
Shimizu, J., and Kaga, T. (1972). U.S. Patent 3,664,927.
Shimizu, S., Morioka, H., Tani, Y., and Ogata, K. (1975). *J. Ferment. Technol.* **53,** 77.
Shukla, T. P. (1975). *CRC Crit. Rev. Food Technol.* **5,** 325.
Sitton, O. C., and Gaddy, J. L. (1980). *Biotechnol. Bioeng.* **22,** 1735.
Sitton, O. C., Magruder, B. C., and Gaddy, J. L. (1981). *Adv. Biotechnol.* **2,** 231–237.
Skryabin, G. K., Koscheenko, K. A., Surovtsev, V. I., Mogilnitskii, G. M., Fikhte, B. A., and Tyurin, V. S. (1974). *J. Steroid Biochem.* **5,** 397.
Skryabin, G. K., Koscheenko, K. A., Mogil'nitskii, G. M., Surovtsev, V. I., Tyurin, V. S., and Fikhte, B. A. (1975). *Izv. Akad. Nauk SSSR, Ser. Biol.* **6,** 857.
Skryabin, G. K., Koscheenko, K. A., Sukhodolskaya, G. V., and Arinbasarova, A. Y. (1976). *Abstr. Pap., Int. Ferment. Symp., 5th, 1976* p. 326.
Slowinski, W., and Charm, S. E. (1973). *Biotechnol. Bioeng.* **15,** 973.
Snell, R. L. (1976). U.S. Patent 3,974,036.
Somerville, H. J., Mason, J. R., and Ruffell, R. N. (1977). *Eur. J. Appl. Microbiol.* **4,** 75.
Sonaer, A. H., and Çağlar, M. A. (1981). *Abstr. Commun., Eur. Congr. Biotechnol., 2nd, 1981* p. 157.
Sonomoto, K., Tanaka, A., Omata, T., Yamane, T., and Fukui, S. (1979). *Eur. J. Appl. Microbiol. Biotechnol.* **6,** 325.
Spaeth, R. W. (1980). *F. O. Licht's Int. Sweetener Rep.* **3,** 37.
Spivey, M. J. (1978). *Process Biochem.* **13**(11), 2.
Stenroos, S.-L., Linko, Y.-Y., Harju, N., Heikonen, M., and Linko, P. (1982). *Enzyme Eng.* **6,** 299–301.
Stieber, R. W., Coulman, G. M., and Gerhardt, P. (1977). *Appl. Environ. Microbiol.* **34,** 733.
Stottmeister, U. (1979). *Z. Allg. Mikrobiol.* **19,** 763.
Sun, W.-R., Wang, Z.-X., Zhang, Y.-Y., Zhang, Q.-X., and Wang, X.-Q. (1980). *Wei Sheng Wu Hsueh Pao* **20,** 407.
Suzuki, H., Ozawa, Y., and Tanabe, O. (1963). *Nippon Nogei Kagaku Kaishi* **37,** 637.
Suzuki, H., Ozawa, Y., Oota, H., and Koshida, H. (1969). *Agric. Biol. Chem.* **33,** 506.
Suzuki, H., Ozawa, Y., and Tanabe, O. (1972). U.S. Patent 3,647,625.
Suzuki, H., Yoshida, H., Ozawa, Y., Kamibayashi, A., Sato, M., Mori, S., and Endo, M. (1973). U.S. Patent 3,767,526.
Suzuki, S. (1977). Japan Kokai 77/47,983.
Suzuki, S., and Karube, I. (1979a). *ACS Symp. Ser.* **106,** 59–72.
Suzuki, S., and Karube, I. (1979b). Japan Kokai 79/132,294.
Suzuki, S., Ozawa, Y., and Tanabe, O. (1964). *Nippon Nogei Kagaku Kaishi* **38,** 334.
Suzuki, S., Hirano, K., Hiraga, K., and Takagi, K. (1977). Japan Kokai 77/03,891.
Suzuki, S., Karube, Y., and Morikawa, Y. (1979). Japan Kokai 79/138,194.
Suzuki, S., Karube, I., Matsunaga, T., Kuriyama, S., Suzuki, N., Shirogami, T., and Takamura, T. (1980). *Biochimie* **62,** 353.
Taguchi, H., Suga, K., Yoshida, T., and Yuda, S. (1975). *In* "Immobilized Enzyme Technology" (H. H. Weetall and S. Suzuki, eds.), pp. 151–168. Plenum, New York.
Takahashi, Y. (1978). Japan Kokai 78/79,088.
Takamatsu, S., Yamashita, K., and Sumi, A. (1980). *J. Ferment. Technol.* **58,** 129.

Takasaki, Y. (1976). U.S. Patent 3,950,222.
Takasaki, Y., and Kanbayashi, A. (1973). U.S. Patent 3,753,858.
Takasaki, Y., and Tanabe, O. (1971). U.S. Patent 3,616,221.
Takasaki, Y., Kosugi, Y., and Kanbayashi, A. (1969). *In* "Fermentation Advances" (D. Perlman, ed.), pp. 561–589. Academic Press, New York.
Takata, I., Yamamoto, K., Tosa, T., and Chibata, I. (1979). *Eur. J. Appl. Microbiol. Biotechnol.* **7,** 161.
Takata, I., Yamamoto, K., Tosa, T., and Chibata, I. (1980). *Enzyme Microb. Technol.* **2,** 30.
Tanaka, A., Yasuhara, S., Gellf, G., Osumi, M., and Fukui, S. (1978). *Eur. J. Appl. Microbiol. Biotechnol.* **5,** 17.
Tanaka, A., Jin, I.-N., Kawamoto, S., and Fukui, S. (1979). *Eur. J. Appl. Microbiol. Biotechnol.* **7,** 351.
Thomas, G. V., Kalra, M. S., and Ajit, S. (1980). *Indian J. Exp. Biol.* **18,** 1020.
Toda, K. (1975). *Biotechnol. Bioeng.* **17,** 1729.
Toda, K., and Shoda, M. (1975). *Biotechnol. Bioeng.* **17,** 481.
Tolls, T., Showyrs, J., Sandine, W., and Elliker, F. (1970). *Appl. Microbiol.* **19,** 649.
Tosa, T., Sato, T., Mori, T., Yushi, M., and Chibata, I. (1973). *Biotechnol. Bioeng.* **15,** 69.
Tosa, T., Sato, T., Mori, T., and Chibata, I. (1974). *Appl. Microbiol.* **27,** 886.
Tosa, T., Sato, T., Nishida, Y., and Chibata, I. (1977). *Biochim. Biophys. Acta* **483,** 193.
Tosa, T., Sato, T., Mori, T., Yamamoto, K., Takata, I., Nishida, Y., and Chibata, I. (1979). *Biotechnol. Bioeng.* **21,** 169.
Toyo Jozo Company, Ltd. (1974). British Patent 1,347,665.
Tramper, J., and van den Tweel, W. J. J. (1981). *Abstr. Commun., Eur. Congr. Biotechnol., 2nd, 1981* p. 161.
Tsumura, N., and Kasumi, T. (1976). *Abstr. Pap., Int. Ferment. Symp., 5th, 1976* p. 291.
Tsumura, N., and Kasumi, T. (1977). U.S. Patent 4,001,082.
Tsumura, N., Hagi, M., and Sato, T. (1967). *Agric. Biol. Chem.* **31,** 902.
Tsumura, N., Kasumi, T., and Ishikawa, M. (1978). *Stäerke* **30,** 420.
Uchida, T., Watanabe, T., Kato, T., and Chibata, I. (1978). *Biotechnol. Bioeng.* **20,** 255.
Vandamme, E. J. (1976). *Chem. Ind.* (*London*) p. 1070.
Vandamme, E. J., and Voets, J. P. (1974). *Adv. Appl. Microbiol.* **17,** 311–369.
van den Berg, L., and Lentz, C. F., and Armstrong, D. W. (1981). *Adv. Biotechnol.* **2,** 251–256.
van Velzen, A. G. (1974). U.S. Patent 3,838,007.
Venkatasubramanian, K., ed. (1979). "Immobilized Microbial Cells," ACS Symp. Ser. No. 106. Am. Chem. Soc., Washington, D.C.
Venkatasubramanian, K., and Toda, Y. (1980). *Biotechnol. Bioeng. Symp.* **10,** 237.
Venkatasubramanian, K., and Vieth, W. R. (1979). *Prog. Ind. Microbiol.* **15,** 61–86.
Venkatasubramanian, K., Constantinides, A., and Vieth, W. R. (1978). *Enzyme Eng.* **3,** 29–41.
Vieth, W. R., and Venkatasubramanian, K. (1976). *In* "Methods in Enzymology" (K. Mosbach, ed.), Vol. 44, pp. 768–775. Academic Press, New York.
Vieth, W. R., and Venkatasubramanian, K. (1978). *Enzyme Eng.* **4,** 307–316.
Vieth, W. R., and Venkatasubramanian, K. (1979). *ACS Symp. Ser.* **106,** 1–12.
Vieth, W. R., Wang, S. S., and Saini, R. (1973). *Biotechnol. Bioeng.* **15,** 565.
Vieth, W. R., Venkatasubramanian, K., and Constantinides, A. (1976a). *Abstr. Pap., Int. Ferment. Symp., 5th, 1976* p. 298.
Vieth, W. R., Wang, S. S., and Saini, R. (1976b). U.S. Patent 3,972,776.

Villet, R., Dillon, J., and Manderson, G. (1979). *Sun 2 [Two], Proc. Int. Sol. Energy Soc. Silver Jubilee Congr., 1979* Vol. 1, p. 78.
Vogelmann, H., Ghahremani, B., and Wagner, F. (1975). *Eur. J. Appl. Microbiol. Biotechnol.* **2,** 19.
Voishvillo, N. E., Kamernitskii, N. E., Khaikova, A. V., Leontev, I. G., Paukov, V. N., and Nahkapetyan, L. A. (1976). *Izv. Akad. Nauk SSSR, Ser. Chem. Sci.* **25,** 1303.
Vojtiset, V., Zeman, R., Barta, M., and Culik, K. (1980). Ger. Offen. 2,950,985.
Vorlop, K.-D., and Klein, J. (1981). *Biotechnol. Lett.* **3,** 9.
Wada, M., Kato, J., and Chibata, I. (1979). *Eur. J. Appl. Microbiol. Biotechnol.* **8,** 241.
Wada, M., Kato, J., and Chibata, I. (1980a). *J. Ferment. Technol.* **58,** 327.
Wada, M., Kato, J., and Chibata, I. (1980b). *Eur. J. Appl. Microbiol. Biotechnol.* **10,** 275.
Wada, M., Uchida, T., Kato, J., and Chibata, I. (1980c). *Biotechnol. Bioeng.* **22,** 1175.
Wada, M., Kato, J., and Chibata, I. (1981). *Eur. J. Appl. Microbiol. Biotechnol.* **11,** 67.
Walton, M. T., and Martin, J. L. (1979). *In* "Microbial Technology" (H. J. Peppler and D. Perlman, eds.), 2nd ed., Vol. 1, pp. 187–209. Academic Press, New York.
Wang, D., Shinsky, A., and Butterworth, T. (1970). *In* "Membrane Science and Technology" (J. Flinn, ed.). Plenum, New York.
Wang, Q.-C., Ye, X.-L., Zhao, F.-X., Liu, G.-R., Liu, G.-J., Ou, G.-Q., Shen, C.-Q., and Xu, J.-D. (1980). *Acta Biochim. Biophys. Sin.* **12,** 307.
Weber, P. (1976). Ger. Offen. 2,619,322.
Weckström, L., Linko, Y.-Y., and Linko, P. (1980). *In* "Food Process Engineering" (P. Linko and J. Larinkari, eds.), pps. 148–151. Applied Science Publishers, London.
Weetall, H. H., and Bennett, M. A. (1976). *Abstr. Pap., Int. Ferment. Symp., 5th, 1976* p. 299.
Weetall, H. H., and Krampitz, L. O. (1980). *J. Solid-Phase Biochem.* **5,** 115.
Wheatley, M. A., and Phillips, C. R. (1981a). *Adv. Biotechnol* **2,** 47–54.
Wheatley, M. A., and Phillips, C. R. (1981b). *Abstr. Commun., Eur. Congr. Biotechnol., 2nd, 1981* p. 152.
White, F. H., and Portno, A. D. (1978). *J. Inst. Brew.* **84,** 278.
Whittier, E. C., and Rogers, L. A. (1931). *Ind. Eng. Chem.* **23,** 532.
Wick, E., and Popper, K. (1977). *Biotechnol. Bioeng.* **19,** 235.
Wiersma, M., Lucas, C. C. H., Kok, J. J., and Olijve, W. (1981). *Abstr. Commun., Eur. Congr. Biotechnol., 2nd, 1981* p. 50.
Wiseman, A. (1979). *Top. Enzyme Ferment. Technol.* **3,** 267–288.
Yakovleva, V. I. (1980). *In* "Food Process Engineering" (P. Linko and J. Larinkari, eds.), pps. 158–161. Applied Science Publishers, London.
Yamada, H., Okamura, S., Kojima, H., Okamoto, Y., and Ito, Y. (1976). Japan Kokai 76/144,779.
Yamada, H., Yamada, K., Kumagai, H., Hino, T., and Okamura, S. (1978). *Enzyme Eng.* **3,** 57–62.
Yamada, H., Shimizu, S., and Tani, Y. (1980a). *Enzyme Eng.* **5,** 405–411.
Yamada, H., Shimizu, S., Shimida, H., Tani, Y., Takahashi, S., and Ohashi, T. (1980b). *Biochimie* **62,** 395.
Yamamoto, K., Sato, T., Tosa, T., and Chibata, I. (1974a). *Biotechnol. Bioeng.* **16,** 1589.
Yamamoto, K., Sato, T., and Chibata, I. (1974b). *Biotechnol. Bioeng.* **16,** 1601.
Yamamoto, K., Tosa, T., Yamashita, K., and Chibata, I. (1976). *Eur. J. Appl. Microbiol. Biotechnol.* **3,** 169.
Yamamoto, K., Tosa, T., Yamashita, K., and Chibata, I. (1977). *Biotechnol. Bioeng.* **19,** 1101.

Yamamoto, K., Tosa, T., and Chibata, I. (1980). *Biotechnol. Bioeng.* **22,** 2045.
Yamane, T. (1971). *Sucr. Belge* **90,** 345.
Yamanouchi Pharmaceutical Co., Ltd. (1979). Japan Kokai 79/154,592.
Yang, H. S., and Studebaker, J. F. (1978). *Biotechnol. Bioeng.* **22,** 17.
Yang, L.-W., and Zhong, L.-C. (1980). *Wei Sheng Wu Hsueh Pao* **20,** 296.
Yoshimura, S., Danno, G., and Natake, M. (1966). *Agric. Biol. Chem.* **30,** 1015.
Young, J. C., and McCarthy, P. L. (1976). *Eng. Ext. Ser.* (*Purdue Univ.*) **129**(2), 559.
Zedan, H., and El-Tayeb, O. (1977). *Planta Med.* **31,** 163.
Zittan, L., Oulsen, P. B., and Hemmingsen, S. H. (1975). *Stärke* **27,** 236.
Zueva, N. N., Shcherbakova, V. N., Yakovleva, V. I., Avsyuk, I. V., Chan, T. M., and Berezin, I. V. (1980a). *Biokhimiya* **45,** 2206.
Zueva, N. N., Shcherbakova, V. N., Yakovleva, V. I., Nikitin, Y. S., Avsyuk, I. V., Chan, T. M., and Berezin, I. V. (1980b). *Prikl. Biokhim. Mikrobiol.* **16,** 918.

Immobilized Organelles

Hideo Ochiai

Laboratory of Biochemistry
College of Agriculture
Shimane University
Shimane, Japan

Atsuo Tanaka and Saburo Fukui

Laboratory of Industrial Biochemistry
Department of Industrial Chemistry
Kyoto University
Kyoto, Japan

I. INTRODUCTION

In living cells, enzymes usually associate with different classes of biomolecules to form supramolecular systems: membranes, lipoproteins, and ribosomes. Furthermore, especially in eukaryotes these supramolecular systems are assembled into cellular organelles and other

APPLIED BIOCHEMISTRY AND BIOENGINEERING
Volume 4

ISBN 0-12-041104-0

cellular structures, such as nuclei, mitochondria, chloroplasts, microbodies (peroxisomes), endoplasmic reticulum (microsome fraction), lysosomes, and vacuoles. Many of the cellular metabolic reactions, such as ATP generation and the synthesis and degradation of biomolecules, are segregated and organized within these structural components surrounded by membranes or on such membranes. Thus, the cellular organelles isolated will become important, excellent biocatalysts for performing sequential bioreactions when they are immobilized and stabilized.

The chloroplast photosystems of higher plants are well known as unique photoconverters using water molecules as primary electron donors. Such chloroplast photoconverters are able to use solar energy with great efficiency, to induce (1) charge separation of water molecules, (2) $NADP^+$ reduction to form NADPH, (3) ATP synthesis through photophosphorylation, and (4) carbon dioxide assimilation. Hence, stabilization of the chloroplast photosystems and application of the stable photosystems have been desirable for projects using the chloroplasts as photoconverters. Indeed, a variety of recent experiments have attempted, for instance, to produce hydrogen via biophotolysis of water (Yagi and Ochiai, 1978), to generate photocurrent from the chloroplast electrode (Ochiai *et al.*, 1979), and to regenerate ATP photosynthetically from ADP plus orthophosphate (Sawa *et al.*, 1980).

However, the chloroplast electron-flow systems, which are supposed to be composed of two systems—photosystem I (PS-I) and photosystem II (PS-II)—are notoriously unstable *in vitro* to be used as the bioreactor. Preservation of PS-II activity has been an especially refractory problem in the project. Therefore, considerable attention has been focused on immobilization technique as one of the most important approaches aiming at the stabilization of the chloroplast photosystems. A number of current approaches to chloroplast immobilization are listed in Table I. The chloroplast photosystems are located in thylakoid membranes, which are functional associations composed of a variety of lipids and proteins. Degradation of these components through the action of "lytic" enzymes or agents must lead to a marked reduction in the photochemical activities. From such a point of view, entrapment of chloroplasts with semipermeable supports would be most appropriate to stabilize or protect the functional association from the degradation. In this chapter, some recent attempts at chloroplast entrapment are reviewed. Aspects of chloroplasts immobilized by covalent cross-linking with bifunctional compounds as well as the application of the immobilized chloroplasts for hydrogen production have

TABLE I
Techniques and Properties of Chloroplast Immobilization

Methods and materials	Activity retention[a] (% PS-II or O_2 evolution)	Reference
I. Adsorption		
A. DEAE-Cellulose	95–100	Shioi and Sasa (1979)
II. Covalent cross-linking[b]		
A. Glutaraldehyde	30–35	West and Packer (1970)
	46	Oku *et al.* (1973)
	(H_2 Production)	Rao *et al.* (1976)
	45	Papageorgiou and Isaakidou (1977)
B. Dimethyl suberimidate	44	Papageorgiou and Isaakidou (1977)
III. Microcapsulation		
A. Gelatin, protamine in dibutyl phthalate and toluene diisocyanate	(PS-I, 33%)	Kitajima and Butler (1976)
IV. Entrapment		
A. Polyacrylamide gel	11	Ochiai *et al.* (1977)
	(CO_2 Fixation)	Karube *et al.* (1979)
B. Poly(vinyl alcohols)		
1. Instant chloroplast	80–100	Ochiai *et al.* (1978b)
2. Insoluble film	35	Ochiai *et al.* (1979)
C. Proteins (plus glutaraldehyde cross-linked)		
1. Collagen	52	Vieth and Venkatasubramanian (1978)
2. Bovine serum albumin	71	Cocquempot *et al.* (1980)
D. Marine algal polysaccharides		
1. Ca alginate	—	Kierstan and Bucke (1977)
	100	Gisby and Hall (1980)
2. Agar (3%)	61	Berezin and Varfolomeev (1979)
3. Agar (2%)	($NADP^+$ Reduction)	Karube *et al.* (1980)
4. Carrageenan	3	Cocquempot *et al.* (1980)
E. γ Irradiation of 2-hydroxyethyl acrylate	65	Fujimura *et al.* (1980)
F. Photo-cross-linkable resins	19	Cocquempot *et al.* (1981)
G. Urethane polymer	32	Cocquempot *et al.* (1981)

[a] Activity retention, percentage expressed activity = [(activity of immobilized chloroplasts/freshly prepared chloroplasts) × 100] %.
[b] For further details, see Papageorgiou (1979), and Rao and Hall (1979).

TABLE II

IMMOBILIZATION OF MITOCHONDRIA, MICROSOMES, AND PEROXISOMES

Organelles and source	Immobilization method	Application	Reference
Mitochondria			
Rat liver	Adsorption on alkylsilylated glass beads	Physiological study	Arkles and Brinigar (1975)
Avocado pears	Entrapment in Ca alginate gels	Production of ATP	Kierstan and Bucke (1977)
Acetate-grown yeast	Entrapment in photo-cross-linked resin and urethane polymer	Adenylate kinase	Tanaka *et al.* (1980)
Electron-transport particles (beef heart)	Entrapment in agar gels	Assays of NADH and succinate	Aizawa *et al.* (1980)
Microsomes (rat liver)	Entrapment in hollow fiber	Detoxification of drugs	Kastl *et al.* (1978)
Peroxisomes (methanol-grown yeast)	Entrapment in photo-cross-linked resin, urethane polymer, and proteinic polymer	Physiological study and multifunctional biocatalyst	Tanaka *et al.* (1977, 1978, 1979)

been the exciting subjects in excellent publications (Papageorgiou, 1979; Rao and Hall, 1979).

In contrast, chromatophores isolated from photosynthetic bacteria, vesicles 50 times smaller than the plant chloroplasts, are relatively stable under ambient temperature and illumination conditions, and are able to synthesize ATP via a cyclic photophosphorylation system. This situation prompts us to use the chromatophores as the possible ATP-regeneration system. The immobilization of chromatophores and its applications are also described in this chapter.

In addition to the photosynthetic organelles and chromatophores, other cellular organelles also have their own distinctly organized functions for metabolizing different cellular substances. This means that these organelles will become excellent biocatalysts to carry out multistep reactions, if they are isolated intact and immobilized without the loss of their native characteristics. However, it is not easy to immobilize these organelles intact, because their functions are closely associated with the membrane systems, which will be easily disordered during immobilization processes. In fact, only a few reports have been published concerning the immobilization of organelles (Table II) other than photosynthetic ones (Table I).

In this chapter, the results on the immobilization of mitochondria, microsomes, and peroxisomes are described with several examples of reconstituted models of these organelles.

II. CHLOROPLASTS

A. Entrapment within Polyacrylamide Gel (PAG)

Inclusion inside polyacrylamide gel has been hitherto most widely applied for the preparation of immobilized enzymes or microbial cells. To make polyacrylamide gel, the following reagents are usually used: acrylamide monomer, *N,N'*-methylene bisacrylamide (BIS), persulfate as a polymerization initiator, and β-dimethylaminopropionitrile as a polymerization accelerator. Addition of any one of the reagents to the chloroplast preparation (Type C, mainly lamellae fraction; Hall, 1972) exerted little influence on the PS-II activity. However, gelation on mixing them with the chloroplasts usually resulted in complete loss in activity of PS-II. This suggests that "radicals" formed by mixing the reagents are harmful to the PS-II component(s) (Ochiai *et al.*, 1977, 1978a).

However, the immobilization of active photosystems can be achieved by another polymerization technique called "redox polymerization": acrylamide monomer (15% in final concentration), BIS (0.8%), the polymerization initiator (0.02%), ascorbate (0.06%), bovine serum albumin (BSA, 1%), D-mannitol (0.01%) as protectants, and the chloroplast suspension (about 1 mg Chl/ml) were incubated in 0.4 *M* sucrose–0.05 *M* Tris–0.01 *M* NaCl (STN buffer, pH 7.2) at 10°C for 2 h under anaerobic conditions (Ochiai *et al.*, 1977, 1978a). The green gel beads, obtained after crushing and washing with cold STN buffer, showed specific activities: PS-I, 131 μmol oxygen uptake/mg Chl/h; and PS-II, 20.1 μmol 2,6-dichlorophenolindophenol (DPIP) reduction/mg Chl/h (11% of those of the native chloroplasts). The PAG-immobilized chloroplasts retained the activities even after being allowed to stand for several weeks in the STN buffer at 4°C. In one experiment, a conventional two-electrode photocell was constructed: one tank contained the chloroplast gel, methyl viologen (MV) as an electron acceptor, glucose plus glucose oxidase as an oxygen trap, and sodium azide in the STN buffer; the other tank contained the same reagents but lacked the gel. On illumination to the former tank, a weak photo-induced electric current was generated. This was the first demonstration of conversion of light energy to electrical energy by the use of the immobilized-chloroplast gel (Ochiai *et al.*, 1978a).

Recently Karube *et al.* (1979) reported the PAG immobilization of "whole chloroplasts" prepared by the method of Jensen and Bassham (1966). To the whole chloroplasts suspended in medium C were added acrylamide monomer (8% in final concentration) and BIS, and the system was blanketed with nitrogen. The polymerization was then initiated with β-dimethylaminopropionitrile and potassium persulfate solution, and allowed to proceed for 30 min at 25°C. Thus immobilized chloroplasts maintained the activity of the carbon dioxide fixation at around 65% of that of free counterparts. Probably the envelopes of the whole chloroplasts protected the photosynthetic functions inside the envelopes against injury from radicals. These workers asserted that light penetration of the gel was not a limiting parameter of the carbon dioxide fixation, and the lifetime of the immobilized chloroplasts was found to be 90 min, three times longer than that of free chloroplasts.

B. Entrapment within Poly(Vinyl Alcohols) (PVA)

In most studies on chloroplast preparations, polyol compounds such as sucrose or sorbitol have been used as stabilizing reagents. Polyols are also known experimentally to stabilize many other enzymes. Poly(vinyl alcohol) (PVA) is a polyol and, in fact, has a protective effect similar to sucrose on chloroplast photosystems (Ochiai *et al.*, 1978b,c). Furthermore, poly(vinyl alcohol) is able to polymerize to form film under dehydrating conditions. Therefore, entrapment inside the PVA matrix provides extremely mild conditions for immobilization. Ochiai *et al.* prepared two forms of PVA-immobilized chloroplasts: one is soluble in buffer, rendering the chloroplast suspension almost identical in activities to the freshly prepared chloroplasts; the other is an insoluble film that is available for the purpose of immobilizing chloroplasts.

1. Instant Chloroplasts

Freshly prepared chloroplasts from sources such as spinach and higher plants have variable activity and readily deteriorate on aging. Hence, chloroplast preparations having constant activity, if possible in a ready-to-use form, have been widely desired for laboratory use. Immobilization with low molecular weight PVA provides the instant chloroplasts, which can be stored with minimal precaution. The standard immobilization procedure is as follows: the chloroplast suspension (Type C), in the STN buffer containing 80 m*M* sodium borate plus 1% BSA, is mixed with 20% PVA-105 (an average degree of polymerization, 550) solution buffered by the STN. The resulting viscous solution is gently kneaded and then dried under reduced pressure in the darkness to give a foam-structured substance. The substance is crushed to flakes for storage. This dry and flaky solid can be dissolved easily in the buffer when needed, to give a suspension as native as the original one. The recovery of PS-I activity is almost quantitative, PS-II 80% or more of the native chloroplasts.

As shown in Fig. 1, the activities of PS-I and PS-II in the flake were retained for at least 5 weeks with a loss of less than 20%. Cyclic photophosphorylation mediated by phenazine methosulfate (PMS, 5-methylphenazinium methyl sulfate) was also performed with the resuspension prepared from the PVA–chloroplast flaky solid. The activity, as much as 117 μmol ATP formed/mg Chl/h (44% of the control) was observed even after storage for 2 weeks in cold, dark conditions, as shown in Table III. Thus the chloroplast preparation stabilized through the immobilization with PVA-105, in soluble form, is widely useful as a source of portable and instant chloroplasts (Ochiai *et al.*, 1978b,c). Such an outstanding effect of PVA on activity retention may be explained by the fact that the oxygen permeability of PVA film is found to be extremely low in comparison with other natural or synthetic polymers (Ochiai *et al.*, 1978c).

2. Chloroplast Electrode

Excitation of the chloroplast photosystems I and II by light leads to the pumping of an electron by at least 1.2 V from a level of water oxidation ($E_0' = +0.82$ V versus a normal hydrogen electrode at pH 7)

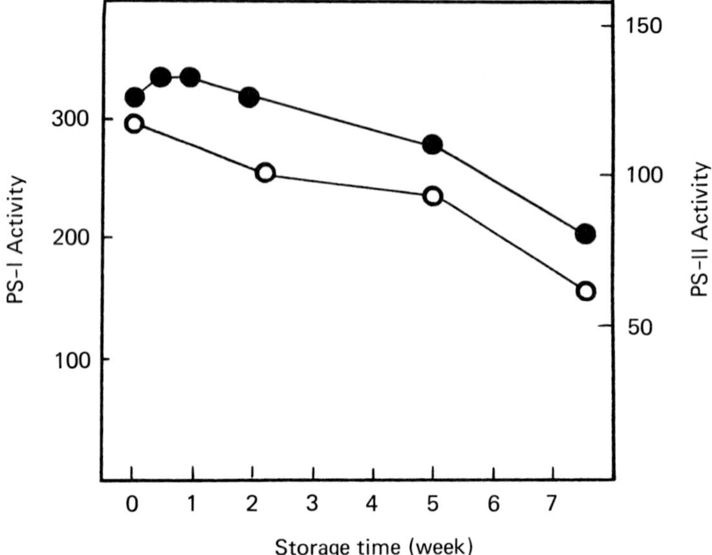

Fig. 1. Changes in the photosystem activities during storage of the chloroplasts entrapped with poly(vinyl alcohol) as soluble flakes. ●——●, Photosystem I (PS-I) activity (μmol O_2 absorbed/mg Chl/h); ○——○, Photosystem II (PS-II) activity (μmol 2,6-dichlorophenolindophenol reduced/mg Chl/h). The flake was dissolved in the 0.4 *M* sucrose–0.05 *M* Tris–0.01 *M* NaCl buffer (pH 7.2) at indicated time during storage under cold, dark conditions, then the activities of PS-I and PS-II were determined. [Adapted from Ochiai *et al.* (1978b, p. 684), with permission from The Agricultural Chemical Society of Japan.]

to that of ferredoxin reduction ($E_0' = -0.43$ V). In fact, light-driven electric cells have been extensively studied making use of chloroplast membranes (Allen, 1977), chloroplasts (Haehnel *et al.*, 1978), and PS-I particles (Gross *et al.*, 1978). Recently, Ochiai and co-workers (1979) prepared the PVA-immobilized chloroplast electrode, as described here.

The freshly prepared chloroplasts (Type C) were gently mixed with 0.05 *M* Tris–0.01 *M* NaCl buffer (pH 7.2) containing 18% PVA-117 (average degree of polymerization, 1750; degree of saponification, 98.5% or more) and 1% BSA. The resulting sticky mixture was spread uniformly on the surface of an SnO_2 optically transparent electrode (SnO_2 OTE) glass plate and dried under reduced pressure over phosphorus pentoxide. The PVA–chloroplast film thus deposited on the SnO_2 OTE, ~25 μm in thickness, was covered and fixed with a monolayer of nylon cloth in order to prevent the PVA film from swelling and coming off the SnO_2 glass plate, and was used as a working

TABLE III
ACTIVITY FOR CYCLIC PHOTOPHOSPHORYLATION MEDIATED BY PMS OF FRESH CHLOROPLAST SUSPENSION, AND OF SOLUBLE-FLAKE AND INSOLUBLE-FILM CHLOROPLAST PREPARATIONS[a]

Sample	Activity[b] (μmol ATP formed/mg Chl/h)
Fresh suspension	265
Flake[c]	
After 2 weeks storage	117
After 4 weeks storage	43
Film[d]	
Fresh[e]	22
After 1 week storage	11

[a] From Ochiai *et al.* (1978b), p. 684, with permission from The Agricultural Chemical Society of Japan.

[b] The activity was assayed according to the method of Asada *et al.* (1972), with a slight modification. After a 5-min illumination to the reaction mixture at 25°C with continuous stirring, esterified ^{32}P was separated from unreacted ^{32}P phosphate and determined with a liquid scintillation counter, without the scintillator, using a Čerenkov radiation of ^{32}P.

[c] Chloroplasts entrapped with PVA-105 as soluble flake.

[d] Chloroplasts immobilized with PVA-117 as insoluble film.

[e] The fresh film was obtained after drying *in vacuo* for 24 h.

electrode. The film obtained had apparent PS-I activities of as much as 72.5 μmol oxygen uptake/mg Chl/h (42% of the control), and PS-II activities of as much as 65 μmol DPIP reduction/mg Chl/h (35% of the control). The comparatively low activities should be interpreted as the result of diffusion resistance of PVA film to substrate mass transfer. The PVA–chloroplast film on the SnO_2 OTE was mounted as the window of an electrochemical cell with the film side facing the electrolyte solution, as shown in Fig. 2. A platinum plate and saturated calomel electrode (SCE) served as the counterelectrode and the reference electrode, respectively. On illumination, the chloroplast electrode functioned as a photoanode with concomitant oxidation of water molecules, though the magnitude of the observed photocurrent depends on the electrode potential imposed, the light intensity used, the density of the chlorophyll deposited on the SnO_2 OTE, the illumination area, the efficiency of the cut-off filter employed, the nature of electrolyte solu-

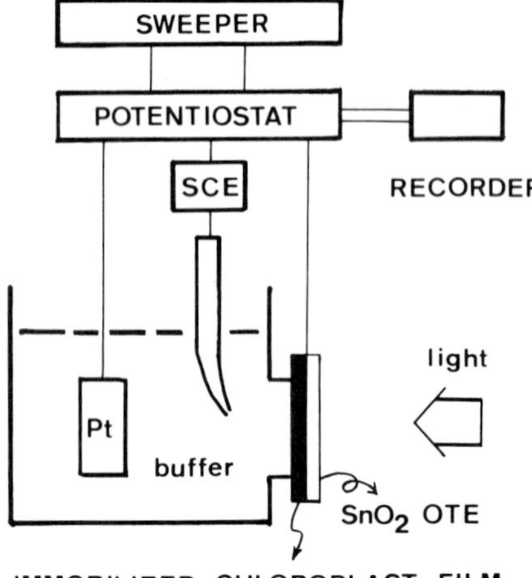

Fig. 2. Diagram showing the experimental setup for photocurrent measurement. Buffer used as electrolyte: 0.05 *M* Tris–0.01 *M* NaCl (pH 7.2). Temperature, 25°C. SCE, Saturated calomel electrode; SnO_2 OTE, SnO_2-coated, optically transparent electrode. [Reproduced from Ochiai *et al.* (1979, p. 881), with permission from The Agricultural Chemical Society of Japan.]

tion, and especially on the surviving activity of the chloroplast preparations. Practical outputs of around 1 μA have been observed under laboratory conditions. In general, one could measure the photocurrent output immediately after the illumination under the potentiostatic condition, whereas under the short-circuit condition one could determine the maximum photocurrent 10 min after the onset of the light. On turning the light off the photocurrent returned to zero with a half-time of approximately 1 min.

Figure 3 shows the potential dependence of photocurrents at the PVA-chloroplast electrode on illumination of 10×10^4 erg/s/cm^2 in contact with a 0.05 *M* Tris–0.01 *M* NaCl buffer (pH 7.2) as an electrolyte at 25°C. The dark current is also illustrated in the figure. The anodic photocurrent resulting from the excitation of the chloroplast photosystems arose from around +0.2 V versus the SCE and increased linearly with the potential imposed. As can be seen in the figure, about 50 nA with a Y-46 filter (which eliminates the photoresponse of SnO_2) and about 90 nA without the filter were obtained as a "specific" photocurrent, with the working electrode poised at +0.7 V versus the SCE.

The potential was determined as a function of the illuminated time and potentials as high as 150–300 mV have been observed in the open-circuit condition several minutes after the onset of illumination.

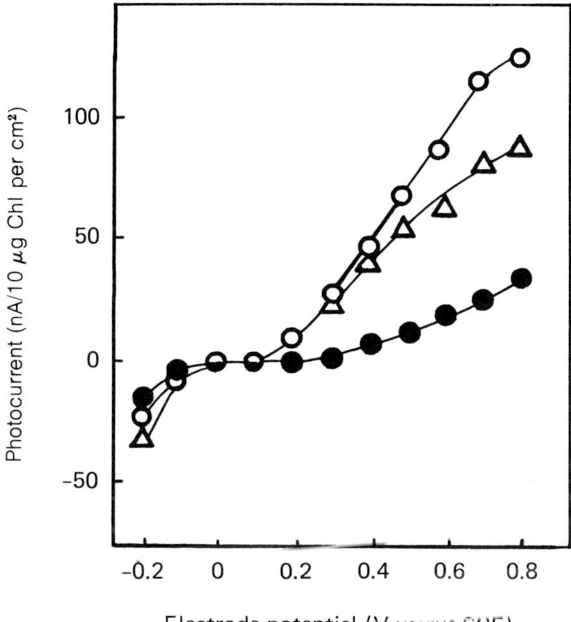

Fig. 3. Photocurrent-potential curves at PVA–chloroplasts fixed on the SnO_2 OTE. ○——○, Light current, without the filter; △——△, light current, with Y-46 filter; ●——●, dark current (nA/cm^2 of SnO_2 OTE). The immobilized-chloroplast electrode was illuminated in 0.05 *M* Tris–0.01 *M* NaCl buffer (pH 7.2) at 25°C. Light intensity was 10×10^4 $erg/s/cm^2$, and the light below 460 nm was eliminated by a Y-46 filter in one of the experiments. (From Ochiai *et al.*, 1982.)

A maximum photovoltage of about 1.2 V is expected, but electrical leaks from the electron-flow components in the thylakoid membrane, as well as mismatching of the energy levels of the photosystems with the energy levels of the SnO_2, could account for the small potentials observed. Janzen and Seibert (1980) described the similar phenomenon of the electrode coated with reaction centers isolated from a photosynthetic bacterium, *Rhodopseudomonas spheroides* R-26.

Unquestionably a cut-off filter is unnecessary when practical output of the photocurrent is needed. In such a case the output could be about twice as high as that with the filter. Addition of 1-methoxy-5-methylphenazinium methyl sulfate (methoxy-PMS, 30 μM in final concentration) to the electrolyte solution greatly enhanced the photocurrent yield, as much as 60-fold under the short-circuit condition. Methoxy-PMS, a photochemically stable derivative of PMS, is known as a cyclic electron carrier, diverting electrons from *P*-430 and cycling them around PS-I (Hisada and Yagi, 1977). Hence, the SnO_2 OTE could accept electrons much more efficiently by means of the carrier. At the condition of +0.5 V versus the SCE, DPIP (30 μM) also exerted an enhancement effect on the photocurrent yield of as much as 13-fold.

A typical photocurrent spectrum is illustrated in Fig. 4., together with the absorption spectrum of an immobilized-chloroplast film deposited on the SnO_2 OTE. As is evident in the figure, both spectra coincide fairly well, showing the maxima in the wavelength regions of around 450 and 680 nm, which are considered to correspond with carotenoids and chlorophylls, respectively. Kinetic studies on the effects of heat treatment, photosynthetic inhibitors, and electrolyte pH on the chloroplast electrode also proved that current output depends on the presence of active chloroplast photosystems using water molecules as primary electron donors (Ochiai *et al.*, 1979, 1982).

Thus, the PVA–chloroplast film coating the SnO_2 OTE in a three-electrode configuration functions precisely as a photoconverter of radiant energy into electric energy. Therefore, such a technique of

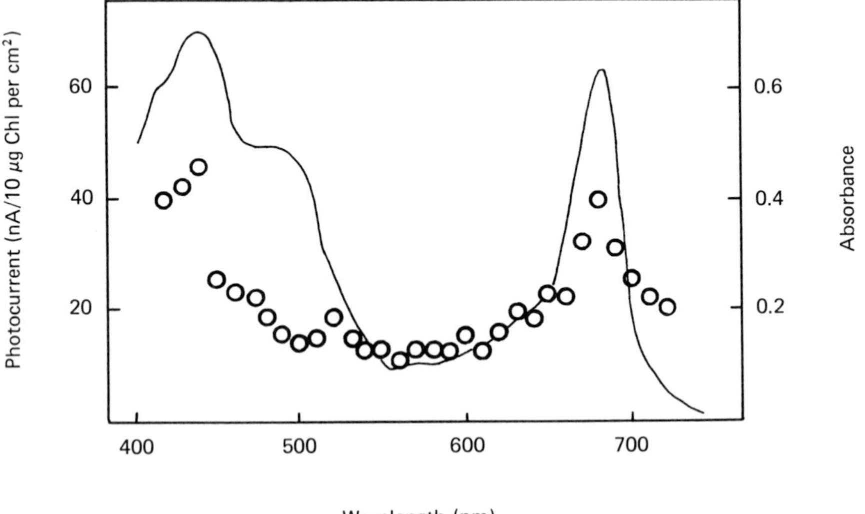

Fig. 4. A comparison of action spectrum with absorption spectrum. Open circles (○) show the action spectrum from the anodic photocurrent at the PVA–chloroplasts immobilized on the SnO_2 OTE. Electrolyte: 0.05 M Tris–0.01 M NaCl buffer (pH 7.2), at 25°C. Light intensity: 20×10^4 erg/s/cm². Electrode potential: +0.8 V versus the SCE. Solid line (—) is the absorption spectrum of the chloroplasts deposited on the SnO_2 OTE. (From Ochiai *et al.*, 1982.)

electrically coupling the chloroplast photosystems to semiconductor electrodes in a photoelectrochemical cell provides new methods of probing the primary photochemistry of photosynthesis. For instance, these electrodes could be useful for the direct determination of time courses of the photosystem activities under continuous illumination, which has not been possible previously. Moreover, photoconverters of this type have been suggested as systems to donate electron to hydrogenase in the production of hydrogen through water photolysis (Benemann *et al.*, 1973), especially for schemes using a two-stage system (Yagi, 1976).

C. Entrapment within Proteins

Vieth and Venkatasubramanian (1978) mixed the stripped chloroplasts with a 4% (w/v) collagen dispersion at pH 8.5. The mixture was cast to form a thin membrane after drying, and then, if necessary, this was followed by treatment with glutaraldehyde to a desired level of mechanical strength. The chloroplasts thus immobilized retained 18.5 μmol oxygen evolution/mg Chl/h, about half the activity of free chloroplasts. An important factor in the activity retention is the tanning conditions, that is, concentration of glutaraldehyde used, and treatment duration. After 2 weeks storage, the collagen-entrapped chloroplasts still showed 60% of their initial activity.

Very recently, Cocquempot *et al.* (1980) have succeeded in preparing other protein-entrapped chloroplasts by using a co-cross-linking method with BSA and glutaraldehyde at subzero temperatures. They mixed the following solution in the order shown: (1) 3 ml of 0.02 *M* (pH 6.8) phosphate buffer solution, (2) 2.5 ml of 20% BSA, (3) 1 ml of thylakoid suspension (equivalent to 4 mg chlorophyll), and (4) 2 ml of 1.5% glutaraldehyde. The resulting glutaraldehyde concentration was 0.35%. The mixture was frozen at −30°C and after 2 h slowly thawed to 4°C to obtain the foam structure with good mechanical properties. The activity yields of this porous spongelike chloroplast compound are presented in Table IV.

In general, the photophosphorylation systems of thylakoids are known to be sensitive to chemical treatment, especially to covalent cross-linking with bifunctional compounds (Papageorgiou, 1979). However, by this immobilization process under the subzero temperature, the activity was not completely denatured. It is rather interesting that PS-I and the cyclic photophosphorylation system were found more sensitive to this processing than "unstable" PS-II and the noncyclic photophosphorylation system. A half-life of oxygen-evolution activity of this preparation under continuous working conditions with light was

TABLE IV
ACTIVITY OF DIFFERENT SYSTEMS OF NATIVE AND BSA-IMMOBILIZED THYLAKOIDS[a,b]

Thylakoids	PS-II (FeCN)	PS-I + II (MV)	PS-I + II ($NADP^+$)	PS-I (Ascorbate + DPIP → MV)	ATP	
					Noncyclic	Cyclic
Native	180	70	50	165	80	234
Immobilized	127	25	6	56	21	22
Yield (%)	71	35	12	34	26	9.5

[a] From Cocquempot *et al.*, (1980, p. 617), by courtesy of Societe de Chimie Biologique, France.

[b] Oxygen evolution (PS-I and PS-II) given as μmol O_2 produced/mg Chl/h. ATP regeneration expressed as μmol ATP regenerated/mg Chl/h from ADP and P_i. Abbreviations: FeCN, Ferricyanide; MV, methyl viologen.

shown to be 60 min, about four times longer than that of native chloroplasts. In one experiment, hydrogenase from *Clostridium pastorianum* and thylakoids from lettuce were immobilized within two different foam structures. On illumination, hydrogen evolution was observed and the yield was 9% with an absolute activity of 1.2 μmol hydrogen/mg Chl/h (Cocquempot *et al.*, 1980). The yield is low when compared with the native systems, but equal to 75% when activity is compared to the results obtained with ferredoxin for oxygen evolution (see Table IV). The use of better hydrogenase could be a way to overcome the difficulty.

D. Marine Algal Polysaccharides

Calcium alginate gel also provides mild conditions for immobilization. Kierstan and Bucke (1977) first reported the calcium alginate-immobilized chloroplasts that showed the Hill reaction. Most recently, Gisby and Hall (1980) have entrapped chloroplasts within calcium alginate gel on reinforcing grids of nylon and stainless steel to form a strengthened film. Chloroplasts thus immobilized showed oxygen-evolution rates almost the same as nonimmobilized ones. Using such strengthened films, they prepared a variety of hydrogen-production systems, which consisted of the calcium alginate–chloroplasts coimmobilized with either hydrogenase or PVA–platinum, and of a biological or synthetic electron carrier. On illumination in the presence of O_2 trap, a certain system produced hydrogen gas with a maximum amount of 5 μmol hydrogen/mg Chl/h. The longevity of the hydrogen-production system was prolonged twofold, although the initial apparent rate was less than half of the free counterpart systems.

In 1979, Berezin and Varfolomeev described in detail the study of chloroplasts entrapped within agar gel in an excellent chapter of Volume 2 in this series. They stated that the irreversible inactivation of agar-entrapped chloroplasts proceeds as a function of agar gel strength; the chloroplasts are more stable in gel of lower agar concentration.

Karube *et al.* (1980) have tried to regenerate NADPH from $NADP^+$ as one application of 2% agar-entrapped chloroplasts, which were coimmobilized with ferredoxin. For the inclusion of chloroplasts inside the agar gel, for example, 0.24 g agar was dissolved in 10 ml physiological saline in a flask at 100°C and cooled to 50°C. Then, 2 ml of 0.1 *M* Tris buffer (pH 8) containing 0.2 g of chloroplasts (0.6–0.8 mg Chl), crude ferredoxin (8 μM), and $NADP^+$ (0.8 m*M*) were added to the flask, mixed, and immediately cooled to 30°C. Crude ferredoxin and $NADP^+$-ferredoxin oxidoreductase isolated from spinach were used as electron carriers. On illumination, $NADP^+$ was continuously reduced for 2 h with a hollow-fiber reactor containing immobilized chloroplasts, and $NADP^+$ and NADPH were separated with a hollow-fiber dialyzer from ferredoxin and $NADP^+$-ferredoxin oxidoreductase. The percentage conversion of $NADP^+$ to NADPH was found to range from 40 to 80%.

In another experiment, Kayano *et al.* (1981) used the 2% agar-immobilized chloroplasts and 2% agar-entrapped *Clostridium butyricum* for the purpose of water-splitting hydrogen production. Ferredoxin, as a circulating electron carrier, was reduced by the chloroplast photosystems in a light bioreactor under nitrogen bubbling. Hydrogen was produced through oxidation of this reduced ferredoxin by the hydrogenase of *Cl. butyricum* in a dark reactor, and was applied to a hydrogen–oxygen fuel cell, as shown in Fig. 5. Thus they observed photoinduced current of 0.4–1.5 mA for 4 h. The percentage conversion from hydrogen to current was found to range from 80 to 100%. However, the current gradually decreased with reaction time. After a 4-h run, the immobilized *Cl. butyricum* retained the initial hydrogenase activity, whereas the activity of immobilized chloroplasts decreased to 30% of the initial activity. Hence, the decrease in current might be caused by the disintegration of chloroplast functions.

E. Synthetic Polymers

Practically speaking, the chloroplast preparations from higher plants are so unstable on warming that it is preferable to treat them at temperatures as low as possible. With this in mind, Fujimura *et al.* (1980) tried to immobilize the chloroplasts at low temperature (below −24°C)

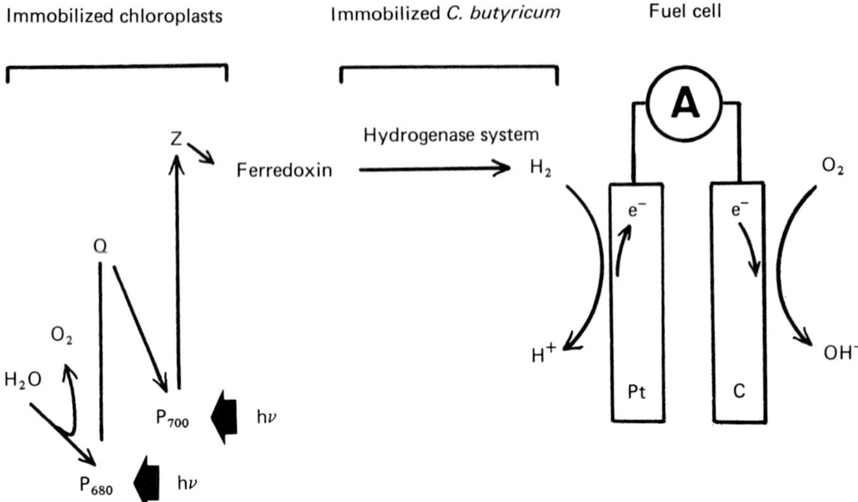

Fig. 5. Principle of the photoelectrochemical fuel cell. (From Kayano *et al.*, 1981. Adapted with permission from *Biotechnol. Bioeng.* Copyright by John Wiley & Sons, Inc.)

TABLE V

ACTIVITY OF NATIVE AND IMMOBILIZED THYLAKOID MEASURED WITH POTASSIUM FERRICYANIDE AS ELECTRON ACCEPTOR IN THE ABSENCE AND PRESENCE OF NH_4Cl[a]

	Oxygen evolution[b] (μmol O_2 evolved/mg Chl/h, 20°C)	
	No NH_4Cl	5 m*M* NH_4Cl
Native thylakoids	51	112
Cross-linked albumin polymer	68	76
Polyurethane–BSA	38	56
Cross-linked gelatin polymer	49	50
Urethane polymer	24	31
Alginate gels	25	25
Photo-cross-linkable resins	19	21
Carrageenan gels	3	3

[a] Adapted from Cocquempot *et al.* (1981, p. 194), with permission from Springer-Verlag, Inc.

[b] Without ammonium chloride, the results should represent the sum of two phenomena: the partial denaturation and the partial uncoupling of the electron-transfer chain after immobilization. With 5 m*M* ammonium chloride, a well-defined activity yield should express the integrity of the electron-transfer chain.

through γ irradiation in the presence of BSA, D-mannitol, and sodium ascorbate as protectants, and 2-hydroxyethyl acrylate as a prepolymer. Thus immobilized chloroplasts in a hydrophilic-polymer matrix showed the stable duration of oxygen-evolution activity of more than 700 h at 4°C. Thermostability of chloroplasts was also improved greatly through immobilization. The active center of PS-II in the immobilized chloroplasts was retained even after 60 min standing at 50°C.

Very recently, Cocquempot *et al.* (1981) compared oxygen-evolution activities of chloroplasts immobilized by various methods selected to provide the chloroplast membrane with different environments. These included proteins (albumin and gelatin), polysaccharides (carrageenan and alginate gels), and synthetic polymers such as photo-cross-linkable resins (Fukui *et al.*, 1976; Tanaka *et al.*, 1978) and polyurethane (Fukushima *et al.*, 1978). Table V lists their results, which show large variations in the activity yield after immobilization. The immobilization technique at subzero temperature using the cross-linked albumin polymer has resulted in the most favorable outcome overall, as described in Section II,C.

III. IMMOBILIZATION OF CHROMATOPHORES AND ITS APPLICATION

Bacterial chromatophores are membrane fractions that can be isolated from photosynthetic bacteria grown in the light. They are roughly spherical particles, with a diameter of approximately 0.1 μm, and they contain bacteriochlorophyll and carotenoid pigments, phospholipids, and all the enzymes necessary for both light-induced electron transport and phosphorylation of ADP. In fact, chromatophores from *Rhodospirillum rubrum* have a cyclic electron-transport system that allows photophosphorylation to occur without the net production of reducing equivalents. Therefore, much attention has been focused on the light-induced ATP-regeneration system using the bacterial chromatophores.

Pace *et al.* (1976) demonstrated the availability of the free chromatophores isolated from *R. rubrum* for the continuous photosynthetic regeneration of ATP from ADP in an ultrafiltration reactor in which the chromatophores were completely retained by a PM10 membrane. They observed the production of 0.8 m*M* ATP under illumination of 4.4×10^4 erg/s/cm^2 at 30°C, in the presence of 1 m*M* ADP, 5 m*M* inorganic phosphate, 1 m*M* fumarate, 10 m*M* $MgCl_2$, 17 m*M* ascorbic acid, and 50 m*M* Tricine (pH 7.8). However, decline in ATP production occurred with time due to the decrease in the activity of chromatophores. A considerable retention of the chromatophore activity was achieved through immobilization inside polyacrylamide gel (Yang *et al.*, 1976). It was found that the gel, which resulted from a monomer concentration of 7.5%, a cross-linking level of 12.5%, and chromatophore concentration of 5.0–7.5 mg/ml of gel, was the most appropriate to keep the high activity. Apparent photophosphorylating activity was around 40% of free chromatophores, but kinetic parameters showed that the rate of photophosphorylation in PAG particles was controlled by the diffusion of substrate into the gel particles. Light penetration of the gel particles was not a limiting parameter. The temperature and pH optima for the PAG-immobilized chromatophores were similar to the native photosynthetic apparatus. Figure 6 shows time courses of both the immobilized and free chromatophores stored in 0.1 *M* Tricine (pH 7.8) at 3 and 25°C, respectively. Evidently the decrease in activity followed first-order kinetics. At 3°C, for instance, immobilized chromatophores lost only 5% of their initial activity after 3 days storage, in contrast with 50% of the free counterparts. Immobilization increased the stability of the chromatophores toward denaturation.

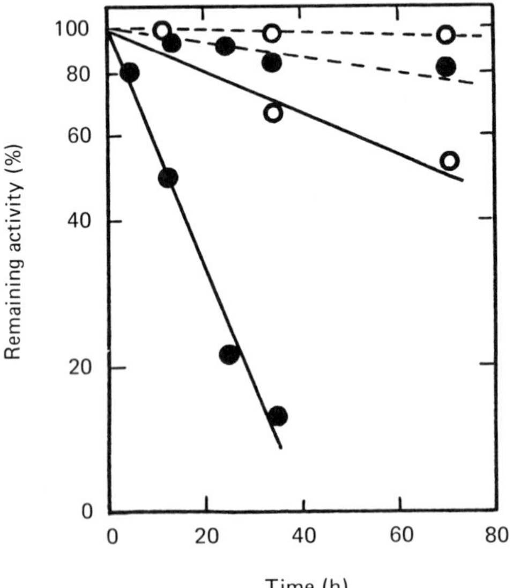

Fig. 6. Stability of polyacrylamide-immobilized (○) and native (●) chromatophores stored in 0.1 *M* Tricine buffer (pH 7.8) at 3°C (---) and 25°C (——). [From Yang *et al.* (1976, p. 1431). Reprinted with permission from *Biotechnol. Bioeng.* Copyright by John Wiley and Sons, Inc.

In general, the alginate gel formed by calcium binding was rapidly destroyed by millimolar concentrations of inorganic phosphate such as are necessary for ATP synthesis. Paul and Vignais (1980) compared alginate beads prepared with barium, strontium, and calcium cations as gel-forming agents for their mechanical strength, chemical resistance against disruption by phosphate-induced swelling, and yield of photophosphorylation activity. Barium alginate beads could maintain their structure for relatively long periods of time, even in the presence of 10 m*M* orthophosphate. Chromatophores of *Rhodopseudomonas capsulata* within the barium alginate gel retained a high yield of up to 70% of their phosphorylating capacity with an absolute activity of 3 μmol ATP formed/mg protein/h, whereas activity yields of only 40 and 20% were obtained with strontium and calcium alginate gels, respectively. A large increase in the Michaelis constant for ADP and orthophosphate was not caused by the alginate entrapment. However, embedding gave no additional protection against rapid inactivation of chromatophores on storage at 3°C.

Very recently, Garde *et al.* (1981) presented comparative studies on ATP regeneration by bacterial chromatophores (*R. capsulata*) immobilized within various supports. The results are shown in Table VI. The chromatophores entrapped with foam-structured BSA, the most favorable medium, were prepared in a manner similar to the case of the chloroplasts: a solution was made by mixing 1 ml of 0.02 *M* (pH 6.8) phosphate buffer, 1.25 ml of 20% BSA, 1 ml of 1.5% glutaral-

TABLE VI
ATP PRODUCTION BY THE NATIVE AND IMMOBILIZED CHROMATOPHORES[a]

Immobilization method	Support	ATP (μmol/mg bacteriochlorophyll/h)	Activity yield (%)
Native chromatophores		560	100
Entrapment	κ-Carrageenan gel	16	3
	Ca^{2+} alginate gel	91	16
	Photo-cross-linkable resin	95	17
	Urethane polymer	263	47
Co-cross-linking process	BSA film	160	29
	Gelatin film	219	39
	Foam-structured BSA	369	66
	Foam-structured gelatin	235	42

[a] Adapted from Garde *et al.* (1981, p. 136), with permission from Springer-Verlag, Inc.

dehyde solution, and 1 ml of chromatophore suspension. The resulting concentration of glutaraldehyde was 0.35%. This mixture was frozen at −20°C for several hours and then slowly thawed at 4°C and rinsed. The porous, spongelike chromatophores thus obtained retained their photophosphorylating activity at about 66% of that of the native controls. It is important to optimize the glutaraldehyde concentration in order to obtain entrapped chromatophores capable of the high ATP production. The half-life under storage in the dark increased from 20 h to 10 days for chromatophores after immobilization with cross-linked BSA.

IV. UTILIZATION OF THERMOSTABLE LIVING CHLOROPLASTS

As already mentioned, through the entrapment of chloroplasts as well as chromatophores, their photosynthetic systems are generally stabilized and the activities can be maintained for much longer periods on storage under cold and dark conditions. However, many authors always finish their reports with the statement that "further stabilization of chloroplasts is required for practical utilization," or "stability is a major requirement for future photoconverting devices." Indeed, unfortunately, no one has described success in a long, continuous run of the chloroplast apparatus under illumination.

The chloroplast preparation, especially PS-II, readily deteriorates on warming. This point suggests that the use of the chloroplast photoconverter must meet with serious difficulty because the illumination is naturally accompanied by heat. In addition, the chloroplast photosystems liberate molecular oxygen on illumination. In the presence of light and oxygen at the pigmented organelles, there usually appears "active oxygen," which could cause damage to the functional membrane, for example through peroxidation of lipids. Some free fatty acids, which are known to exert an inhibitory effect on the photosystem activities, would be released from the damaged lipids. Furthermore, photodecomposition of chlorophylls could result from the active oxygen poisoning. These situations may induce serious limitations in using chloroplasts, even the immobilized ones, as practical photoconverters.

Alternatively, Ochiai *et al.* have considered the use of "thermostable living chloroplasts" in place of labile chloroplasts prepared from higher plants. In general, chloroplast organelles in plant cells are stabilized by cellular dynamic processes. Therefore, the problems of stability and longevity that occur with the chloroplast photoconverters must be avoided by the use of thermostable, live chloroplasts. Indeed,

Ochiai *et al.* (1980) have reported some favorable work using a strain of thermophilic blue-green alga, *Mastigocladus laminosus,* which was isolated from Matsue hot springs, Matsue, Shimane, Japan. The strain of this alga was easily cultured at 47 ± 2°C in the light under laboratory conditions without any living contaminants. It should be emphasized that one can assay both PS-I and PS-II activities by the use of the *intact* algal cells, a very unique circumstance as compared to the general experience with other algae. Electrons are available from the intact cells directly to electron carriers such as DPIP and methyl viologen. Moreover, the photochemical activities of the intact alga are comparable to those of the freshly prepared chloroplasts from higher plants. Such an ability to take up electrons from the intact cells suggests the availability of the intact alga as a source of "live chloroplasts," a suitable replacement for unstable chloroplast preparations.

A. Living Electrode

The typical procedure for preparing the immobilized-algae electrode is as follows: the cells of the algae are suspended in an 8% sodium alginate solution of the hot spring water. The suspension is deposited and spread on the surface of the SnO_2 OTE, and the electrode plate is dipped into 50 m*M* $CaCl_2$ solution buffered at pH 8 for 5 min. The electrode is washed with the hot spring water. Thus a living-algae electrode of ~250 μm in thickness is obtained. On illumination, this electrode functions as a working photoanode in the conventional three-electrode photoelectrochemical cell, as illustrated in Section I,B,2, in a manner similar to the chloroplast electrode. Increasing output of the photocurrent was observed with increase of potential. The specific output value on the basis of chlorophyll concentration per square centimeter was similar to that of the chloroplast electrode. Figure 7 shows time courses of the photocurrent output at the calcium alginate–algae electrode illuminated continuously in the culture medium and the chlorophyll accumulation in the matrix. When the algae plate is illuminated in the culture medium, the algae remain alive and grow in the alginate matrix, as shown by the chlorophyll-accumulation curve. In fact, the algae electrode is able to generate steady photocurrent on continuous illumination for 20 days or more, indicating that the algae electrode could function as a long-lived photoconverter. An algae electrode of this kind is called a "living electrode." The living electrode shows promise as a suitable replacement for the unstable chloroplast electrode (Ochiai *et al.*, 1980).

Most recently, Ochiai *et al.* (1983) have made the exciting discovery that drying of a "wet" algal electrode at 50°C for 60 min increased the photocurrent yield by 100-fold, probably because of better contact

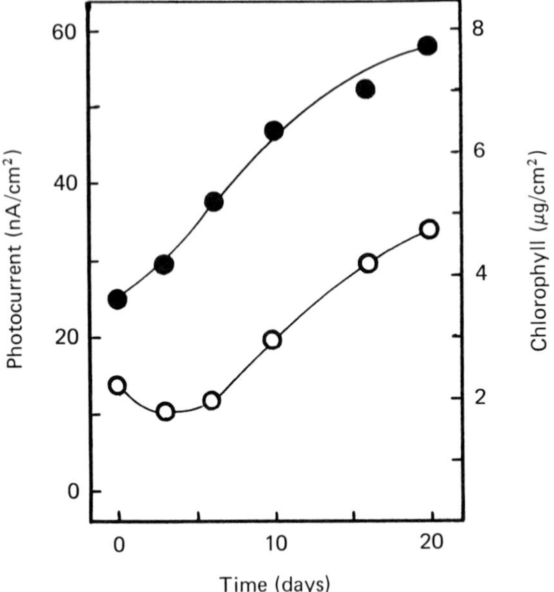

Fig. 7. Time courses of photocurrent output at the calcium alginate–algae electrode (versus the SCE) under continuous illumination, and of chlorophyll accumulation in the calcium alginate matrix. Electrode potential: +0.7 V versus the SCE. ○——○, Photocurrent; ●——●, chlorophyll. (From Ochiai *et al.*, 1980.)

of the algal cells with the SnO_2 OTE. Moreover, the photocurrent yield was more than doubled on addition of an electron carrier such as methyl viologen (20 μM in final concentration) to the electrolyte. All of the kinetic studies done hitherto indicate that water molecules contribute to the light-induced current. Figure 8 shows a profile of the photocurrent output with the working electrode poised at +0.6 V versus the SCE. So far, a practical output up to 1 mA has been achieved using the immobilized-algae electrode with an illumination area of 50 cm². These results prompted us to set up a practical hydrogen-production system by water photolysis using the two-stage apparatus as shown in Fig. 9 (Ochiai *et al.*, 1979).

B. ATP-Regeneration System

ATP regeneration is a challenging problem in the field of enzyme technology. The blue-green alga *Mastigocladus laminosus* can also accumulate ATP in the culture medium through its photophosphorylating function in the presence of exogenous ADP plus orthophosphate. The addition of methoxy-PMS, a photochemically sta-

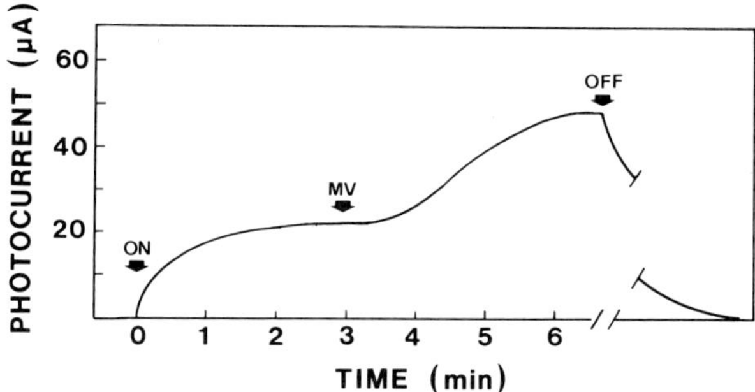

Fig. 8. Anodic photocurrent output at the calcium alginate–algae electrode (versus the SCE) as a function of illumination time. The algal cells, equivalent to 4.3 μg chlorophyll *a*/cm², were immobilized on the SnO_2 OTE and allowed to dry at 50°C for 60 min, then used as a working electrode, with a Y-46 filter that cuts off the light below 460 nm. Electrolyte: 0.05 *M* H_3BO_3–KCl–Na_2CO_3 (pH 8) at 40°C. Electrode potential: +0.6 V versus the SCE. Light intensity: 2.5×10^5 erg/s/cm². Illumination area: 7 cm². Methyl viologen (MV) was added at indicated position to be 20 μM in final concentration. (From Ochiai *et al.*, 1983.)

ble electron mediator around PS-I, to the reaction mixture increased the rate of ATP synthesis about threefold, indicating that ATP was synthesized mainly through the cyclic photophosphorylation system (Sawa *et al.*, 1980). Table VII summarizes the effects of various parameters on ATP accumulation. The optimal temperature was around 45°C, which is very close to the growth temperature of the algal cells. Clearly, the ATP formation in this alga was thermophilic. A sharp decline in ATP accumulation was observed above 50°C. This optimal temperature was similar to that of hydrogen evolution in the blue-

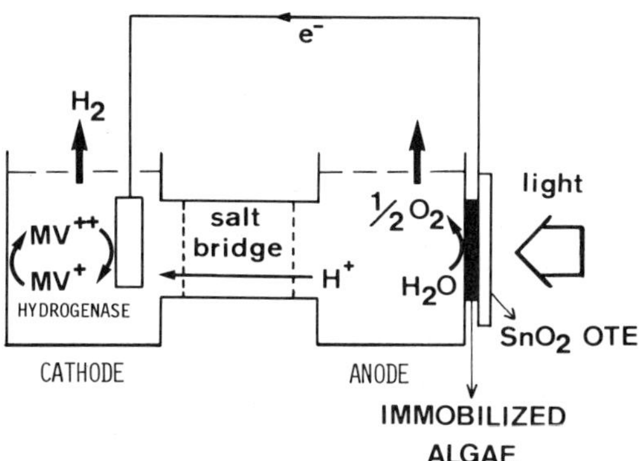

Fig. 9. Possible setup for hydrogen production by water photolysis. Hydrogenase may be replaced by PVA–platinum (Gisby and Hall, 1980). [Adapted from Ochiai *et al.* (1979, p. 883), with permission from The Agricultural Chemical Society of Japan.]

TABLE VII
PARAMETERS AFFECTING THE PHOTOSYNTHETIC ATP ACCUMULATION BY INTACT BLUE-GREEN ALGAL CELLS[a]

Variable	Condition required for full ATP accumulation
Temperature	45°C
pH	7.4
Light	15 mW/cm^2
Methoxy-PMS	3 μM
ADP	2 mM
PO_4^{3-}	1 mM
Mg^{2+}	7 mM
Algal concentration	Equivalent to 65 μg Chl/ml

[a] From Sawa *et al.* (1982, p. 311). Reprinted with permission from *Biotechnol. Bioeng.* Copyright by John Wiley & Sons, Inc.

TABLE VIII
EFFECTS OF VARIOUS BUFFER REAGENTS ON PHOTOSYNTHETIC ATP ACCUMULATION BY INTACT BLUE-GREEN ALGAL CELLS[a,b]

Buffer	Specific activity (μmol ATP accumulated/mg Chl/h)	Relative value
None (culture medium)	53.5	1.00
Tris-HCl	78.6	1.47
Tricine-NaOH	103.0	1.93
Tris malate-NaOH	141.5	2.64
Phosphate	72.6	1.36
Borate	161.4	3.02

[a] From Sawa *et al.* (1982, p. 313). Reprinted with permission from *Biotechnol. Bioeng.* Copyright by John Wiley & Sons, Inc.

[b] The reaction mixture was preincubated for 10 min in the dark before 10 min illumination. All buffers were used at 50 mM (pH 7.5).

green alga *Mastigocladus laminosus,* employed by Miura *et al.* (1980). Table VIII shows the dependence of ATP synthesis on various buffer reagents. Notably, borate buffer (50 m*M*, pH 7.5) exerted a stimulating effect on the photophosphorylation. However, the oxygen evolution and DPIP photoreduction by the algal cells were not affected in the presence of borate buffer. Therefore, the stimulating effects may have resulted from the enhancement of permeability of such compounds as ATP and ADP through the algal cell membrane and cell wall. With a continuous reactor at a flow rate of 1 ml/h, efficient photoconversion of ADP (2 m*M*, at substrate reservoir) to ATP (1 m*M*, at product outlet) was maintained from 4 to 60 h after the onset of illumination, though the efficiency has decreased in a further 2-day period (Sawa *et al.*, 1982). Thus the ATP-regeneration system using the living blue-green alga could function with much more prolonged stability in comparison with immobilized chloroplasts (Cocquempot *et al.*, 1980) and chromatophores (Yang *et al.*, 1976). An avenue of great promise would be to design an ATP-regeneration device composed of immobilized living algae using light as the sole external energy source.

V. MITOCHONDRIA

The first study on the immobilization of mitochondria was reported by Arkles and Brinigar (1975). They immobilized rat liver mitochondria by adsorption on alkylsilalated glass beads essentially in a monolayer as evidenced by electron microscopical observation. Chain length of the alkyl moiety and amount of hydrocarbon on the glass beads affected markedly the amount of bound mitochondria. That is, maximum mitochondrial binding occurred when the beads were coated with at least 3% alkylsilane and the length of the alkyl group exceeded 8 carbons (maximum 18 carbons). It is interesting that mitochondria bind tightly on the beads at 27°C, whereas the organelles immediately dissociate from the beads at 0°C. In a flow system, immobilized mitochondria exhibited normal characteristics: (*a*) respiratory control, that is, stimulation of oxygen uptake by ADP; (*b*) P : O ratio of 2.7, essentially the same as the value obtained for free mitochondria; (*c*) uncoupling by 2,4-dinitrophenol or other agents; and (*d*) inhibition by cyanide, azide, rotenone, oligomycin, and antimycin. Mitochondrial respiratory was recovered rapidly and completely from the inhibition by cyanide and azide after the inhibitor was removed from the reaction system, whereas inhibition by rotenone, oligomycin, and antimycin was essentially irreversible. The immobilized mitochondria retained the activity up to 4 h at 27°C.

The use of immobilized mitochondria appears to offer a new technique for the study of the dynamic behavior of the organelles *in vivo* where the concentrations of substrates and other components are continuously changing. Immobilized mitochondria will also be useful for the synthesis of ATP via oxidative phosphorylation, as well as for the assay of various compounds, if the organelles can be stabilized extensively.

Kierstan and Bucke (1977) immobilized avocado pear mitochondria by entrapment in calcium alginate gels. Although the entrapped mitochondria were demonstrated to convert ADP to ATP, no evidence was given that ATP formation arose from oxidative phosphorylation. Entrapment of mitochondria in calcium alginate gels is not promising because partial solubilization of the gels was observed in the presence of phosphate ion.

Mitochondria isolated from acetate-grown *Candida tropicalis* were entrapped with photo-cross-linkable resin prepolymer and urethane prepolymer by the present authors (Tanaka *et al.*, 1980). Although the entrapped mitochondria showed a low but distinct respiratory activity—that is, about 9% of that of the free counterparts—respiratory control by ADP and oxidative phosphorylation of ADP were not observed. However, the preparation exhibited a relatively high activity of adenylate kinase, converting ADP to ATP and AMP. Immobilization greatly enhanced the stability of mitochondrial adenylate kinase during storage at 4°C or incubation at 30°C. The immobilized mitochondria retained nearly the original enzyme activity even after 19 batches of the reaction (total operational period, 380 min) (Fig. 10), indicating their utility as an immobilized preparation of adenylate kinase to be used for the regeneration of ATP from AMP.

Apart from immobilization of whole mitochondria, immobilization of submitochondrial particles and various mitochondrial enzymes has also been tried from fundamental and practical viewpoints. Two typical examples are here described.

Aizawa *et al.* (1980) immobilized nonphosphorylating electron-transport particles (ETP) prepared from beef heart mitochondria. Of various immobilization methods examined, entrapment in agar gels gave the highest activity of NADH oxidase by ETP. Not only NADH but also succinate were oxidized by the immobilized ETP. The membrane-shaped immobilized ETP were mounted on an oxygen probe, and used to assay NADH or succinate. This organelle sensor can be applied to either turbid or nonhomogeneous samples, for which the spectrophotometric method and others are not useful. NADH or succinate can be measured in a range of 0 to 200 μM with the response time

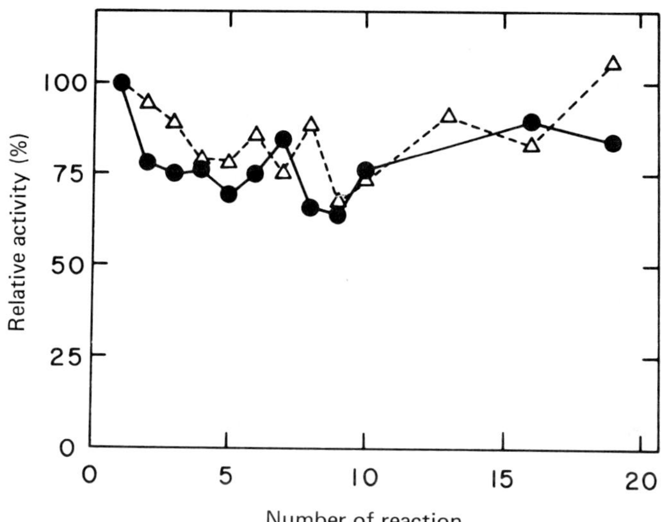

Fig. 10. Stability of adenylate kinase in entrapped mitochondria on repeated use. Each reaction was carried out for 20 min. ●——●, Photo-cross-linked resin-entrapped mitochondria; △-----△, urethane resin-entrapped mitochondria.

of 6–8 min. The immobilized ETP were reusable and found to retain their function for at least 2 weeks.

An immobilized three-enzyme system consisting of malate dehydrogenase, citrate synthase, and lactate dehydrogenase was constructed as a model for mitochondria, and the rates of oxaloacetate production and utilization were investigated (Srere *et al.*, 1973). In this case, lactate dehydrogenase was introduced to mimic the NADH-utilizing system of mitochondria. The rate of citrate production from malate, NAD, and acetyl-CoA was enhanced about twofold by coimmobilization of the three enzymes compared with that of the mixed free enzyme system. Addition of pyruvate to reoxidize NADH stimulated the reaction about fivefold over that of the free system. These results indicate a possible organization of enzymes by coimmobilization. Among the immobilization systems, the system entrapped in polyacrylamide gels was more effective than that coupled covalently on Sephadex G-50 or Sepharose 4B. Such studies—that is, coimmobilization of plural enzymes relating to the same organelle—will be useful for a variety of purposes, including fundamental investigation of the functions of organelles and practical applications such as the production of useful compounds and assays of various substances.

VI. MICROSOMES

Although microsomes contain important detoxification enzymes for xenobiotics, such as cytochrome *P*-450 and NADPH–cytochrome *P*-450 reductase, few attempts have been made to immobilize the organelles intact. This is likely due to the instability of highly organized membrane-bound enzyme systems during the isolation and immobilization processes.

Kastl *et al.* (1978) entrapped microsomes from phenobarbital-induced rat liver in hollow fiber to develop an extracorporeal drug detoxifier. The entrapped microsomes could metabolize hexobarbital in the presence of molecular oxygen and an NADPH-regenerating system (e.g., glucose 6-phosphate and glucose 6-phosphate dehydrogenase) for at least 3 h. They calculated a hexobarbital clearance of 9 ml/min for 10,000 cm^2 hollow fiber, and suggested the usefulness of the entrapped microsomes for detoxification of serum.

In general, membrane-bound enzymes become unstable when solubilized and purified. However, immobilizing them on or in a suitable support will increase their stability, giving a similar circumstance *in vivo.* Thus, many solubilized and purified microsomal enzymes were immobilized to study their functions *in vivo* and also to apply to practical purposes. Some typical examples are described.

The synergism of hydroxylation and glucuronidation was observed when the enzymes involved in the combined reactions (UDP-glucuronyltransferase, NADPH-cytochrome *c* reductase, and cytochrome *P*-450) were coimmobilized on BrCN-activated Sepharose 4B, and the necessary cofactors (UDPGA, NADPH, and the phospholipid) were placed inside the enzyme beads (Brunner and Lösgen, 1978). The activities of the transferase and the reductase in the multienzyme complex were sufficiently high. Hydroxylation activity on hexane as substrate was found to be low but distinct. In addition, binding of UDP-glucuronyltransferase to another support, acrylamide–acrylic acid imide copolymer, brought about a 30-fold increase in stability compared to native microsomes.

Reconstitution of liver microsomal system within liposomes was also reported (Ingelman-Sundberg and Glaumann, 1977). Cytochrome *P*-450LM_2 and NADPH–cytochrome *P*-450 reductase were entrapped in vesicles of phosphatidylcholine. The recoveries of cytochrome *P*-450 and the reductase in the liposomes were 65–75% and 80–90%, respectively. The liposomes showed about eightfold higher apparent V_{max} for 6β-hydroxylation of androstenedione than that of nonvesicular preparations. The high activity may be ascribable to maintenance of active conformation of the enzymes incorporated into the lipophilic circumstance, or to favored interactions between cytochrome *P*-450 and the reductase in the microenvironment of liposomes. It was found that the catalytic properties, such as substrate specificity and reaction kinetics, of the hydroxylation system were markedly dependent on the type of phospholipid used for the reconstitution, namely, the properties of the membrane (Ingelman-Sundberg and Johansson, 1980).

VII. PEROXISOMES (MICROBODIES)

Yeast peroxisomes containing alcohol oxidase, catalase, and D-amino acid oxidase (Fukui *et al.*, 1975; Fukui and Tanaka, 1979) were isolated from methanol-grown cells of *Kloeckera* sp., and immobilized intact in matrices of photo-cross-linked resin (Tanaka *et al.*, 1977, 1978) (Fig. 11), proteinic polymer (Tanaka *et al.*, 1978) (Fig. 12), urethane polymer (Tanaka *et al.*, 1979), and polyacrylamide (Tanaka *et al.*, 1978). These entrapped preparations of peroxisomes showed relatively high activities of alcohol oxidase, catalase, and D-amino acid oxidase, as seen in Table IX (Fukui *et al.*, 1978; Tanaka *et al.*, 1978, 1979).

The immobilized peroxisomes oxidized 2 mol of methanol to form 2 mol of formaldehyde with consumption of 1 mol of molecular oxygen in both the presence and absence of 0.65 *M* sucrose (Table X). Addition of 3-amino-1,2,4-triazole, an inhibitor for catalase, decreased the formation of formaldehyde from methanol by half, but not the oxygen consumption. These results clearly reveal the synergic reaction of alcohol oxidase and catalase in the peroxisomes; that is, oxidation of 1 mol of methanol by alcohol oxidase forming formaldehyde and hydro-

TABLE IX
ACTIVITY YIELDS OF IMMOBILIZED *Kloeckera* PEROXISOMES

Peroxisome condition	Relative enzyme activity (%)		
	Catalase	Alcohol oxidase	D-Amino acid oxidase
Free	100	100	100
ENT-4000-entrapped[a]	46–55	79–80	49
PEGM-2000-entrapped[a]	36	76	31
PU-9-entrapped[b]	46	58	—
Albumin-entrapped	48	73	—
Polyacrylamide-entrapped	74	49	—

[a] Photo-cross-linked resins.
[b] Urethane resin.

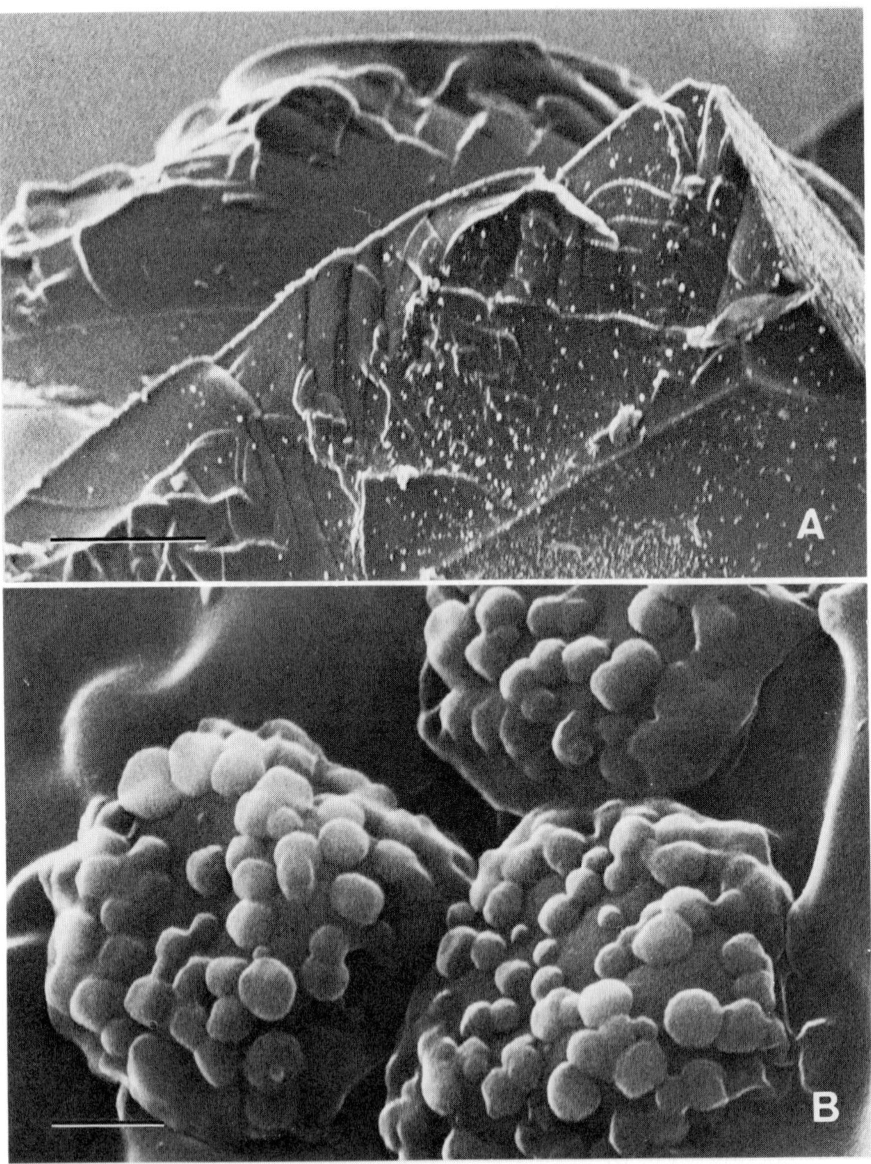

Fig. 11. (A) Electron micrograph of peroxisome-entrapped photo-cross-linked resin film. Aggregated peroxisomes are observed as small white dots. Bar = 100 μm. (B) Enlarged photograph of immobilized *Kloeckera* peroxisomes. The immobilized peroxisomes remained intact. Bar = 500 nm.

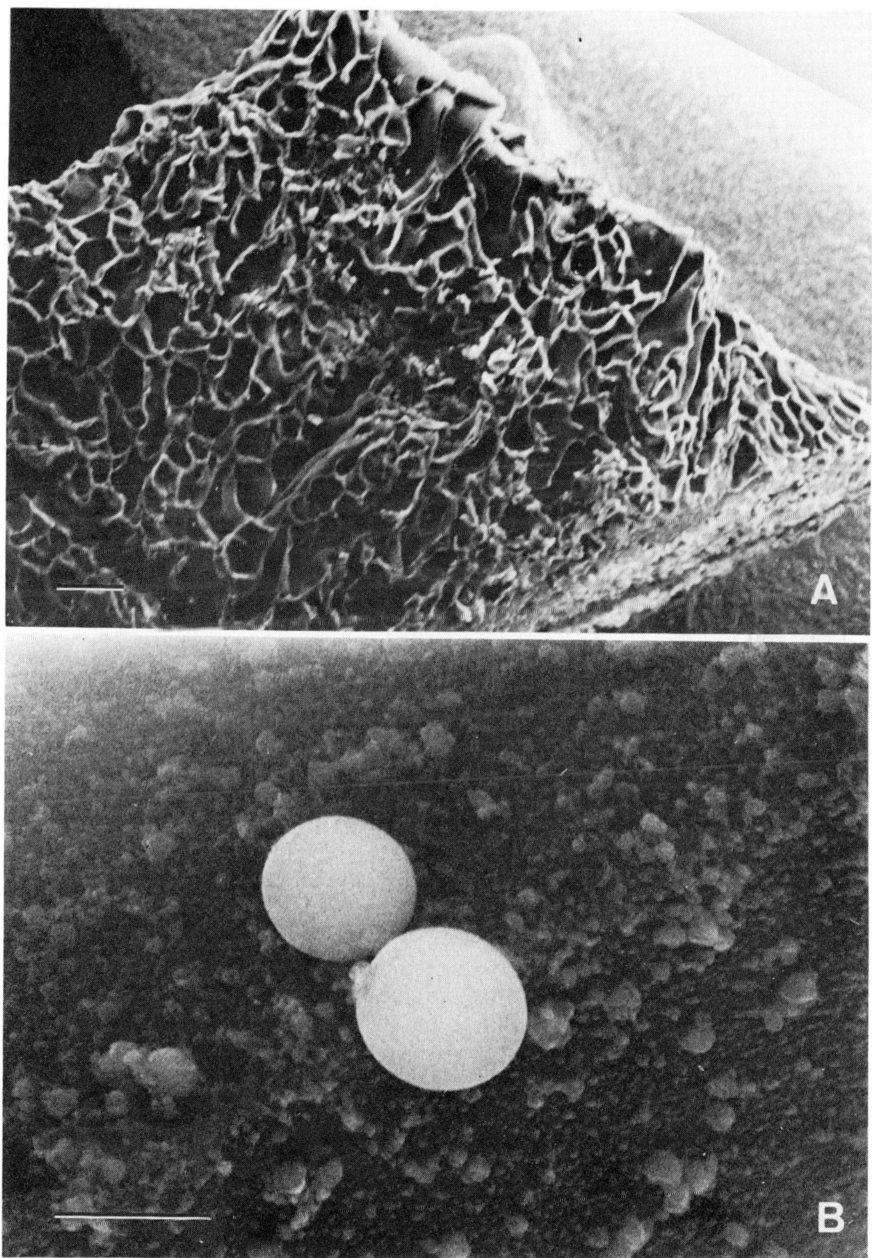

Fig. 12. (A) Electron micrograph of peroxisome-entrapped proteinic polymer. Bar = 100 μm. (B) Enlarged photograph of immobilized *Kloeckera* peroxisomes. Bar = 500 nm.

TABLE X
STOICHIOMETRY OF METHANOL OXIDATION BY PHOTO-CROSS-LINKED RESIN-ENTRAPPED PEROXISOMES OF METHANOL-GROWN *Kloeckera* sp.[a]

Addition to the assay mixture	O_2 consumed (*A*) (nmol/min/mg protein)	Formaldehyde formed (*B*) (nmol/min/mg protein)	*B*/*A*
0.65 *M* Sucrose	259	526	2.0
None	215	401	1.9
0.15 *M* 3-Amino-1,2,4-triazole	203	207	1.0

[a] From Tanaka *et al.* (1977, p. 196) and Fukui and Tanaka (1979, p. 193), with permission from Springer-Verlag, Inc., and Academic Press, respectively.

gen peroxide, and oxidation of yet another mole of methanol by peroxidative action of catalase using hydrogen peroxide thus formed (Fig. 13). Such synergic action of these enzymes of peroxisomes located in the living cells could be observed even when the entrapped peroxisomes were burst inside gel matrices, indicating the utility of immobilized organelles for the study of their functions *in vivo*.

Alcohol oxidase of the entrapped peroxisomes showed substrate specificity similar to the free counterparts (Table XI). Not only methanol but also primary alcohols of two to five carbons were oxidized at significant rates. Stability of alcohol oxidase and catalase were improved to some extent, but not satisfactorily so far as tested. However, immobilized peroxisomes will be applicable to continuous assays or treatments of hydrogen peroxide, alcohols, and D-amino acids as a multifunctional biocatalyst.

TABLE XI
SUBSTRATE SPECIFICITY OF ALCOHOL OXIDASE IN FREE AND IMMOBILIZED YEAST PEROXISOMES[a]

Alcohol	Relative enzyme activity (%)		
	Free peroxisomes	Photo-cross-linked resin-entrapped peroxisomes	Albumin-entrapped peroxisomes
Methanol	100	100	100
Ethanol	80	96	120
n-Propanol	56	60	50
n-Butanol	43	49	47
n-Amyl alcohol	39	35	14
Benzyl alcohol	28	11	8

[a] From Tanaka *et al.* (1978, p. 25), with permission from Springer-Verlag, Inc.

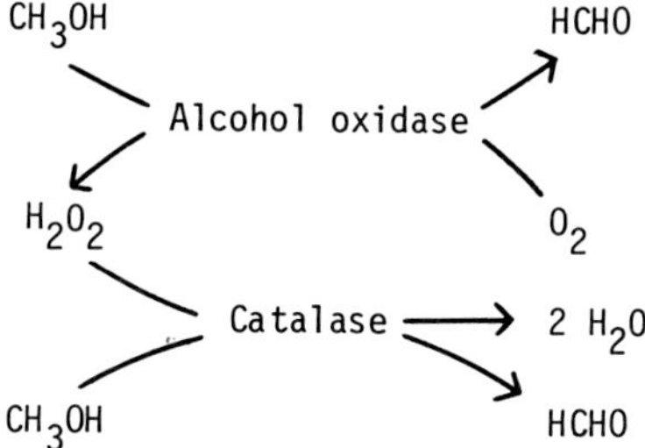

Fig. 13. Synergic action of alcohol oxidase and catalase in peroxisomes of methanol-grown yeast cells.

VIII. CONCLUSIONS

In the last few years, considerable progress has been made in chloroplast immobilization. In most cases the longevity of chloroplast functions *in vitro* is much improved through immobilization, as compared with the labile nature of fresh chloroplast preparations. However, workers have always had the annoying problem of the decline in the photochemical activities that occurs during the course of light runs. Thermostable living chloroplasts could overcome such a cumbersome problem. Immobilization of living algal cells such as *Mastigocladus laminosus* and its application would lead to development of an efficient apparatus for a biological solar energy-conversion system in the near future. Another approach to improve the instability of the system should be establishment of a synthetic photoconverting device that resists the light and oxygen poisoning. In fact, considerable investigation aiming at this goal has been carried out, attracting worldwide interest. Knowledge of chloroplast functions and hence progress in biomimetic technology should be required for the establishment of practical photoconverters.

In contrast, reports concerning the immobilization of cellular organelles are very few, with the exception of photosynthetic organelles. However, organelles are sophisticated and well-organized intracellular organs in which many important biological reactions are mediated by the conjugated actions of multiple enzymes. For example, microsomes contain the enzyme system applicable to detoxification of drugs, assays of various compounds, and production of valuable compounds. Reconstitution of microsomal functions is also useful for these purposes. Yeast peroxisomes, especially those from alkane-grown cells having more complex functions than those from methanol-grown cells (Fukui and Tanaka, 1979), will be utilized as a multifunctional and multistep biocatalyst. In any case, it is necessary to immobilize these organelles without injury to their membranous systems, in which the catalytic activities of synergic multistep reactions are embedded.

We hope to see the further utilization of immobilized organelles for fundamental studies and for practical purposes.

REFERENCES

Aizawa, M., Wada, M., Kato, S., and Suzuki, S. (1980). *Biotechnol. Bioeng.* **22,** 1769–1783.

Allen, M. J. (1977). *In* "Living Systems as Energy Converters" (R. Buvet, M. J. Allen, and J.-P. Massué, eds.), pp. 271–274. Elsevier/North-Holland Biomedical Press, Amsterdam.

Arkles, B., and Brinigar, W. S. (1975). *J. Biol. Chem.* **250,** 8856–8862.

Asada, K., Takahashi, M., and Urano, M. (1972). *Anal. Biochem.* **48,** 311–315.

Benemann, J. R., Berenson, J. A., Kaplan, N. O., and Kamen, M. D. (1973). *Proc. Natl. Acad. Sci. U.S.A.* **70,** 2317–2320.

Berezin, I. V., and Varfolomeev, S. D. (1979). *Appl. Biochem. Bioeng.* **2,** 259–289.

Brunner, G., and Lösgen, H. (1978). *Enzyme Eng.* **3,** 391–396.

Cocquempot, M. F., Garde, V. L., and Thomas, D. (1980). *Biochimie (Paris)* **62,** 615–621.

Cocquempot, M. F., Thomasset, B., Barbotin, J. N., Gellf, G., and Thomas, D. (1981). *Eur. J. Appl. Microbiol. Biotechnol.* **11,** 193–198.

Fujimura, T., Yoshii, F., Kaetsu, I., Inoue, Y., and Shibata, K. (1980). *Z. Naturforsch., C: Biosci.* **35C,** 477–481.

Fukui, S., and Tanaka, A. (1979). *J. Appl. Biochem.* **1,** 171–201.

Fukui, S., Kawamoto, S., Yasuhara, S., Tanaka, A., Osumi, M., and Imaizumi, F. (1975). *Eur. J. Biochem.* **59,** 561–566.

Fukui, S., Tanaka, A., Iida, T., and Hasegawa, E. (1976). *FEBS Lett.* **66,** 179–182.

Fukui, S., Tanaka, A., and Gellf, G. (1978). *Enzyme Eng.* **4,** 299–306.

Fukui, S., Sonomoto, K., Itoh, N., and Tanaka, A. (1980). *Biochimie (Paris)* **62,** 381–386.

Fukushima, S., Nagai, T., Fujita, K., Tanaka, A., and Fukui, S. (1978). *Biotechnol. Bioeng.* **20,** 1465–1469.

Garde, V. L., Thomasset, B., Tanaka, A., Gellf, G., and Thomas, D. (1981). *Eur. J. Appl. Microbiol. Biotechnol.* **11,** 133–138.

Gisby, P. E., and Hall, D. O. (1980). *Nature (London)* **287,** 251–253.

Gross, E. L., Youngman, D. R., and Winemiller, S. L. (1978). *Photochem. Photobiol.* **28,** 249–256.

Haehnel, W., Heupel, A., and Hengstermann, D. (1978). *Z. Naturforsch., C: Biosci.* **33C,** 392–401.

Hall, D. O. (1972). *Nature (London), New Biol.* **235,** 125–126.

Hisada, R., and Yagi, T. (1977). *J. Biochem. (Tokyo)* **82,** 1469–1473.

Ingelman-Sundberg, M., and Glaumann, H. (1977). *FEBS Lett.* **78,** 72–76.

Ingelman-Sundberg, M., and Johansson, I. (1980). *Biochemistry* **19,** 4004–4011.

Janzen, A. F., and Seibert, M. (1980). *Nature (London)* **286,** 584–585.

Jensen, R. G., and Bassham, J. A. (1966). *Proc. Natl. Acad. Sci. U.S.A.* **56,** 1095–1101.

Karube, I., Aizawa, K., Ikeda, S., and Suzuki, S. (1979). *Biotechnol. Bioeng.* **21,** 253–260.

Karube, I., Otsuka, T., Kayano, H., Matsunaga, T., and Suzuki, S. (1980). *Biotechnol. Bioeng.* **22,** 2655–2665.

Kastl, P. R., Baricos, W. H., Chambers, R. P., and Cohen, W. (1978). *Enzyme Eng.* **4,** 199–206.

Kayano, H., Matsunaga, T., Karube, I., and Suzuki, S. (1981). *Biotechnol. Bioeng.* **23,** 2283–2291.

Kierstan, M., and Bucke, C. (1977). *Biotechnol. Bioeng.* **19,** 387–397.

Kitajima, M., and Butler, W. L. (1976). *Plant Physiol.* **57,** 746–750.

Miura, Y., Yokoyama, H., Kanaoka, K., Saito, S., Iwasa, K., Okazaki, M., and Komemushi, S. (1980). *Plant Cell Physiol.* **21,** 149–156.

Ochiai, H., Shibata, H., Matsuo, T., Hashinokuchi, K., and Yukawa, M. (1977). *Agric. Biol. Chem.* **41,** 721–722.

Ochiai, H., Shibata, H., Matsuo, T., Hashinokuchi, K., and Yukawa, M. (1978a). *Nippon Nogei Kagaku Kaishi* **52,** 31–36.

Ochiai, H., Shibata, H., Matsuo, T., Hashinokuchi, K., and Inamura, I. (1978b). *Agric. Biol. Chem.* **42,** 683–685.

Ochiai, H., Shibata, H., Matsuo, T., Hashinokuchi, K., Yukawa, M., and Inamura, I. (1978c). *Amino Acid Nucleic Acid (Tokyo)* **37,** 54–63.

Ochiai, H., Shibata, H., Fujishima, A., and Honda, K. (1979). *Agric. Biol. Chem.* **43,** 881–883.

Ochiai, H., Shibata, H., Sawa, Y., and Katoh, T. (1980). *Proc. Natl. Acad. Sci. U.S.A.* **77,** 2442–2444.

Ochiai, H., Shibata, H., Sawa, Y., and Katoh, T. (1982). *Photochem. Photobiol.* **35,** 149–155.

Ochiai, H., Shibata, H., Sawa, Y., Shoga, M., and Ohta, S. (1983). *Appl. Biochem. Biotechnol.* in press.

Oku, T., Sugahara, K., and Tomita, G. (1973). *Plant Cell Physiol.* **14,** 385–396.

Pace, W. G., Yang, H. S., Tannenbaum, S. R., and Archer, M. C. (1976). *Biotechnol. Bioeng.* **18,** 1413–1423.

Papageorgiou, G. C. (1979). *In* "Photosynthesis in Relation to Model Systems" (J. Barber, ed.), pp. 211–241. Elsevier/North-Holland Biomedical Press, Amsterdam.

Papageorgiou, G. C., and Isaakidou, J. (1977). *In* "Bioenergetics of Membranes" (L. Packer, G. C. Papageorgiou, and A. Trebst, eds.), pp. 257–268. Elsevier/North-Holland, Amsterdam.

Paul, F., and Vignais, P. M. (1980). *Enzyme Microb. Technol.* **2,** 281–287.

Rao, K. K., and Hall, D. O. (1979). *In* "Photosynthesis in Relation to Model Systems" (J. Barber, ed.), pp. 299–329. Elsevier/North-Holland Biomedical Press, Amsterdam.

Rao, K. K., Rosa, L., and Hall, D. O. (1976). *Biochem. Biophys. Res. Commun.* **68,** 21–28.

Sawa, Y., Kanayama, K., and Ochiai, H. (1980). *Agric. Biol. Chem.* **44,** 1967–1969.

Sawa, Y., Kanayama, K., and Ochiai, H. (1982). *Biotechnol. Bioeng.* **24,** 305–315.

Shioi, Y., and Sasa, T. (1979). *FEBS Lett.* **101,** 311–315.

Srere, P. A., Mattiasson, B., and Mosbach, K. (1973). *Proc. Natl. Acad. Sci. U.S.A.* **70,** 2534–2538.

Tanaka, A., Yasuhara, S., Osumi, M., and Fukui, S. (1977). *Eur. J. Biochem.* **80,** 193–197.

Tanaka, A., Yasuhara, S., Gellf, G., Osumi, M., and Fukui, S. (1978). *Eur. J. Appl. Microbiol. Biotechnol.* **5,** 17–27.

Tanaka, A., Jin, I.-N., Kawamoto, S., and Fukui, S. (1979). *Eur. J. Appl. Microbiol. Biotechnol.* **7,** 351–354.

Tanaka, A., Hagi, N., Gellf, G., and Fukui, S. (1980). *Agric. Biol. Chem.* **44,** 2399–2405.

Vieth, W. R., and Venkatasubramanian, K. (1978). *Enzyme Eng.* **4,** 307–316.

West, J., and Packer, L. (1970). *Bioenergetics* **1,** 405–412.

Yagi, T. (1976). *Proc. Natl. Acad. Sci. U.S.A.* **73,** 2947–2949.

Yagi, T., and Ochiai, H. (1978). *Adv. Hydrogen Energy* **3,** 1293–1307.

Yang, H. S., Leung, K.-H., and Archer, M. C. (1976). *Biotechnol. Bioeng.* **18,** 1425–1432.

Immobilized Living Cells and Their Applications

John F. Kennedy

Research Laboratory for the Chemistry of Bioactive Carbohydrates and Proteins
Department of Chemistry
University of Birmingham
Birmingham, England

Joaquim M. S. Cabral

Laboratório de Engenharia Bioquímica
Departamento de Engenharia Química, Instituto Superior Técnico
Universidade Técnica de Lisboa
Lisbon, Portugal

APPLIED BIOCHEMISTRY AND BIOENGINEERING
Volume 4

ISBN 0-12-041104-0

I. INTRODUCTION

For many years the dominant catalysts in the biological industry have been metabolizing microorganisms, whose great versatility has led to the development of the present fermentation industry. During the fermentation process large numbers of microorganisms are formed that, either during or after the growth phase, act as catalysts for the synthesis of products. For product formation, not only must the enzymes involved in the synthesis be present, but also the microorganisms must still be metabolizing nutrients to form the energy required for product synthesis. In some cases, however, only one or a few enzymes in the microorganisms are required for the desired transformation, and nonmetabolizing cells, which may no longer be viable, are adequate.

Over the past two decades there have been rapid developments in the use of enzymes as catalysts for industrial, analytical, and medical purposes, and a new field of research, called enzyme technology, has appeared. In this area enzymes have been immobilized in order to make their use more convenient, in such a way that they resemble ordinary solid-phase catalysts used conventionally in the synthetic chemical reactions (Kennedy and Cabral, 1983).

Enzymes can be classified into two categories: extracellular if they are excreted from the cells into the growth medium, and intracellular if the enzyme is retained in the cell during cultivation. In order to immobilize intracellular enzymes, they have to be extracted and isolated from the cell. However, such extracted enzymes are unstable in many cases and not suitable for practical use; additionally the operations of extraction are normally expensive. Thus, in order to avoid the expensive, lengthy, and tedious extraction and purification of enzymes, often resulting in only low yields, and by procedures that are difficult to scale up, and mainly to avoid the problems of instability of the active enzyme, whole cells have been immobilized directly and used as a solid catalyst, with the inherent advantages of enzyme immobilization. Thus current and industrial applications of continuous single-enzyme reactions are carried out using immobilized microbial cells (Chibata *et al.*, 1974).

In practice, for single-enzyme reactions, there is no clear boundary between enzyme and cell immobilization. The choice to be made between the use of an immobilized enzyme or an immobilized cell is similar in many respects to the more familiar choice to be made between the use of purified or crude soluble enzymes as biocatalysts.

However, the potential disadvantages associated with the immobilization of cells must be considered. Cells contain numerous catalytically active enzymes, which may catalyze unwanted side reactions in some processes, so those interfering enzymes should be easily inactivated by simple methods, such as heat or pH treatment. Other disadvantages are the possible loss of some of the desirable catalytic activity due to enzyme inactivation during the immobilization process or to diffusional barriers that hinder substrate access to the active sites of the enzyme. The cell membrane itself serves as a diffusion barrier and in some single-step reactions must be made permeable.

However, many useful compounds are usually produced, especially in fermentative processes, by the joint action of several kinds of enzymes (multienzyme systems). Although it is no doubt possible to coimmobilize individually purified enzymes or to mix together separately immobilized enzymes, so as to reconstitute multienzyme pathways, by far the most satisfactory way of reusing multienzyme reactions in the form of convenient reactors is to immobilize whole living cells.

The advantages of such a whole-living-cell approach are immediately obvious. The tedious and time-consuming procedures for enzyme extraction and purification are instantly eliminated, the cellular enzymes are often already organized into the requisite metabolic pathways, the problem associated with enzyme instability may also be automatically avoided, and for single- and multienzyme reactions—which are likely to involve enzymes requiring cofactors and coenzymes—the latter are readily at hand in the cell, obviating the alternative but inadequate method of coimmobilization of cofactors with enzymes and their regeneration, a successful route for which has not yet been devised. Thus immobilized living cells should be preferred to immobilized enzymes for degradative and synthetic reactions, which require energy and expensive cofactors and the use of complete metabolic pathways or can effectively and efficiently use the complete metabolism of the whole cells. Another major advantage of immobilized cells is that the operational stability of the immobilized living cells may often be greatly enhanced by regeneration of the enzyme activities of the immobilized cells; this regeneration may be achieved either by reinduction or by causing the immobilized cells to divide *in situ*.

Like enzyme immobilization, the immobilization of cells has the same advantages, when comparing the immobilized cells with the free cells. Thus the immobilization process makes possible the reuse or the continuous use of this type of biocatalyst; immobilized cells are often more stable than the equivalent free cells; immobilized cell processes are easier to automate and enable exploitation of the advantages of various reactor configurations.

When compared with the versatile fermentation method, which is the dominant form of industrial biological catalysis, immobilized living cells present several potential advantages. Immobilized cells are convenient to handle, appear to be less susceptible to microbial contamination, and permit easy separation of products from the biocatalyst. However, contamination by cells produced by multiplications of the immobilized living cells or materials derived from the cell lysis may occur. The use of immobilized cells enables greater control throughout the reaction. Furthermore, nondividing immobilized cells require only maintenance energy, and yields of product will be greater than with fermentation methods. Immobilization also facilitates the use of dense cell populations by altering the rheological properties of the suspending medium. The fluid viscosity is lower than that appertaining when comparable numbers of cells are freely suspended in solution. Lower viscosities contribute to better mixing and mass-transfer properties in the reactor.

Furthermore, the applications of immobilized living cells are not limited to the production of chemical fermentation products, but have been extended to the production of viral particles or synchronous cells, the chromatographic separation of special cells, the culturing of animal tissue, and more recently, the immobilization and use of plant cells in the production of alkaloids.

The purpose of this chapter is to illustrate the exploitation of the aforementioned advantages of immobilized cells, and to discuss the methods used for whole-cell immobilization and their applications, with particular attention to immobilized living cells.

A. Historical Perspective

Although living-cell immobilization has been very recently receiving widespread attention (Venkatasubramanian, 1979; Larsson *et al.*, 1979; Kennedy, 1980; Klein *et al.*, 1976, 1978; Navarro, 1978; Chibata, 1979; Linko, 1981), immobilized viable cells have in fact been successfully exploited for many years, but without the workers involved realizing the importance and relevance of their processes.

One of the earliest examples of the use of immobilized living cells was in the vinegar-manufacturing industry. The "quick" process, invented in 1823 by Scheutzenbach, utilized wooden vats with perforated bottoms; these vats were packed with wood shavings on which microbial films developed. Alcoholic solutions were trickled through the vats and were oxidized to acetic acid by the microbial films on the shavings.

However, this conversion of wine to vinegar was believed to be a mere chemical reaction until 1864, when Pasteur (1868) showed the microbial dependence of this transformation. Pasteur developed vinegar manufacturing through the now well-known "Pasteur" or "Orleans" process, which is the first rationally designed and recognized immobilized-cell system. In this process a barrel was partially filled with a wine–vinegar mixture and a microbial film was allowed to develop on the surface of a mechanical support into a thick gelatinous mat. The support immobilized the microorganisms and was used repeatedly from one fermentation to another, the vinegar being periodically harvested and replaced with wine.

The effect of solid surfaces on microbial growth has long been recognized (Russell, 1891; Whipple, 1901; ZoBell, 1943; Heukelekian and Heller, 1940). Under natural conditions, microbial films develop on a wide variety of biotic and abiotic supports, including such diverse materials as sand grains, mineral faces, metal surfaces, poly(vinyl chloride) tubing, dental plaque, and soil and marine environments. For example, studies of the microbial population in rivers have shown that very few microorganisms are in free suspension, but are rather associated with solid surfaces, such as silt particles (Heukelekian and Dondero, 1964).

Immobilized living cells are principal components of activated sludge and trickling-filter waste-treatment systems. Immobilized living cells are also responsible for effective leaching of low-grade mineral ores. This leaching occurs through the oxidation of mineral sulfides by chemoautotrophic bacteria attached to the mineral surfaces (Malouf and Prater, 1961; Muir and Berry, 1976). In continuous culture, the trivial washout state often cannot be attained (Larsen and Dimmick, 1964) due to the adhesion of some part of the microbial population to the internal surfaces of the reaction vessel. Animal tissue culture and its attendant area of virus production are routinely operated so that the cells are grown as a surface monolayer (for review, see Van Wezel, 1973).

Thus the immobilization of whole cells, viable or nonviable, is not a novel concept, but rather a refinement of a phenomenon observed in nature and in those industrial microbiological processes in which the surface growth of cells is favored.

However, the study of immobilized microbes is a rapidly expanding area. During the 1970s new commercial immobilized-cell processes have been developed. The first industrial application of immobilized cells is attributable to Chibata in 1973, for the production of L-aspartic acid from fumaric acid, using immobilized *Escherichia coli* with L-aspartase ammonia lyase (aspartase, EC 4.3.1.1) activity (Chibata *et al.*, 1974). Following the L-aspartic acid production, Chibata and co-workers succeeded, in 1974, in the second application of immobilized cells, the production of L-malic acid from fumaric acid using immobilized *Brevibacterium ammoniagenes* with fumarate hydratase (fumarase, EC 4.2.1.2) activity (Yamamoto *et al.*, 1977). Another very important industrial application of immobilized cells is the D-glucose isomerization reaction, using microbial cells with D-glucose isomerase (D-xylose isomerase EC 5.3.1.5) activity. However, the processes just described, using immobilized microbial cells, are primarily catalyzed by a single enzyme and the immobilized cells are in dead state, though the enzyme is in an active state. These immobilized microbial cells are used advantageously instead of immobilized intracellular enzymes, so they can be more accurately described as immobilized nonpurified enzymes.

The early reports (Updike *et al.*, 1969; Franks, 1971, 1972) in the field of immobilized whole cells deal with the viability problem of the cells. However, with the main emphasis on single-enzyme reactions during the following years, the viability problem was left aside; the problem regained interest and attention only recently in connection with multienzyme reactions (Klein *et al.*, 1976, 1978) and with fermentations (Kennedy *et al.*, 1976b), mainly ethanol production (Kierstan and Bucke, 1977; Navarro, 1978; Chibata, 1979; Ghose and Bandyopadhay, 1980; Linko, 1981; Linko and Linko, 1981).

Comparatively little work has been carried out on the immobilization of cells compared with the voluminous literature on fermentation and enzyme immobilization, but the potential of immobilized cells has recently begun to be recognized, as shown by the reports presented on the occasions of the Enzyme Engineering Conferences of 1979 and 1981, and of the First and Second European Congresses on Biotechnology. A flood of publications is to be expected in the very near future, with emphasis on fermentation and on animal- and plant-cell technologies.

Cell-immobilization methods have been reviewed thoroughly by several authors (Chibata, 1977; Abbott, 1976, 1977, 1979; Jack and Zajic, 1977a; Vandamme, 1976; Klein and Wagner, 1978; Durand and Navarro, 1978; Cheetam, 1980; Mosbach, 1981).

B. Classification of Immobilized Cells

For the purpose of this chapter, cell immobilization is defined similarly to immobilized enzymes as drafted at the First Enzyme Engineering Conference. Immobilized cells are whole cells that are physically confined or localized in a certain defined region of space with retention of their catalytic activities, and which can be used repeatedly and continuously.

The term *immobilized cell* includes (*a*) cells attached to solid surfaces (microbial films); (*b*) entrapped cells; (*c*) free suspended cells used in reactors equipped with semipermeable membranes, allowing the passage of reaction products, but retaining the free cells inside the reactor; and (*d*) pelletized or flocculated cells that can be used continuously in a column reactor or can be easily and nondestructively recovered from batch reactors by sedimentation or filtration devices, allowing their reuse.

With this definition, certain kinds of continuous-fermentation processes can be considered as immobilized-living-cell systems, such as in activated sludge wastewater treatment, single-cell protein fermentations, winemaking, brewing, citric acid fermentation, vinegar fermentation, and any fermentation that requires a biomass recovery and reuse for product isolation.

As with the case of enzyme immobilization, there are several ways of classifying the various types of immobilized cells that fall within the foregoing definition. One possible classification system is based on the nature of the interaction responsible for immobilization; that is, the immobilization can be achieved either through (*a*) chemical means, which includes any method that involves the formation of covalent or partially covalent bonds between any reactive component of the cell surface and the support, or between two or more whole cells, or (*b*) by physical means, which includes any method that involves localizing the whole cell in any manner whatsoever that is not dependent on covalent bond formation. However, the immobilization of living cells by chemical means has as the main disadvantage the problem of the toxicity to the cells of the chemical agents usually used. It is preferable, therefore, to classify the immobilized living cells according to other classifications based on both the nature of the support and the type of reaction between cells and support.

The classification presented here is very similar to that suggested by Kennedy and Cabral (1983), in their review of immobilized enzymes, and it is not fundamentally different from others proposed by several authors (Jack and Zajic, 1977a; Klein and Wagner, 1978; Cheetam,

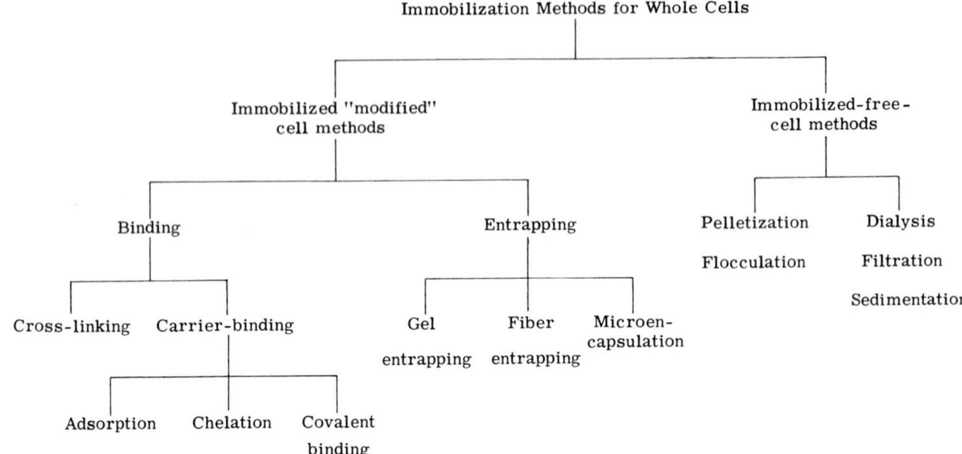

Fig. 1. Classification of immobilization methods for whole cells.

1980). It attempts to combine the nature of interaction responsible for immobilization and the nature of the support. Figure 1 illustrates this classification system and lists the various individual methods that are discussed in detail in subsequent sections.

II. TECHNIQUES OF WHOLE-CELL IMMOBILIZATION

In recent years a number of methods to immobilize cells have been developed. Some recent reviews (Jack and Zajic, 1977a; Klein and Wagner, 1978; Durand and Navarro, 1978; Cheetam, 1980) discuss these methods in detail. The aim of this chapter is to review the more recently published data on the preparation of immobilized living cells, according to the classification suggested by us.

A. Cross-Linking Method of Immobilization

This immobilization method is a carrierless technique in which whole cells are attached to each other by chemical or physical means.

Chemical cross-linking consists of the covalent attachment of cells to each other with bi- or multifunctional reagents such as aldehydes (Chibata *et al.*, 1974; Poulson and Zittan, 1976; Hughes and Thurman, 1970) or amines (Lartigue and Weetall, 1976). However, the toxicity of these chemicals (Murton and Russell, 1973) obviously limits their applicability in immobilizing living cells.

The cross-linking method has been used only for microbial cells that catalyze single reactions. Chibata *et al.* (1974) used this method to

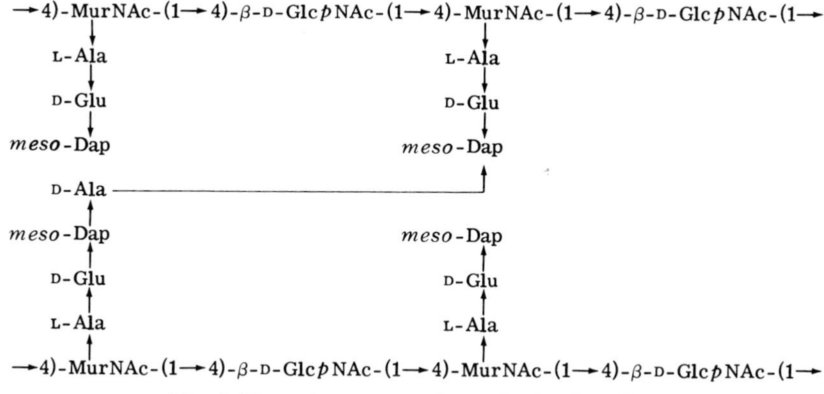

Fig. 2. Typical structure of microbial cell wall.

immobilize *Escherichia coli* cells having high L-aspartate ammonia lyase (aspartase, EC 4.3.1.1) activity, by strengthening the cell walls or cell membranes and cross-linking with the bifunctional reagents, glutaraldehyde and toluene diisocyanate. A 34% retention of L-aspartate ammonia lyase activity was obtained by using glutaraldehyde, but inactivated-immobilized cell preparations were obtained with toluene diisocyanate and toluene diisocyanate plus 1,6-diaminohexane. However, the immobilization of *E. coli* cells by entrapment in polyacrylamide led to more active preparations, with about 73% retention of aspartate ammonia lyase activity.

In order to describe the underlying chemistry of this cross-linking method, the structure of the cell wall material must be understood. Bacterial walls are composed of mucopeptides, which are peptide chains linked together (Fig. 2) with some free amino groups. The linkage is through the α-carboxyl terminal of a residue of D-alanine of one chain to the ε-amino group of a diaminopimelic acid residue in a second chain (Hughes, 1968; Warth and Strominger, 1968), forming heptapeptide dimer units. The cross-linking of microbial cells with glutaraldehyde involves the reaction between this bifunctional reagent and the residual free amino groups of the microbial mucopeptides (Fig. 3). The linkages formed between the amino groups and glutaraldehyde are irreversible and survive extremes of pH and temperature. However, the straightforward participation of the reagent's aldehyde groups to form an aldimine bond (Schiff's base) with protein amino groups has been questioned. It has been suggested (Richards and Knowles, 1968) that the glutaraldehyde reaction most probably involves conjugate addition of amino groups to ethylenic double bonds of α,β-unsaturated oligomers contained in the commercial

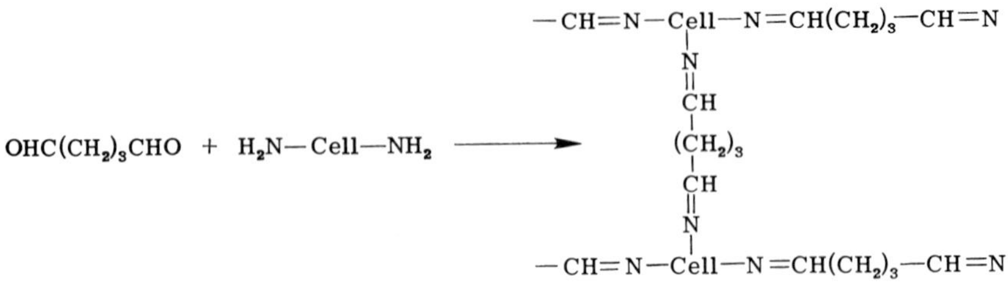

OHC(CH2)3CHO + H2N—Cell—NH2
—CH=N—Cell—N=CH(CH2)3—CH=N
N
CH
(CH2)3
CH
N
—CH=N—Cell—N=CH(CH2)3—CH=N

Fig. 3. Cross-linking of cell wall free amino groups with glutaraldehyde.

aqueous glutaraldehyde solutions usually used (Fig. 4). According to Richards and Knowles, this mechanism explains both the stability of the bond—which cannot be due to a single, simple Schiff's base formation—and also the lower reactivity on protein of solutions of freshly distilled glutaraldehyde.

Very few articles, and these only for single-enzyme reactions, concerning the chemical cross-linking of microbial cells have appeared. If suitable cross-linking agents can be found, there is some greater potential for this method.

In contrast, physical cross-linking of cells by flocculation has a high potential for application in the immobilization of cells (living or dead), as in the light of industrial technology this procedure leads to high cell concentrations per unit volume of reactor. A variety of flocculating agents can be used in promoting the cell aggregation. Such agents include cationic polyelectrolytes such as polyamines, polyethyleneimine, and cationic polyacrylamides; anionic polyelectrolytes such as carboxyl-substituted polyacrylamides, polystyrene sulfonates, and polycarboxyl acids (Lee and Long, 1974); and metallic

OHC(CH2)3CHO
CHO
CHO
CHO
—CH=C—(CH2)2—CH=C—(CH2)2—CH=C—(CH2)2—
H2N—Protein
CHO
CHO
CHO
—CH—CH—(CH2)2—CH=C—(CH2)2—CH—CH—(CH2)2—
NH—Protein
NH—Protein

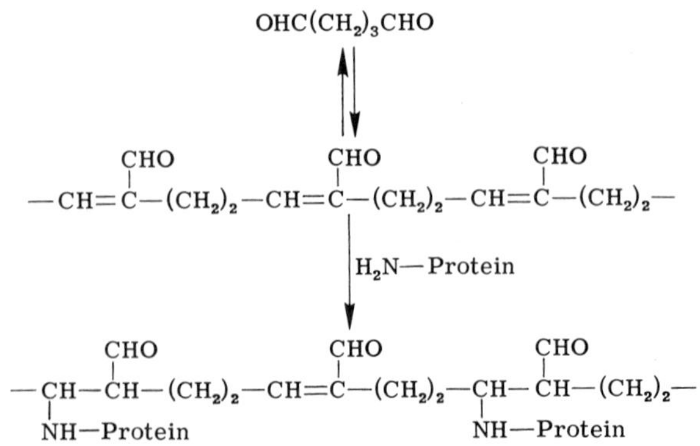

Fig. 4. Suggested glutaraldehyde reaction mechanism.

compounds, namely the oxides, hydroxides, sulfates, and phosphates of Mg^{2+}, Ca^{2+}, Fe^{3+}, and Mn^{2+} (Long, 1976). The use of filter aids such as diatomaceous earth and asbestos, and polymeric adsorbents such as acrylic esters may be advantageous in some instances, when added to the cell suspension at the time of flocculation. Because of the lack of mechanical characteristics for use in column reactors, cell aggregates may be used in conjunction with inert support materials.

Some immobilized-whole-cell materials that have been prepared by chemical or physical cross-linking are listed in Table I.

TABLE I
PRINCIPAL EXAMPLES OF WHOLE CELLS IMMOBILIZED BY INTERCELLULAR CROSS-LINKING

Cross-linking agent	Cells	Reaction (substrate/ product)	Reference
Chemical cross-linking			
Diazotized diamines	*Streptomyces* sp.	Isomerization (D-Glucose/ D-fructose)	Lartigue and Weetall (1976)
Glutaraldehyde	*Bacillus coagulans*	Isomerization (D-Glucose/ D-fructose)	Poulsen and Zittan (1976)
	Escherichia coli	Molecular lysis (Fumaric acid/ L-aspartic acid)	Chibata *et al.* (1974)
	Mold	Removal of raffinose	Nishimaru *et al.* (1975)
Toluene diisocyanate	*E. coli*	Molecular lysis (Fumaric acid/ L-aspartic acid)	Chibata *et al.* (1974)
Toluene diisocyanate + 1,6-diaminohexane	*E. coli*	Molecular lysis (Fumaric acid/ L-aspartic acid)	Chibata *et al.* (1974)
Physical cross-linking			
Cationic polyelectrolyte (Primafloc C-7)	*Aspergillus niger*	Oxidation (D-glucose/ D-gluconic acid)	Lee and Long (1974)
	Streptomyces olivaceous	Isomerization (D-Glucose/ D-fructose)	Lee and Long (1974)
Cationic polyelectrolyte (Primafloc C-7) + anionic polyelectrolyte (Primafloc A-10)	*Arthrobacter* sp.	Isomerization (D-Glucose/ D-fructose)	Lee and Long (1974); Long (1976)

B. Carrier-Binding Method of Immobilization

The carrier-binding method is based on linking whole cells to solid supports and can be further divided into three categories, according to the binding mode of the whole cell: adsorption, chelation or metal binding, and covalent binding.

When cells are immobilized in this way, care must be exercised in the selection of carriers as well as of the binding procedure. A carrier judiciously chosen can enhance the operational stability of the immobilized-cell system. As in the situation with immobilized enzymes, it is recognized that there is no universally successful or acceptable carrier, but a number of desirable characteristics should be common to any material considered for cell immobilization. The chosen carrier should be stable in solution and should not deteriorate under operational solution conditions of composition and pH. It should also be mechanically rigid and show little compaction in high-flow-rate, continuous operations using fixed-bed reactors.

Of the organic and inorganic, natural and synthetic carriers that may be used for immobilization of cells, it appears that the inorganic supports can best fulfill the requirements for use—that is, mainly in industry. This is due to their stabilities against physical, chemical, thermal, and microbial degradation, their mechanical strength and structural stability to avoid compaction and large pressure drops in flow processes, and their relatively simple regenerability.

According to their composition, carriers can be classified as organic or inorganic. Organic carriers can be divided into natural and synthetic polymers. Some examples of natural polymers that have been used as matrices for cell immobilization are polysaccharides such as cellulose, dextran, and agarose; proteins such as gelatin and collagen; and carbon materials such as anthracite and wood. Examples of synthetic polymers are copolymers of ethylene with maleic anhydride, glutaraldehyde–glycidyl methacrylate copolymers, synthetic ion-exchange materials, and plastic materials. Some of the inorganic carriers used are glasses, ceramics, hydrous metal oxides, Kieselguhr, and cordierite.

Although this classification into organic and inorganic carriers has been used to categorize the nature of the support, according to Messing (1975), this description is not adequate for the full characterization of very pertinent carrier parameters, such as surface area and pore diameter, both of which in turn will affect the loading of the biocatalyst. Therefore, the carriers can also be classified based on morphology, as nonporous and porous carriers.

Although some advantages of this latter classification are recognizable, the former classification based on the chemical nature of matrices will be used in this chapter for each specific carrier-binding method. It is more appropriate to consider any relevant morphology within such divisions.

1. Adsorption Method of Immobilization

It is well known that in natural environments, particularly in soil, rivers, rumen, mouth, and skin, microbial cells exist in a sorbed state. Also, the growth of microorganisms is observed on the walls and impellers of the fermenters, and it is assumed that any surface in contact with a microbial suspension will, in time, become biologically active due to adhesion of microorganisms.

The formation of natural biological films by adhesion to surfaces has long been recognized (Russell, 1891), as has already been stated. The adhesion of certain varieties of microorganisms is attributed to the formation of "stalks," or holdfasts, which are responsible for the attachment of the cell to a surface. Once the microbial film is established, adhesion may be automatically increased in various ways (ZoBell, 1943), as, for example, by the formation of a slime film or the secretion of cement-like materials.

Besides inorganic carriers, several other materials have been used for immobilization purposes, among them wood, plastic materials, and mainly ion-exchange organic materials, both natural (e.g., cellulose and dextran) and synthetic resins. One of the very first attempts to immobilize cells was performed by Hattori and Furusaka (1960, 1961), who adsorbed *Escherichia coli* and *Azotobacter agile* to Dowex-1 and measured the oxidation of succinic acid. Among the obvious applications of such specific linkage to charged supports is the chromatographic separation of microbe mixtures or the concentration and isolation of a given microorganism.

Various aspects of the microbial-adsorption phenomenon have been considered in several reviews (Jack and Zajic, 1977a; Klein and Wagner, 1978; Durand and Navarro, 1978; Atkinson and Fowler, 1974; Kolot, 1980; Cheetam, 1980). Recently, a review about the biophysics of cellular adhesion to solid surfaces and substrates (Gerson and Zajic, 1979) discussed the influence of the surface free energy, a parameter that must dominate any explanation of the adhesion between different phases (microorganisms and support) that are not mechanically linked.

Adsorption of cells to supports is dependent on the characteristics of the environment. Because the adsorption or adhesion phenomenon is mainly based on electrostatic interactions—that is, van der Waals forces, and ionic and hydrogen bonds—between the cell surface and the support material, the actual zeta potential on both of them plays an important role in cell-support interaction. From the zeta potential it is possible to obtain an approximate value for the surface-charge density (Abramson *et al.*, 1942). Clearly cells will be attracted to surfaces of opposite zeta potential. If cells have the same zeta potential as a surface, attachment is still possible provided the electrostatic barrier can be penetrated by small surface projections (Van Oss *et al.*, 1975; Grinnell, 1978). Zeta potential and the adhesion of microorganisms to surfaces have been reviewed by several authors (Daniels, 1971; Marshall, 1976; Martin, 1978).

Another factor that influences the adsorption and is strictly dependent on the properties of the microbe is the cell wall composition. The charged nature of the cell wall is chiefly determined by the distribution of carboxyl and amino groups of the peptide amino acid, diaminopimelic acid, and 2-amino-2-deoxyhexose residues of the cell wall surfaces, which may directly interact with the solid surface. For example, it is a well-known fact that yeast cells are negatively charged, therefore it is preferable to choose a positively charged support for their immobilization.

Carrier properties other than zeta potential will also influence the adsorption of the cells to solid supports. One of the most important properties is the carrier composition. So all glasses and ceramics consist of varying proportions of aluminum, silicon, magnesium, zirconium, titanium, and other oxides that in solution act as ion-exchange materials. The corresponding hydroxides (hydrous oxides) can be formed; this allows the replacing of the hydroxyl groups on the carrier surface by suitable amino or carboxyl groups on the cell surface by a chelation mechanism.

A further carrier property that may influence the immobilized-cell preparation is the morphology of the support. The advantage of using porous supports is well established for enzyme immobilization. Messing *et al.* (1979) studied the relationship between the accumulation of stable and viable biomass, and the pore morphology of a dimensionally stable inorganic carrier. They found that this relationship is dependent on the mode of reproduction of the specific microorganism. In the case of microbes that reproduce by fission (*Escherichia coli, Serratia marcescens, Bacillus subtilis*), the maximum accumulation of stable and viable biomass occurs when the pore diameter is between 1 and 5 times the major dimension of the microorganisms. In the case of reproduction by budding, exemplified by yeasts (*Saccharomyces cerevisiae* and *Saccharamyces amurcae*), pores of approximately 4 times the maximum dimension of yeast cells must be employed in order to allow for the increase of yeast length. When microbes exhibit mycelial growth and produce spores (*Aspergillus niger*, *Streptomyces olivochromogenes*, *Penicillium chrysogenum*), a high biomass accumulation occurs at about 16 times the largest dimension of the spores.

The characteristics of environment also influence adsorption. Thus, in the case of ion-exchange materials, any factors that can affect surface charge can also influence the ionic-based adsorption phenomena. The presence of metal ions (Zvyagintsev, 1971) or anions (Daniels and Kempe, 1966) can determine the degree of binding of cells to a cationic or anionic ion-exchange resin by competition for the resin-binding sites and by the neutralization or creation of cell surface charges. Also, pH changes may affect adsorption, by the modification of zeta potentials on support surfaces (Navarro, 1975; Messing *et al.*, 1979; Marcipar *et al.*, 1979).

As an immobilization method, adsorption is a mild, simple, and nonspecific process that has as the major advantage the fact that cells remain alive and their enzymic activities are not affected.

An important and practical advantage, as in the case of enzyme adsorption, is the ability to regenerate the immobilized cell preparation by desorption of inactive cells, followed by readjustment of the pH or ionic strength and adsorption of fresh cells. But a main disadvantage, in terms of preparation stability, arises from the rate of desorption of cells from the support (Marcipar *et al.*, 1979). This problem is especially severe when changes in pH or ionic strength occur, when multiplication of the adsorbed cells occurs (Hattori, 1972), or when cells are sheared from the surfaces of the support as a result of fluid velocity or contact with gas bubbles or with particles. A further disadvantage of adsorption method, common to the other carrier-binding methods, is the relatively low cell concentration per unit volume of reactor, when compared with other immobilization methods, such as flocculation.

Some principal examples of microbial living cells that have been immobilized by the adsorption method are presented in Table II.

2. Chelation or Metal-Binding Method of Immobilization

This method of immobilization was initially developed for enzyme immobilization at the University of Birmingham, England (Novais, 1971; Emery *et al.*, 1972). Subsequent independent work in which the mechanism of the process was elucidated was the development of the use of the hydrous oxides of the transition-metal activators, not only as ligand but also as carrier (Kennedy *et al.*, 1976a,b, 1977, 1981; Kennedy and Kay, 1976; Kennedy and Pike, 1979).

TABLE II
PRINCIPAL EXAMPLES OF IMMOBILIZED LIVING CELLS PREPARED BY ADSORPTION

Carrier	Cells	Reaction (substrate/product)	Reference
Organic support			
Anthracite	*Pseudomonas* sp.	Degradation of phenol	Scott and Hancher (1976)
Bio Rex	*Azotobacter vinelandii*	Nitrogen fixation	Gainer *et al.* (1981)
Coal particles	Waste-treatment bacteria	Biological denitrification	Scott and Hancher (1976)
			Taylor (1978); Hancher *et al.* (1979); Ngian and Martin (1980)
		Biodegradation of phenolic waste liquors	Holladay *et al.* (1976)
DEAE-Sephadex	Primary and finite life-span cell strains	Growth of cells on microcarriers and production of poliomyelitis virus	A. L. Van Wezel (1967); A. N. Van Wezel (1972, 1973)
	BHK 21 Cells	Growth of cells on microcarriers	Van Hemert *et al.* (1968); Spier *et al.* (1977)
		Production of foot-and-mouth disease virus	Spier and Whiteside (1976)
	Cell line 292 (skin cells)	Production of interferon	Giard *et al.* (1979)
	Cell line 316 (skin cells)	Production of interferon	Giard *et al.* (1979)
	Chicken fibroblast	Production of sindbis virus	Levine *et al.* (1979)
	CHO Cells	Production of sindbis virus	Levine *et al.* (1979)
	FS-4 Cells	Production of interferon	Giard *et al.* (1979); Levine *et al.* (1979)
	HEL 299 Cells (human fibroblast)	Growth of cells on microcarriers and production of interferon	Giard *et al.* (1979); Levine *et al.* (1979)
	JLSV 9 Cells	Production of murine leukemia virus	Levine *et al.* (1979)
	MDBK Cells	Growth of cells on microcarriers	Levine *et al.* (1979)

TABLE II (*Continued*)

Carrier	Cells	Reaction (substrate/product)	Reference
	NIH 3T3, Clone 1 (mouse fibroblast)	Production of murine leukemia virus	Levine *et al.* (1979)
Dowex-1	*Azotobacter agilis*	Oxidation of D-glucose and succinic acid	Hattori and Furusaka (1961)
	Escherichia coli	Oxidation of D-glucose and succinic acid	Hattori and Furusaka (1960)
ECTEOLA-Cellulose	*A. vinelandii*	Nitrogen fixation	Seyhan and Kirwan (1979); DeNicola and Kirwan (1980); Gainer *et al.* (1981)
Polypropylene	*Pseudomonas aeruginosa*	Denitrification and heavy-metals recovery	Holló *et al.* (1979)
Poly(vinyl chloride)	Denitrifying microorganisms	Wastewater treatment	Holló *et al.* (1980)
	Ps. aeruginosa	Denitrification and heavy-metals recovery	Holló *et al.* (1979)
	Saccharomyces carlbergensis	Production of beer	Navarro (1975); Corrieu *et al.* (1976)
	Saccharomyces cerevisiae	(Molasses/ethanol)	Ghose and Bandyopadhyay (1980)
Wood chips	*Bacterium schuetzenbachii*	Oxidation (Ethanol/acetic acid)	Fetzer (1930); Prescott and Dunn (1959)
	S. carlbergensis	Production of beer	Navarro (1975)
	S. cerevisiae	(D-Glucose/ethanol)	Lamptey *et al.* (1980); Moo-Young *et al.* 1980, 1981)
	Saccharomyces lipolytica	(D-Glucose/citric acid)	Briffaud and Engasser (1978, 1979)
Inorganic support			
Glass	*Anabaena cylindrica*	Hydrogen production	Lambert *et al.* (1979)
	BHK 21 Cells	Production of foot-and-mouth disease virus	Spier and Whiteside (1976)
	Denitrifying mixed bacteria	Biological denitrification	Mulcahy and La Motta (1978)
	Fusarium moniliform	Oxidation of *n*-alkanes	Heinrich and Rehm (1981a,b)

(*Continued*)

TABLE II (*Continued*)

Carrier	Cells	Reaction (substrate/product)	Reference
	S. cereviisae	Growth of cells and evidence of metabolic activity	Rouxhet *et al.* (1981)
Fritted glass	*Aspergillus niger*	Spore bioaccumulation and mycelial growth	Messing and Oppermann (1979); Messing *et al.* (1979
	Bacillus subtilis	Bioaccumulation	Messing and Oppermann (1979); Messing *et al.* (1979)
	E. coli	Bioaccumulation	Messing and Oppermann (1979); Messing *et al.* (1979
	Penicillium chrysogenum	Mycelial growth	Messing and Oppermann (1979); Messing *et al.* (1979
	S. cerevisiae	Bioaccumulation	Messing and Oppermann (1979); Messing *et al.* (1979
	Streptomyces olivo-chromogenes	Mycetial growth	Messing and Oppermann (1979); Messing *et al.* (1979
Fritted glass activated with silane	*E. coli*	Bioaccumulation	Messing and Oppermann (1979); Messing *et al.* (1979
Borosilicate glass	*A. niger*	Spore bioaccumulation and mycelial growth	Messing and Oppermann (1979); Messing *et al.* (1979
	B. subtilis	Bioaccumulation	Messing and Oppermann (1979); Messing *et al.* (1979
	E. coli	Bioaccumulation	Messing and Oppermann (1979); Messing *et al.* (1979
	P. chrysogenum	Mycetial growth	Messing and Oppermann (1979); Messing *et al.* (1979
	S. cerevisiae	Bioaccumulation	Messing and Oppermann (1979); Messing *et al.* (1979
	S. olivo-chromogenes	Mycetial growth	Messing and Oppermann (1979); Messing *et al.* (1979
Borosilicate glass activated with silane	*E. coli*	Bioaccumulation	Messing and Oppermann (1979); Messing *et al.* (1979

TABLE II (*Continued*)

Carrier	Cells	Reaction (substrate/product)	Reference
Zirconia ceramics	*A. niger*	Spore bioaccumulation and mycetial growth	Messing and Oppermann (1979); Messing *et al.* (1979)
	P. chrysogenum	Mycetial growth	Messing and Oppermann (1979); Messing *et al.* (1979)
	S. cerevisiae	Bioaccumulation	Messing and Oppermann (1979); Messing *et al.* (1979)
	S. olivochromogenes	Mycetial growth	Messing and Oppermann (1979); Messing *et al.* (1979)
Cordierite	*A. niger*	Spore bioaccumulation and mycetial growth	Messing and Oppermann (1979); Messing *et al.* (1979)
	P. chrysogenum	Mycetial growth	Messing and Oppermann (1979); Messing *et al.* (1979)
	S. cerevisiae	Bioaccumulation	Messing and Oppermann (1979); Messing *et al.* (1979)
	S. olivochromogenes	Mycetial growth	Messing and Oppermann (1979); Messing *et al.* (1979)
Ceramic raschig rings	*S. cerevisiae*	Oxidation (D-Glucose/ethanol)	Sitton and Gaddy (1980)
	S. cerevisiae	Oxidation (Molasses/ethanol)	Ghose and Bandyopadhyay (1980)
Ceramic support	*Candida tropicalis*	Investigation of strength of adsorption	Marcipar *et al.* (1979)
	Rhodotorula sp.	Investigation of strength of adsorption	Marcipar *et al.* (1979)
	Trichosporon sp.	Investigation of strength of adsorption	Marcipar *et al.* (1979)
Porous brick	*S. carlbergensis*	Production of beer	Navarro (1975); Corrieu *et al.* (1976)
	S. cerevisiae	Oxidation (Molasses/ethanol)	Ghose and Bandyopadhyay (1980)
Diatomaceous earth	*S. carlbergensis*	(Glucose/ethanol)	Grindbergs *et al.* (1977)

(*Continued*)

TABLE II (*Continued*)

Carrier	Cells	Reaction (substrate/product)	Reference
Diatomite	*S. cerevisiae*	(Glucose/ethanol)	Lamptey *et al.* (1981); Moo-Young *et al.* (1980)
Clay	*C. tropicalis*	Behavior of adsorbed cells	Marcipar *et al.* (1978)
	Rhodotorula sp.	Behavior of adsorbed cells	Marcipar *et al.* (1978)
	S. cerevisiae	Behavior of adsorbed cells	Marcipar *et al.* (1978)
	Trichosporon sp.	Behavior of adsorbed cells	Marcipar *et al.* (1978)
Sand	Denitrifying mixed bacteria	Biological denitrification	Jeris and Owens (1975); Stephenson (1978)
Asbestos	Methanogenic bacteria	Production of methane	Romanov-Skaya *et al.* (1981)
Silochrome	Methanogenic bacteria	Production of methane	Romanov-Skaya *et al.* (1981)
Stainless steel	*Aspergillus foetoidus*	Production of citric acid	Perez (1976); Atkinson *et al.* (1979)
	B. subtilis	Production of α-amylase	Atkinson *et al.* (1979)
	S. cerevisiae	Growth of cells	Atkinson *et al.* (1979)
	Streptomyces griseus	Production of streptomycin	Atkinson *et al.* (1979)

Investigation by Kennedy and co-workers of a number of gelatinous hydrous metal oxides (frequently called hydroxides, although their full structures are uncertain) has established that hydrous Ti(IV), Zr(IV), Fe(III), V(III), and Sn(II) oxides at least are capable of forming insoluble complexes with enzymes with retention of enzyme activity. From the practical viewpoint, hydrous titanium(IV) and zirconium(IV) oxides proved the most satisfactory. Comparatively high retentions of enzyme-specific activities have also proved to be suitable for the immobilization of amino acids and peptides (Kennedy and Kay, 1976), antibiotics with retention of antimicrobial activity (Kennedy *et al.*, 1976b), and polysaccharides (Kennedy *et al.*, 1977).

Hydrous titanium(IV) and zirconium(IV) oxides are insoluble over the normal physiological pH range, and because when acting as enzyme-immobilization matrices they give good retention of enzyme

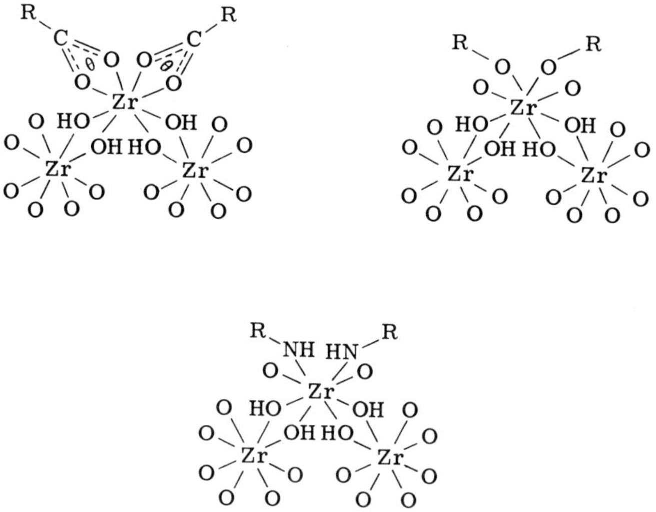

Fig. 5. Immobilization using hydrous zirconium(IV) oxide wherein hydroxyl groups on metal surface are replaced by ligands from enzyme or cell surface (R–C).

activity, they seem to have little or no effect on the function of biologically active molecules. If enzymic activity, which is extremely sensitive to conformational changes in the enzyme molecule, is not seriously affected by immobilization, then there seems to be little reason cell walls should be disrupted or destroyed by this process; thus the cells themselves have a good chance of remaining viable.

The immobilization process for the hydrous metal oxides is envisaged as involving the replacement of hydroxyl groups on the surface of the metal hydrous oxide by suitable ligands from enzyme or cell, resulting in the formation of partial covalent bonds. In the case of enzymes, such ligands could be the side-chain hydroxyl groups of L-serine or L-threonine, the carboxyl groups of L-glutamic acid or L-aspartic acid, and the ε-amino groups of L-lysine residues of the protein chain [illustrated in Fig. 5 using hydrous zirconium(IV) oxide as an example], ligands containing oxygen being preferred to those containing nitrogen. In the case of cells, the structural complexity of the cell wall ensures the availability of a great diversity of suitable ligands from protein as well as from carbohydrate moieties (illustrated in Fig. 6 using hydrous titanium(IV) oxide as an example.

Because the gel formation is pH dependent, the choice of titanium or zirconium reagents is governed by the pH range required for the application of interest, titanium(IV) hydrous oxide being more effective

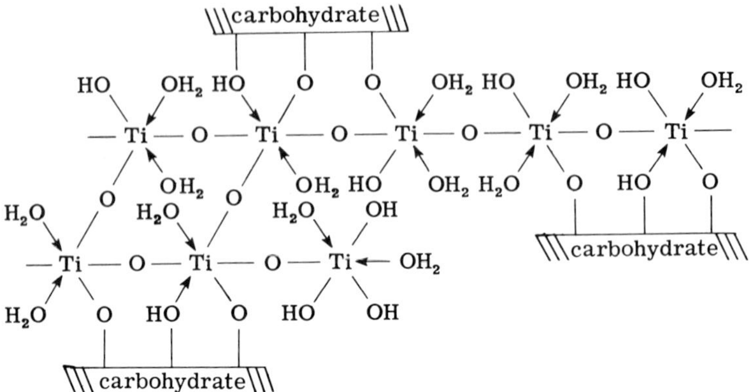

Fig. 6. Coupling of hydrous titanium(IV) oxide to carbohydrate moieties of cells or enzymes.

in acidic conditions and zirconium(IV) hydrous oxide more effective at neutral or higher pH.

Using this method of immobilization, Kennedy and his group (1976b; Kennedy and Pike, 1979, 1980) immobilized several microorganisms, without disruption of their life processes, by conducting the precipitation of the metal hydrous oxides in the suspension of microbial cells. The microorganisms were chelated in the gel-like precipitate formed. *Saccharomyces cerevisiae* and *Escherichia coli* have been immobilized and the retention of the living processes of the cells in the immobilized state was demonstrated by oxygen-uptake experiments. Evidence that the cells became firmly bound to the support was demonstrated by following the course of immobilized colored cells (*Serratia marcescens*), no color being released during continuation of the living processes of the cells after immobilization.

Cells of species of *Acetobacter* have been immobilized on hydrous titanium(IV) oxide and these immobilized living cells were used for the continuous production of malt vinegar from wort in tower fermenters. It was demonstrated that the efficiency of the fermenter was increased well above its normal maximum throughput using only free cells. With other species of *Acetobacter*, prechelation of the titanium species with cellulose further improved the efficiency of the fermenter.

Several advantages of this method of immobilization are recognized: low cost, convenience of preparation (which may be conducted in any location without specialized facilities), the absence of any need for preparation, ability to couple enzymes or cells at neutral pH, high retention of enzymic specific activity.

TABLE III
PRINCIPAL EXAMPLES OF IMMOBILIZED LIVING CELLS PREPARED BY CHELATION

Hydrous metal oxide	Cells	Reaction (substrate/product)	Reference
Hydrous titanium(IV) oxide	*Acetobacter* sp.	Production of vinegar	Kennedy (1979); Kennedy *et al.* (1980)
	Escherichia coli	Evidence of survival and attachment to the matrix	Kennedy *et al.* (1976b); Kennedy (1979)
	Saccharomyces cerevisiae	Evidence of survival and attachment to the matrix	Kennedy *et al.* (1976b); Kennedy (1979)
	Serratia marcescens	Evidence of survival and attachment to the matrix	Kennedy *et al.* (1976b); Kennedy (1979)
Hydrous zirconium(IV) oxide	*E. coli*	Evidence of survival and attachment to the matrix	Kennedy *et al.* (1976b); Kennedy (1979)
	S. marcescens	Evidence of survival and attachment to the matrix	Kennedy *et al.* (1976b); Kennedy (1979)

Some examples of immobilized living cells obtained by chelation are shown in Table III.

3. Covalent-Binding Method of Immobilization

This method of immobilization is one of the most frequently used techniques to insolubilize enzymes, because enzyme preparations with high operational stabilities are obtained. However, for cells this process of immobilization is not widely used because of the toxicity of the coupling agents used, which sometimes results (Chipley, 1974; Shimizu *et al.*, 1975) in the loss of both the enzyme activity and the viability of the cell.

Covalent coupling is the direct linkage of cells to an activated support. Linkage can be to any reactive component of the cell surface; for instance, the amino, carboxyl, thiol, hydroxyl, imidazole, or phenol groups of proteins can be used. To introduce the covalent linkage, the chemical modification of the carrier is usually necessary.

The main advantages of this technique are to produce a system free of the diffusion limitations present in an entrapment process, and cells which are bound to a uniform surface by a bond that is stable for long periods, so that cell leakage is minimized. In the few reports about this method of immobilization of cells, inorganic and organic carriers have

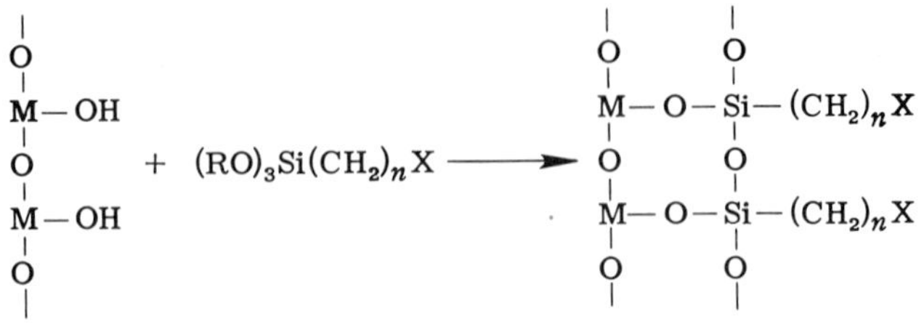

$$
\begin{array}{l} | \\ \mathrm{O} \\ | \\ \mathrm{M-OH} \\ | \\ \mathrm{O} \\ | \\ \mathrm{M-OH} \\ | \\ \mathrm{O} \\ | \end{array}
\quad + \quad \mathrm{(RO)_3Si(CH_2)_{\mathit{n}}X} \longrightarrow
\begin{array}{l} |\qquad\quad | \\ \mathrm{O}\qquad\ \mathrm{O} \\ |\qquad\quad | \\ \mathrm{M-O-Si-(CH_2)_{\mathit{n}}X} \\ |\qquad\quad | \\ \mathrm{O}\qquad\ \mathrm{O} \\ |\qquad\quad | \\ \mathrm{M-O-Si-(CH_2)_{\mathit{n}}X} \\ |\qquad\quad | \\ \mathrm{O}\qquad\ \mathrm{O} \\ |\qquad\quad | \end{array}
$$

Fig. 7. Coupling of alkoxysilane derivatives to oxidized metal support.

been used. In the case of inorganic carriers, mainly silica and ceramics, the support is activated by silanization, as in the case of immobilized enzymes. This method involves the use of trialkoxysilane derivatives containing an organic functional group (Weetall, 1976). These reagents have proved to be the most successful compounds for activation of inorganic carriers for cell-immobilization purposes. Coupling of these reagents to the carrier takes place presumably by displacement of the alkoxy residues on the silane by hydroxy groups on the oxidized surface of the inorganic support to form a Metal—O—Si linkage (Weetall, 1976) (Fig. 7).

A variety of organic functional groups is available in silane coupling agents (Table IV). The most popular and versatile silane compound is the 3-aminopropyltriethoxysilane that produces an alkylamine derivative, and the preparation of its reactive intermediates, for enzyme immobilization, has been discussed extensively in the literature. The pioneering work of silanization was done by Weetall and his collaborators at Corning Glass Works, who introduced aminoalkyl controlled-pore glass as an enzyme support and described methods of preparing reactive intermediates for enzyme immobilization (Weetall, 1976).

Activation of glass can be achieved by either organic or aqueous silanization. With organic silanization, the support is activated by refluxing a 10% (w/v) solution of 3-aminopropyltriethoxysilane in toluene (Weetall, 1976) or xylene (Monsan, 1978), or by an evaporative technique using acetone as solvent (Weetall, 1976). Aqueous silanization at pH 3–4 gives lower amine loading than organic solvent techniques, but experience has shown that greater carrier durability with slightly lower enzyme loading is achieved by the latter technique. Usually after the activation stage there is a heat-treatment (115°C) stage. Once the "covalent" link between the inorganic support and the amino group is formed, the amino glass is activated, usually with glutaraldehyde, and a carbonyl derivative is obtained, which reacts with the amino groups of the cell or enzyme by a Schiff's base formation.

TABLE IV
ORGANIC FUNCTIONAL GROUPS OF SILANE COUPLING AGENTS FOR BIOCATALYST IMMOBILIZATION

Amine (aliphatic)	$—NH_2$
(aromatic)	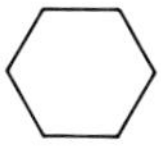 $C_6H_4NH_2$ (benzene ring)—NH_2
Halide	—I; —Br; —Cl
Aldehyde	—CHO
Acetal	$—CH(OC_2H_5)_2$

In spite of the apparent great versatility of this method for the immobilization of enzymes, there are only two reports on its use in cell immobilization (Navarro and Durand, 1977; Messing *et al.*, 1979). Navarro and Durand immobilized successfully *Saccharomyces carlbergensis* to porous silica beads treated with s-aminopropyltriethoxysilane and subsequently activated with glutaraldehyde. The authors found that when glutaraldehyde activation was absent, the amount of cells retained on the carrier decreased with the increase of surface area, and that if the silica was activated, the amount of yeast cells attached increased dramatically with an increase of surface area (Fig. 8). These data suggested that the more silica surface available the more reactive carbonyl groups from glutaraldehyde can be attached to it, and therefore more yeast cells can be linked to the support.

In another report, Messing and co-workers (Messing *et al.*, 1979) coupled cells of *Serratia marcescens, Saccharomyces cerevisiae,* and *Saccharomyces amurcae* by isocyanate coupling agent to different types of carriers (fritted glass, borosilicate glass, zirconia ceramics, and cordierite). In this case the derived carrier is prepared by shaking 0.5 g of the carrier in 10 ml of 0.5% (w/v) polyisocyanate in acetone for 45 min at room temperature. The cells were reacted with the derivatized carrier for 3 h. In all cases, the retention of capacity of carriers declined after treatment, compared with simple adsorption processes.

In the case of organic carriers, carbodiimide reagents have been used. However, in the immobilization of *Escherichia coli, Staphylococcus aureus, Pseudomonas aeruginosa,* and *Bacillus subtilis,* using carbodiimide reagents in the formation of covalent linkages between the cells and agarose hexandioic hydrazide beads, both activity and viability of the cells were lost due to the toxicity of the carbodiimide reagents (Chipley, 1974). More recently, Jack and Zajic

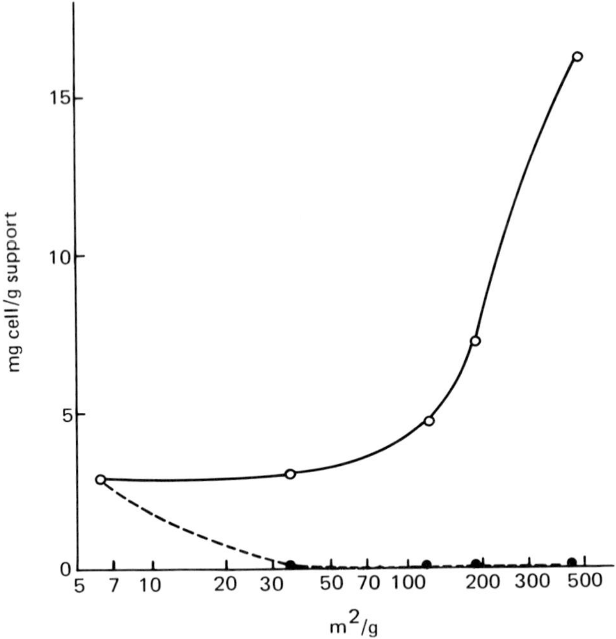

Fig. 8. Effect of silica-specific area on immobilized *Saccharomyces carlbergensis* cell number. ○, Glutaraldehyde-activated silica, ●, nonactivated silica. (From Navarro and Durand, 1977.)

(1977b), by reversion of the procedure and activating the carboxyl groups on the agarose beads with the carbodiimide, immobilized *Micrococcus luteus* cells in a two-step process that avoided exposure of the cells to carbodiimide; thus the urocanic acid-forming activity was still retained after cell viability had been lost.

Shimizu *et al.* (1975) reported an attempt to bind *Brevibacterium ammoniagenes* covalently to a copolymer of ethylene and maleic anhydride, but it results in the loss of coenzyme A-synthetizing activity of the cells.

Another method of covalent immobilization of cells is the linkage of *Zygosaccharomyces lactis* and *Saccharomyces cerevisiae* to hydroxyalkyl methacrylate gel, modified with epichlorhydrin and with spacers of different length, which are activated with glutaraldehyde or carbodiimide. Jirkú *et al.* (1979, 1980) reported the vegetative reproduction of immobilized cells of *S. cerevisiae* on this kind of support, without separation of the daughter cells, and the formation of chainlike filamentous, which indicate a certain polarization of budding. These authors also observed the conversion of the ellipsoid form found with free *S. cerevisiae* cells into rod-shaped forms, which is probably a result of the newly produced cells being immobilized during their longitudinal growth.

In another process, *Aspergillus niger* cells were treated with glutaraldehyde and then with glycidyl methacrylate monomer followed by polymerization (Nelson, 1976).

Principal examples of cells immobilized by covalent attachment are shown in Table V.

C. Entrapment Method of Immobilization

The entrapping method is based on the occlusion of cells within a constraining structure (lattice of a polymer matrix or membrane) tightly enough to prevent the release of cells, while allowing penetration of substrates and diffusion of products. Consequently, only cellular enzyme reactions involving relatively small-size reactants and the synthesis of small products may be carried out successfully by using entrapped-whole-cell preparations.

This method differs from the chemical coupling methods in that the biocatalyst itself does not bind to the gel matrix or membrane. Thus it can be most generally applied to entrap any kind of biocatalyst (i.e., enzymes, whole cells, and organelles of different sizes and properties), with little destruction of their biological activity as compared with the chemical coupling methods. So far, entrapping of whole cells that are dead, resting, or growing, especially into polymers, has been the most extensively used method for cell immobilization. In this chapter the entrapment method is classified into gel entrapment, fiber entrapment, and microencapsulation (Fig. 9).

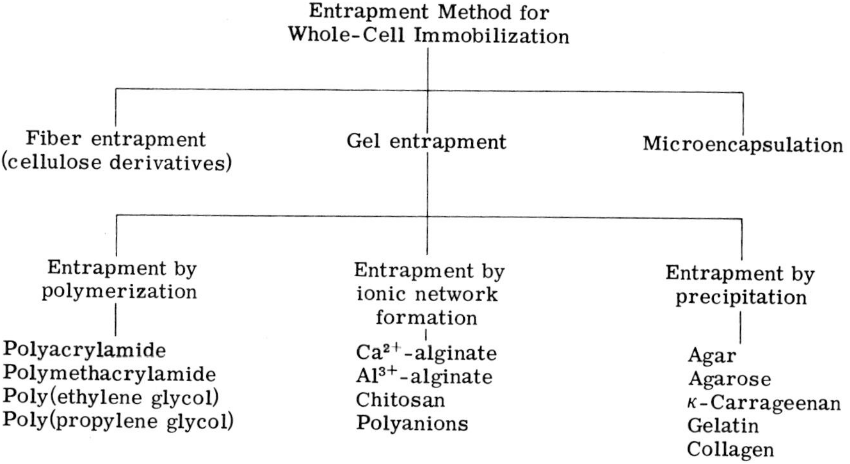

Fig. 9. Entrapment method for whole-cell immobilization.

TABLE V
PRINCIPAL EXAMPLES OF WHOLE CELLS IMMOBILIZED BY COVALENT ATTACHMENT

Carrier	Coupling section	Cells	Reaction (substrate/product)	References
Hydroxyalkyl methacrylate activated with epichlorhydrin + tetramethylenediamine + glutaraldehyde	Schiff's base formation	*Saccharomyces cerevisiae*	Production of biologically active killer toxin and evidence for covalent binding of cells	Jirkú *et al.* (1980)
		Saccharomyces paradoxus	Investigation of optimal method of immobilization	Gulaya *et al.* (1979)
		Zygosaccharomyces lactis	β-D-Galactosidase activity	Jirkú *et al.* (1979)
Porous silica activated with an amine silane reagent + glutaraldehyde	Schiff's base formation	*Saccharomyces carlbergensis*	(D-Glucose/ethanol)	Navarro and Durand (1977)
Cellulose activated with cyanuric chloride	Alkylation	*Azotobacter vinelandii*	Nitrogen fixation	Gainer *et al.* (1981)
		S. cerevisiae	Ethanol production	Gainer *et al.* (1981)
Carboxymethylcellulose activated with carbodiimide	Peptide binding	*Micrococcus luteus*	(L-Histidine/uraconic acid)	Jack and Zajic (1977b)

Agarose hydrazide	Peptide binding	*Bacillus subtilis*	Loss of activity and viability	Chipley (1974)
		Escherichia coli	Loss of activity and viability	Chipley (1974)
		Pseudomonas aeruginosa	Loss of activity and viability	Chipley (1974)
		Staphylococcus aureus	Loss of activity and viability	Chipley (1974)
Ethylene maleic anhydride copolymer	Peptide binding	*Brevibacterium ammoniagenes*	Synthesis of coenzyme A	Shimizu *et al.* (1975)
Fritted glass activated with polyisocyanate	Peptide binding	*Saccharomyces amurcae*	Bioaccumulation	Messing and Oppermann (1979); Messing *et al.* (1979)
		Serratia marcescens	Bioaccumulation	Messing and Oppermann (1979); Messing *et al.* (1979)
Borosilicate glass activated with polyisocyanate	Peptide binding	*S. amurcae*	Bioaccumulation	Messing and Oppermann (1979); Messing *et al.* (1979)
		S. marcescens	Bioaccumulation	Messing and Oppermann (1979); Messing *et al.* (1979)
Zirconia ceramics activated with polyisocyanate	Peptide binding	*S. amurcae*	Bioaccumulation	Messing and Oppermann (1979); Messing *et al.* (1979)
Silochrome activated with toluene isocyanate	Peptide binding	Methane-oxidizing bacteria	Oxidation of methane	Romanov-Skaya *et al.* (1981)

1. Gel-Entrapment Method of Immobilization

The gel-entrapment method involves entrapping of the cell within the interstitial spaces of water-insoluble polymer gels. Cells can be entrapped either by polymerization, by ionic-network formation, or by precipitation by pH, temperature, or solvent changes. In the following sections each one of these methods is described.

a. Entrapment by polymerization: Some of the first examples of cell immobilization were those of entrapment in polyacrylamide gel, which has been the most popular and used method, since the first report of Mosbach and Mosbach (1966). These authors immobilized the powdered thalli of the lichen *Umbilicara pustulata,* whose orsellin decarboxylase (EC 4.1.1.58) activity was retained during periodic tests over 3 months at 20°C.

The retention of viability and enzyme activity of gel-entrapped whole cells were subjects of the very early articles. Updike *et al.* (1969) entrapped *Escherichia coli* and *Tetrahymena pyriformis* cells, which were shown by several methods to retain their viability. The uptake of D-glucose and oxygen across a column packed with the entrapped cells was observed. The protozoa could be observed, under microscope examination, to be swimming inside the gel for several days. Franks (1971, 1972) studied the catabolism of L-arginine using gel-entrapped *Streptococcus faecalis.* The L-arginine-metabolizing ability of the immobilized cells was stable over 11 days, as was the integrity of the cells within the gel as shown by electron microscopy.

Conversion of D-glucose into glutamic acid (L-glutamine) by a multienzyme pathway of polyacrylamide-entrapped *Corynebacterium glutamicum* cells was described by Slowinsky and Charm (1973).

In 1974 the first application of immobilized cells was reported by the Tanabe Seiyaku Co., Japan (Chibata *et al.*, 1974), for the production of L-aspartic acid with immobilized cells of *Escherichia coli* having L-aspartase ammonia lyase (EC 4.3.1.1) activity. Subsequently the same company has developed a similar process to make L-malic acid (Yamamoto *et al.*, 1977). However, in these two industrial cases the immobilized cells are in dead state, though the specific enzyme activity was retained.

The procedure for the preparation of insoluble-gel networks is identical to that employed for the preparation of gel commonly used for disk electrophoresis. This method is based on the free-radical polymerization of acrylamide in an aqueous solution. Because of the solubility in water of the linear polymers, and in order to use them as matrices, they have to be insolubilized by cross-linking with bifunctional compounds, usually *N,N′*-methylenebisacrylamide (BIS).

In the gel-entrapment method of whole cells, the free-radical polymerization of acrylamide is conducted in an aqueous solution containing the whole cell and a cross-linking agent. Polymerization is commonly carried out in the absence of oxygen and at lower temperatures (10°C) in order not to damage the cell during this operation. The polymerizing reaction is initiated by potassium persulfate ($K_2S_2O_8$) or riboflavin and catalyzed by dimethylaminopropionitrile (DMAPN) or *N,N,N′,N′*-tetramethylethylenediamine (TEMED). The resulting gel block can be mechanically dispersed into particles of defined size. However, these gels are quite weak in a mechanical sense and have an open network with a broad distribution of pore sizes. These disadvantages are overcome by optimizing the degree of cross-linking. Chibata *et al.* (1974) had optimized some time ago the most suitable conditions for achievement of highly active, stable, and mechanically strong immobilized-cell preparations, obtaining a polyacrylamide gel with a pore size of 10–40 Å, which permits the transport of small substrate and product molecules, but prevents leaking out of the whole cells, large enzyme molecules, and nucleic acids. Figure 10 describes the structure of whole cells entrapped in polyacrylamide.

However, this entrapping method has a major disadvantage: the toxicity of the acrylamide monomer, the cross-linking agent (BIS), and the

```
—CH2—CH—CH2—CH—CH2—CH—CH2—CH—CH2—CH—CH2—
      |       |       |       |       |
      CO      CO      CO      CO      CO
      |       |       |       |       |
      NH2     NH      NH2     NH2     NH
              |                       |
              CH2                     CH2
              |      (CELLS)          |
              NH                      NH
              |                       |
              CO                      CO
              |                       |
—CH2—CH—CH2—CH—CH2—CH—CH2—CH—CH2—CH—CH2—
      |               |       |
      CO              CO      CO
      |               |       |
      NH              NH2     NH
      |                       |
      CH2                     CH2
      |      (CELLS)          |
      NH                      NH
      |                       |
      CO                      CO
      |                       |
—CH2—CH—CH2—CH—CH2—CH—CH2—CH—CH2—
              |       |
              CO      CO
              |       |
              NH      NH2
              |
              CH2
              |
```

Fig. 10. Structure of polyacrylamide matrix in which cells are entrapped.

initiator (TEMED). Thus the cell exposure to free-radial polymerization results in a decrease not only in the enzymic activity, but also in the viability of the cell. Practically speaking, therefore, microorganisms should be exposed to the reagents for the shortest time possible. As time influences the hardness of the resulting gel, the convenient mechanical strength can be achieved by using a rather high concentration of catalyst with efficient cooling (Larsson *et al.*, 1978).

In a similar process, the use of another monomer, methacrylamide, has been reported, whereby the mechanical strength could be improved with nontoxic suspension-polymerization techniques using butylphthalate as suspension agent (Klein *et al.*, 1978).

Fukui and co-workers (1978; Omata *et al.*, 1979a; Sonomoto *et al.*, 1979) had developed a method to entrap enzymes and whole cells with photo-cross-linkable resin prepolymers and urethane resin prepolymers of either hydrophilic or hydrophobic character. The structures of photo-cross-linkable prepolymers used by these authors are shown in Fig. 11. These oligomers are derivatives of poly(ethylene glycol) (ENT), hydrophilic; poly(propylene glycol) (ENTP), hydrophobic; and polybutadiene (ENTB), hydrophobic, of different chain lengths.

A typical preparation procedure for these prepolymers consists of reacting equimolar amounts of hydroxyethylacrylate and isophorone diisocyanate at 70°C in the presence of a suitable catalyst, such as organic tin compounds or ternary amines. After 2 h a half-molar ratio of the poly(ethylene glycol) is added—the reaction proceeds during 5 h at 70°C. The resulting product is the corresponding photo-cross-linkable resin prepolymer (ENT).

In this process one part of the water-soluble prepolymer is mixed with 0.01 part of an initiator (photosensitizer), usually benzoin ethylether, and the mixture is melted by warming at 60°C. To this molten mixture is added a whole-cell suspension and the resultant mixture is illuminated at 360 nm for 3 min, with a 2-kW high-pressure mercury lamp to initiate a free-radical polymerization reaction that entraps the cell in a nonionic hydrophobic resin. The resin gel thus formed is cut into small pieces and used. In the case of water-insoluble prepolymers, these and the initiator are dissolved in a solvent (benzene : heptane = 1 : 1 by volume), and the cells are added as suspension in the same solvent. The mixture is photo-cross-linked in the same way as the water-soluble prepolymers.

In another polymerization method, Kumakura *et al.* (1978) studied the immobilization of *Streptomyces phaeochromogenes* by radiation-induced polymerization of a glass-forming monomer, such as 2-hydroxyethyl methacrylate, at low temperatures. This polymeriza-

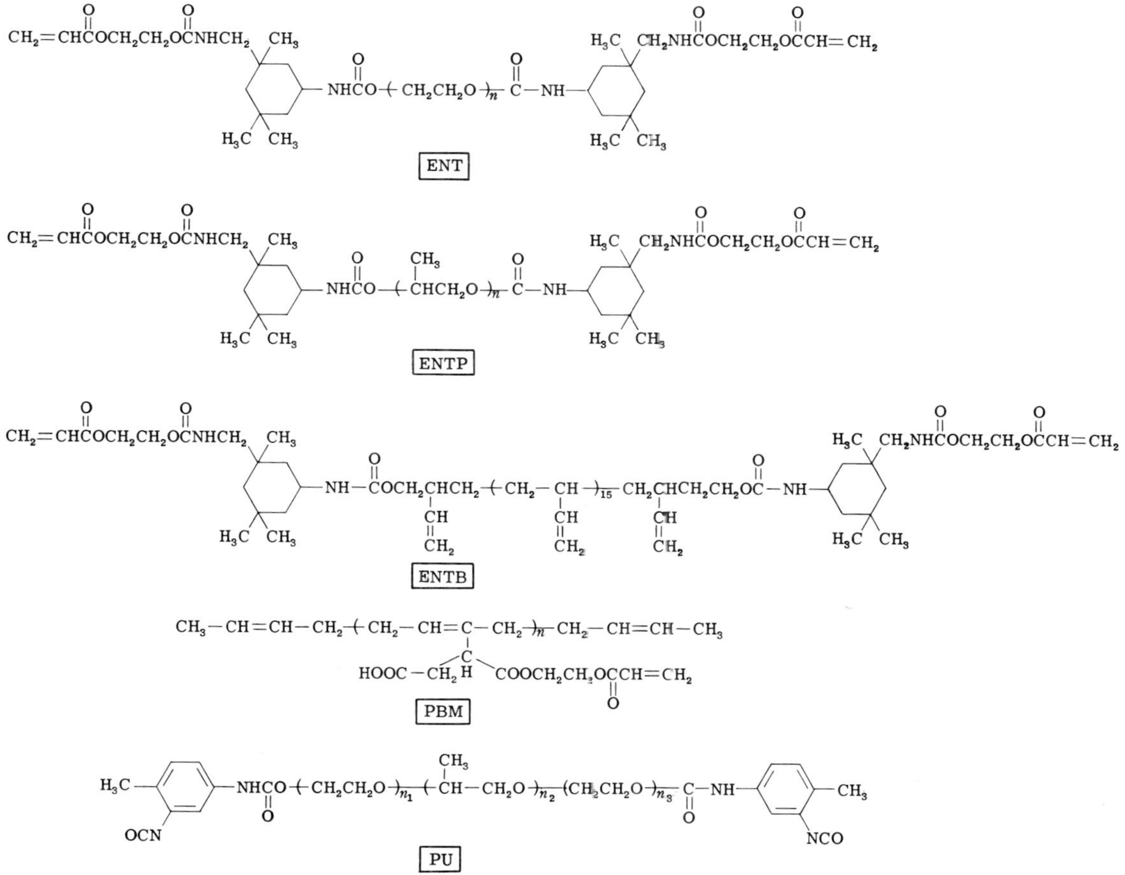

ENT

ENTP

ENTB

PBM

PU

Fig. 11. Structure of photo-cross-linkable resin prepolymers used in cell entrapment. ENT, A water-soluble derivative of poly(ethylene) glycol; ENTP, a water-insoluble derivative of poly(propylene glycol); ENTB, a water-insoluble derivative of polybutadiene; PBM, maleic polybutadiene (water insoluble); PU, polyurethane (water miscible).

tion technique was applied advantageously to the immobilization of enzymes. With this procedure, the authors found that a very high concentration of cells could be immobilized without leakage of the cells from the polymer matrix. However, this process was applied to a single-enzyme reaction, isomerization of D-glucose, where the cells were dead.

Some examples of gel-entrapped living cells are shown in Table VI.

b. Entrapment by ionic-network formation: This technique of entrapment is an example of coarctation, the polymerization of polyelectrolytes by multivalent ions. One of the most popular methods is the immobilization of whole cells by entrapment in calcium alginate. In this very versatile procedure, only nontoxic compounds are used and so it has been preferentially chosen for the immobilization of living cells (Kierstan and Bucke, 1977; Cheetam, 1979) and very sensitive cells such as plant cells (Brodelius *et al.*, 1979, 1980) and protoplasts (Scheurich *et al.*, 1980); such immobilizations have met with success. By this method, reincubation of cells is also possible. The nondenaturing properties of the alginate and the resistance of the pellets to carbon dioxide generation inside the entrapped cells make this method useful for multienzyme conversions, such as ethanol production (Kierstan and Bucke, 1977).

The ionic-network formation procedure was originally developed by Thiele and co-workers (Thiele, 1954; Thiele and Awad, 1969), and the first example of whole-cell immobilization by this process was given by Klein and co-workers (Hackel *et al.*, 1975).

Alginate, which is extracted from seaweed and is a linear block copolymer of D-manuronic and L-guluronic acids, can be gelled by multivalent ions, usually calcium and aluminum.

The process of immobilization includes the preparation of a solution of sodium alginate, the addition of the cell mass, and the dispersion of this mixture into a counterion solution, which results in the formation of a uniform, spherical, highly microporous structure that retains cells, organelles, and enzyme molecules larger than the pores. The pores are large and continuous such that substrate molecules may diffuse throughout the beads.

The mechanical properties of alginate gels are related to the distribution of the D-mannuronic and L-guluronic residues, and to the molecular weight and degree of dispersion within the pellets. The pellets show excellent resistance to hydrostatic pressures, as was reported by Cheetam (1979) when measuring low intrinsic pressure drops of the pellets in packed-bed columns.

However, this mild, simple, rapid, and excellent method has some

TABLE VI
PRINCIPAL EXAMPLES OF GEL-ENTRAPPED LIVING CELLS PREPARED BY POLYMERIZATION

Material	Cells	Reaction (substrate/product)	Reference
Polyacrylamide	*Achromobacter aceris*	NAD-Kinase activity	Uchida *et al.* (1978)
	Achromobacter butyri	Production of D-glucose 6-phosphate	Murata *et al.* (1979b)
	Actinomycetes sp.	Biodegradation of steroids	Germain *et al.* (1981)
	Arthrobacter oxydans	Synthesis of FAD (1,3-Propanediol/lactic acid)	Yamada *et al.* (1980); Yagi *et al.* (1976)
	Arthrobacter simplex	(Cortisol/prednisolone)	Ohlson *et al.* (1978); Larsson *et al.* (1978, 1979)
	Bacillus sp.	Production of bacitracin	Suzuki and Karube (1979); Morikawa *et al.* (1980)
	Bacillus subtilis	Production of α-amylase	Kokubu *et al.* (1978); Suzuki and Karube (1979)
	Brevibacterium ammoniagenes	Production of NADP	Murata *et al.* (1979a)
		Synthesis of CoA	Shimizu *et al.* (1975, 1979); Samejima *et al.* (1978); Yamada *et al.* (1980)
	Candida lipolytica	Production of citric acid	Stottmeister (1980)
	Candida tropicalis	Degradation of phenol	Hackel *et al.* (1975)
	Clostridium butylicum	Production of hydrogen	Karube *et al.* (1976, 1977a,b,c); Suzuki *et al.* (1978); Matsunaga *et al.* (1980)
	Corynebacterium glutamicum	(D-Glucose/L-glutamine)	Slowinski and Charm (1973)
	Escherichia coli	Production of L-tryptophane	Bang *et al.* (1978); Decottignies-Le Maréchal *et al.* (1979); Azerad *et al.* (1980)
		Evidence of survival	Updike *et al.* (1969)

(*Continued*)

TABLE VI (*Continued*)

Material	Cells	Reaction (substrate/product)	Reference
	Gluconobacter melanogenes + *Pseudomonas* sp.	(L-Sorbose/2-ketogulonic acid)	Martin and Perlman (1976)
	Gluconobacter oxydans	(Glycerol/dihydroxyacetone)	Somaer and Caglar (1981)
	Hansenula polymorpha	(Methanol/formaldehyde)	Couderc and Baratti (1980)
	Hyphomicrobium sp.	Nitrate degradation	Divies (1977)
	Lactobacillus bulgaricus + *Streptococcus thermophilus*	(Milk/yogurt)	Divies (1977)
	Lactobacillus casei	Malic acid degradation	Divies and Siess (1976); Divies (1977)
	Lactobacillus delbrueckii	(D-Glucose/lactic acid)	Divies (1977)
	Methanogenic population	(Waste water/methane)	Karube *et al.* (1980)
	Penicillium chrysogenum	Production of penicillin	Morikawa *et al.* (1979); Suzuki and Karube (1979)
	Penicillium cyaneofulvum	Production of erythorbic acid	Kato (1974)
	Pseudomonas aeruginosa	Nitrate degradation	Divies (1977)
	Pseudomonas denitrificans	Biological denitrification	Nilsson *et al.* (1981)
	Pseudomonas fluorescens	Synthesis of pyridoxal 5′-phosphate	Yamada *et al.* (1980)
	Pseudomonas putida	Benzene metabolism	Sommerville *et al.* (1977); Mason *et al.* (1978)
	Pseudomonas testosteroni	Δ'-Dehydrogenation of Reichstein compound S	Yang and Studebaker (1978)
	Saccharomyces cerevisiae	(AMP/ATP)	Samejima *et al.* (1978)
		(CMP/CDP)	Samejima *et al.* (1978)
		(D-Glucose/ethanol)	Marituna and Sinclair (1978)

		(Grape juice/wine)	Divies (1977)
		(Molasses/ethanol)	Divies (1977)
		Production of glutathione	Murata *et al.* (1978, 1981)
	Streptococcus faecalis	L-Arginine catabolism	Franks (1971, 1972)
	Tetrahymena pyriformis	Evidence of survival	Updike *et al.* (1969)
	Thiobacillus ferrooxydans	Ferrons iron oxidation	Kutsal and Caglar (1981)
Copolymer of acrylamide and acrylate	*B. ammoniagenes*	NAD-Kinase activity	Hayashi *et al.* (1979)
Polymethacrylamide	*C. tropicalis*	Oxidative degradation of phenol	Klein *et al.* (1978)
	E. coli	(Penicillin G/6-amino penicillanic acid)	Klein and Wagner (1978, 1980)
Polystryrene	*C. tropicalis*	Degradation of phenol	Hackel *et al.* (1975)
Poly(vinyl alcohol)	*B. ammoniagenes*	Production of CoA	Shimizu *et al.* (1979)
2-Hydroxyethyl methacrylate	*Streptomyces phaeochromogenes*	(D-Glucose/D-fructose)	Kumakura *et al.* (1978, 1979)
Polyurethane	*Kloekera* yeast	—	Tanaka *et al.* (1979)
	Nocardia rhodocrous	Biconversion of steroids	Fukui *et al.* (1980); Omata *et al.* (1979a)
Maleic polybutadiene (PBM)	*A. simplex*	(Hydrocortisone/ prednisolone)	Omata *et al.* (1979b)
	N. rhodocrous	Bioconversion of steroids	Fukui *et al.* (1980); Omata *et al.* (1979a)
Poly(ethylene glycol)	*A. simplex*	(Hydrocortisone/ prednisolone)	Fukui *et al.* (1978); Sonomoto *et al.* (1979)
	N. rhodocrous	Bioconversion of steroids	Omata *et al.* (1979a); Yamane *et al.* (1979); Fukui *et al.* (1980)
Poly(propylene glycol)	*A. simplex*	(Hydrocortisone/ prednisolone)	Fukui *et al.* (1978); Sonomoto *et al.* (1979)
	N. rhodocrous	Bioconversion of steroids	Fukui *et al.* (1980); Omata *et al.* (1979a,b); Yamane *et al.* (1979)

disadvantages. One is the use of calcium alginate beads in a medium containing calcium-chelating agents, such as phosphates, and certain cations such as Mg^{2+} or K^{2+}, which results in the disruption of the gel by solubilizing the bound Ca^{2+}. Very recently, Birnbaum *et al.* (1981) circumvented the instability of alginate in phosphate-containing media by treating the gel with polyamines and cross-linking. This treatment provides a stabilized peripheral layer that prevents the release of cells. Cell leakage from the matrices has also occurred when cell division within the pellets takes place and when the pellets are used in stirred vessels.

A modified form of alginate immobilization has been described by Klein *et al.* (1978). In this process, called entrapment by polycondensation reactions, a copolymer of calcium alginate and epoxide containing the entrapped cells was formed and dried; then the calcium alginate was dissolved away by a phosphate solution, leaving an open-pored and rigid immobilized-cell pellet, irreversibly formed.

Besides calcium alginate, other polymers may be used in this method of entrapping whole cells. Some anionic polymers and possible counterions are shown in Table VII.

Recently Vorlop and Klein (1981) described a method of gel entrapment using the ionotropic gelation of a polycation, chitosan, with different multivalent anionic counterions (Table VIII). They reported that the immobilization under sterile conditions caused no problem, and that a later reincubation of living *Escherichia coli* cells with L-tryptophane synthetase activity was also possible.

Principal examples of immobilized living cells in ionic-network polymers are shown in Table IX.

TABLE VII
ANIONIC POLYMERS AND COUNTERIONS USED IN THE GEL-ENTRAPMENT METHOD BY IONIC-NETWORK FORMATION

Polyanions	Counterions
Sodium alginate	Ca^{2+}
Carboxymethylcellulose	Fe^{2+}
Copoly(styrene-maleic acid)	Zn^{2+}
Copoly(acrylamide-acrylate)	Co^{2+}
Copoly(*trans*-stilbene-maleic acid)	Al^{3+}
Copoly(isoprene-maleic acid)	Fe^{3+}
Copoly(vinyl acetate-maleic acid)	
Copoly(isobutene-maleic acid)	
Copoly(vinylmethylether-maleic acid)	
Copoly(furan-maleic acid)	

c. Entrapment by precipitation: Gels may be formed by precipitation of some natural and synthetic polymers by changing one or more parameters in the solution, such as temperature, salinity, pH, or solvent.

Collagen has been widely used as an enzyme-, cell-, and organelle-immobilization matrix (Vieth and Venkatasubramanian, 1979). Collagen is the most abundant protein constituent of higher vertebrates. It can be readily isolated from a number of biological sources and reconstituted into various forms without losing its native structure. This, in conjunction with its ready availability from a large number of biological species—from fish to cattle—makes it an inexpensive matrix.

The major work with this carrier has been done by Vieth and Venkatasubramanian (1979), who used collagen for immobilization of several enzymes and cells by three different procedures: complexation, electrodeposition, and impregnation.

Although these authors consider these to be individual processes, they can be integrated within a larger classification. Thus, the so-called complexation method can be classified as a gel-entrapment process, as the cells are suspended in an aqueous collagen dispersion and comixed, before casting into a membrane. The membrane-impregnation method and the electrodeposition method can be classified as adsorption, as the major forces responsible for binding whole cells to the collagen matrix are multiple ionic linkages, hydrogen bonds, and van der Waals interactions.

To increase the mechanical strength of the collagen–whole-cell membrane and to maximize the amount of whole cells retained, mainly in conjunction with the complexation method, a tanning step is performed by addition of suitable bifunctional reagents, usually glutaraldehyde, to the dried collagen–whole-cell complex membrane.

Gel entrapment of whole cells is also possible in gelatin. This mate-

TABLE VIII
CATIONIC POLYMERS AND COUNTERIONS USED IN THE GEL-ENTRAPMENT METHOD BY IONIC-NETWORK FORMATION

Polycation	Counterions
Chitosan	$Fe(CN)_6^{4-}$ $Fe(CN)_6^{3-}$ Polyphosphates Poly(aldehydrocarbonic acid) Poly(1-hydroxy-1-sulfonate propen-2)

TABLE IX
PRINCIPAL EXAMPLES OF IMMOBILIZED LIVING CELLS IN IONIC-NETWORK POLYMERS

Material	Cells	Reaction (substrate/product)	Reference
Aluminum alginate	*Candida tropicalis*	Degradation of phenol	Hackel *et al.* (1975); Klein *et al.* (1979)
Calcium alginate	*Actinomycetes* sp.	Biodegradation of steroids	Germain *et al.* (1981)
	Arthrobacter simplex	(Cortisol/prednisolone)	Larsson *et al.* (1979); Ohlson *et al.* (1979)
	Catharanthus roseus	Synthesis of ajmalicine	Brodelius *et al.* (1979, 1981); Brodelius and Nilsson (1981)
	Clostridium acetobutylicum	*n*-Butanol production	Häggström (1981); Häggström and Molin (1980)
	Clostridium butylicum	*n*-Butanol and isopropanol production	Krouwell *et al.* (1980)
	Curvularia lunata	(Reichstein compound S/cortisol)	Ohlson *et al.* (1981)
	Daucus carota	Biotransformation of cardenolides	Jones and Veliky (1981); Veliky and Jones (1981)
	Digitalis lanata	(Digitoxin/digoxin)	Brodelius *et al.* (1979, 1981)
		(Digitoxin/purpurea glycoside A)	Alfermann *et al.* (1981a)
	Digitalis mentha	Biotransformation of glycosides and monoterpenoids	Alfermann *et al.* (1981b)
	Escherichia coli	(Penicillin G/6-aminopenicillanic acid)	Klein and Wagner (1980)

	Gluconobacter oxydans	(Glycerol/dihydroxyacetone)	Holst *et al.* (1981)
	Kluyveromyces maxianus	(Insulin/ethanol)	Kierstan and Bucke (1977)
	Morinda citrifolia	De novo synthesis	Brodelius *et al.* (1979, 1981)
	Penicillium chrysogenum	Production of penicillin	Morikawa *et al.* (1979); Suzuki and Karube (1979)
	Pseudomonas denitrificans	Biological denitrification	Nilsson *et al.* (1981); Nilsson and Molin (1981)
	Rhizopus nigricans	(Reichstein compound S/ 11-epicortisol)	Ohlson *et al.* (1981)
	Rhodopseudomonas capsulata	Evidence of immobilized-living-cell system	Larreta-Garde *et al.* (1981)
	Saccharomyces cerevisiae	(D-Glucose/ethanol)	Kierstan and Bucke (1977); Larsson and Mosbach (1979); Linko (1981); Linko and Linko (1981); Birnbaum *et al.* (1981)
	Saccharomyces urarum	(Sucrose/ethanol)	Cheetham (1979)
	Trigonopsis variabilis	Production of α-keto acids	Brodelius *et al.* (1980)
	Vicia faba	Immobilization of protoplasts	Scheurich *et al.* (1980)
	Zymomonas mobilis	(D-Glucose/ethanol)	Grote *et al.* (1980)
Chitosan + $Fe(CN)_6^{4-}$	*E. coli*	L-Tryptophane synthesis	Vorlop and Klein (1981)
Chitoson + $Fe(CN)_6^{3-}$	*E. coli*	L-Tryptophane synthesis	Vorlop and Klein (1981)
Chitosan + poly(aldehydo-carbonic acid)	*E. coli*	L-Tryptophane synthesis	Vorlop and Klein (1981)
Chitosan + poly(1-hydroxy-1-sulfonate propen-2)	*E. coli*	L-Tryptophane synthesis	Vorlop and Klein (1981)
Chitosan + polyphosphate	*E. coli*	L-Tryptophane synthesis	Vorlop and Klein (1981)
Styrene-maleic acid copolymer + Al^{3+}	*C. tropicalis*	Degradation of phenol	Klein *et al.* (1978)

rial is dissolved in an aqueous medium at 40–50°C and is mixed with the cell suspension and cooled at around 10°C. This gel-entrapped whole cell may be obtained in a particulate form, by stirring the aqueous cell–gelatin suspension into an organic liquid poorly miscible or immiscible in water at 50°C, rapidly cooled to 10°C whereby cell- and gelatin-containing particles are formed. However, because this gel does not possess sufficient physical stability to be used alone, it is necessary to incorporate a cross-linking agent—formaldehyde or glutaraldehyde—to increase its mechanical stability.

A very similar process, known as co-cross-linking and initially applied for enzymes, has been developed by Broun and co-workers (Broun, 1976) at the Technological University of Compiègne in France. By this method of immobilization, suspensions of whole cells are mixed with an enzymically inert L-lysine-rich protein (bovine serum albumin) and the glutaraldehyde is added; cross-linking between the whole cells and the enzymically inert protein is observed and the resultant immobilized whole cell can be obtained in the form of particles, membranes, and so on. The same method was also recently reported (De Rosa *et al.*, 1981), but with a modification in which the bovine serum albumin was replaced by a cheaper, albumin-rich matrix: crude egg white. However, all the reports of cell immobilization by co-cross-linking refer to only single-enzyme reactions, and data as to the viability of the cells have not yet been reported.

A very promising matrix, κ-carrageenan, for immobilization of living cells was reported by Chibata and co-workers (Wada *et al.* 1979; Chibata, 1980). κ-Carrageenan is a polysaccharide from seaweeds, used a nontoxic food additive. Gel entrapment of whole cells based on this matrix has been accomplished by mixing a cell suspension at 45–50°C with a solution of κ-carrageenan at the same temperature. The gel is formed by cooling, as in the case of agar, or by contact with metal ions, ammonium ion, amines, or water-miscible organic solvent such as methanol or acetone. The gel formed can be granulated into particles with a suitable size and shape, and in this case where the gel strength of the particles was not satisfactory, the immobilized cells can be treated with hardening agents, such as glutaraldehyde and 1,6-diaminohexane. Alternatively the gel can be further strengthened by incorporation of locust bean gum, which is a D-galacto-D-mannan extracted from locust bean and used for increasing gel strength of jelly foods (Takata *et al.*, 1978).

The gel-entrapped whole cells in the κ-carrageenan matrix have operational stabilities superior to those obtained with polyacrylamide. Another advantage is the nontoxicity of the mild, cheaper, and simpler method of entrapment, again in contradistinction to the polyacrylamide method. It was proven that yeast cells immobilized into κ-carrageenan are kept alive and multiply in the gel, although some leaking out of cells from gel occurred during operation (Chibata, 1979). Therefore, this method of immobilization does not affect the metabolic activity of cells.

Entrapment of cells in agar gel is an obvious method of cell immobilization but it has been little used, presumably because of the poor mechanical strength of the gel, and the characteristics of the gel being associated with oxygen- and product-diffusion limitations. Cells of the yeast *Saccharomyces pastorianus* have been immobilized by entrapment in agar—by direct injection of a hot 2.5% (w/w) agar solution (50°C), containing the whole cells, into toluene or tetrachloroethylene. Spherical pellets were formed with the cells distributed homogeneously throughout (Toda and Shoda, 1975). An application of this method for fermentative processes was reported recently (Margalith and Holcberg, 1981).

The major disadvantage of the entrapment-by-precipitation method is the unavoidable phase–parameter change which is attributable to the precipitation of the gel, because heat or organic solvent application can damage the cells.

Principal matrices that have been used in the precipitative-entrapment technique and examples of immobilized living cells obtained thereby are shown in Table X.

2. Fiber-Entrapment Method of Immobilization

A method of immobilizing enzymes and whole cells by entrapment within microcavities of synthetic fibers has been developed by Dinelli (1972). This method is a variant procedure of entrapment by precipitation with solvents. Biocatalysts can be entrapped in fibers and continuously produced by the conventional wet-spinning techniques for the manufacture of man-made fibers using apparatus very similar to that used in the textile industry.

With this technique, the physical entrapment of cells is achieved by dissolving a fiber-forming polymer in an organic solvent immiscible with water and emulsifying this solution with the aqueous suspension of cells containing glycerol. The emulsion is extruded through a spinnet into a liquid coagulant (toluene or petroleum ether), which precipitates the polymer in filamentous form, microdroplets of microbial-cell suspension being entrapped in this fiber.

This method shows several advantages. The fibers are resistant to weak acids or alkalis, high ionic strength, and some organic solvents. However, their use is limited to low molecular weight by steric hindrance to access, and by the necessity of using water-immiscible liquid

TABLE X
PRINCIPAL EXAMPLES OF GEL-ENTRAPPED LIVING CELLS PREPARED BY PRECIPITATION

Material	Cells	Reaction (substrate/product)	Reference
Agar	*Azotobacter chroococcum*	Nitrogen fixation	Karube *et al.* (1981)
	Catharanthus roseus	Synthesis of ajmalicine	Brodelius and Nilsson (1981)
	Clostridium butylicum	Production of hydrogen	Matsunaga *et al.* (1980); Suzuki *et al.* (1979)
	Escherichia coli	Synthesis of pantothenic acid	Kawabata and Demain (1979, 1980)
	Methane-oxidizing bacteria	Oxidation of methane	Romanov-Skaya *et al.* (1981)
	Methanogenic bacteria	(Wastewater/methane)	Karube *et al.* (1980)
	Rhodospirillum rubrum	(Malate/hydrogen)	Weetall and Bennett (1976)
	Saccharomyces cerevisiae	Production of ethanol	Margalith and Holcberg (1981)
	Saccharomyces pastorianus	(Sucrose/D-glucose + D-fructose)	Toda and Shoda (1975); Toda (1975)
	Trichospora brassicae	Determination of acetic acid	Hikuma *et al.* (1979a)
Agarose	*C. roseus*	Synthesis of ajmalicine	Brodelius and Nilsson (1981)
Albumin + glutaraldehyde	*Aspergillus ochraceus* + aminoacylase (EC 3.5.1.14)	Aminoacylase activity	Hirano *et al.* (1977)
	A. ochraceus + D-glucose oxidase (EC 1.1.3.4)	D-Glucose oxidase activity	Karube *et al.* (1977c)
	E. coli	(Lactose/D-glucose + D-galactose)	Petre *et al.* (1978)

κ-Carrageenan	*Acetobacter aceti*	Production of vinegar	Mori *et al.* (1981)
	Acetobacter suboxydans	Production of L-sorbose	Wada *et al.* (1979)
	Alcaligenes eutrophus	Hydrogenase activity	Klibanov and Puglisi (1980)
	Brevibacterium ammoniagenes	(Fumarate/L-malic acid)	Takata *et al.* (1979)
	Brevibacterium flavum	(Fumarate/L-malic acid)	Chibata *et al.* (1978); Takata *et al.* (1979, 1980); Tosa *et al.* (1979)
	Brevibacterium helvolum	(Fumarate/L-malic acid)	Takata *et al.* (1979)
	C. roseus	Synthesis of ajmalicine	Brodelius and Nilsson (1981)
	E. coli	(Fumarate/L-aspartic acid)	Chibata *et al.* (1978); Tosa *et al.* (1979); Wada *et al.* (1979)
	Proteus vulgaris	(Fumarate/L-malic acid)	Takata *et al.* (1979)
	Penicillium urticae	Production of patulin	Deo and Gaucher (1983)
	Pseudomonas dacunhae	(L-Aspartic acid/L-alanine)	Yamamoto *et al.* (1980)
	Pseudomonas denitrificans	Biological denitrification	Nilsson *et al.* (1981)
	Pseudomonas fluorescens	(Fumarate/L-malic acid)	Takata *et al.* (1979)
	S. cerevisiae	Production of ethanol	Wada *et al.* (1979, 1981); Chibata (1980)
	Sarcina arauntiaca	(Fumarate/L-malic acid)	Takata *et al.* (1979)
	Sarcina flava	(Fumarate/L-malic acid)	Takata *et al.* (1979)
	Sarcina ureae	(Fumarate/L-malic acid)	Takata *et al.* (1978)
	Sarcina variabilis	(Fumarate/L-malic acid)	Takata *et al.* (1979)
	Serratia marcescens	Production of L-isoleucine	Wada *et al.* (1979, 1980); Chibata (1980)
	Streptomyces phaeochromogenes	(D-Glucose/D-fructose)	Chibata *et al.* (1978); Tosa *et al.* (1979)
	Zymomonas mobilis	Production of ethanol	Grote *et al.* (1980)

(Continued)

TABLE X (*Continued*)

Material	Cells	Reaction (substrate/product)	Reference
Collagen + glutaraldehyde	*Anacystis nidulans*	(H_2O_2/O_2)	Vieth and Venkatasubramanian (1979)
		Biological denitrification	Vieth and Venkatasubramanian (1979)
	Aspergillus niger	Production of citric acid	Vieth and Venkatasubramanian (1978)
	Citrobacter freundii	Cephalosporin sensor	Suzuki and Karube (1978a, 1979)
	Cl. butylicum	Production of hydrogen; BOD sensor	Matsunaga *et al.* (1980); Karube *et al.* (1977b)
	Corynebacterium glutamicum	(D-Glucose/L-glutamic acid)	Browstein *et al.* (1974)
	Corynebacterium lilium	(D-Glucose/L-glutamic acid)	Venkatasubramanian *et al.* (1978)
	Corynebacterium simplex	(Hydrocortisone/ prednisolone)	Constantinides (1980); Venkatasubramanian *et al.* (1978)
	Klebsiella pneumoniae	Nitrogen fixation	Venkatasubramanian and Toda (1981)
	Lactobacillus arabinous	Nicotinic acid sensor	Suzuki and Karube (1978a)
	Methanogenic bacteria	(Wastewater/methane)	Karube *et al.* (1980)
	Mixed bacteria from soil and activated sludge	BOD sensor	Suzuki and Karube (1978a,b); Karube *et al.* (1976)

	Mycobacterium rhodochrous	(Cholesterol/Δ^4-cholestenone)	Vieth and Venkatasubramanian (1979)
	Penicillium chrysogenum	Production of penicillin	Morikawa *et al.* (1979); Suzuki and Karube (1979)
	Pseudomonas aeruginosa	Concentration of plutonium from wastewaters	Vieth and Venkatasubramanian (1979)
	Pseudomonas fluorescens	D-Glucose sensor	Suzuki and Karube (1978a, 1979); Karube *et al.* (1979)
	S. cerevisiae	Bioelectrochemical sensor	Suzuki and Karube (1978)
	S. marcescens	(D-Glucose/2-ketogluconic acid)	Venkatasubramanian *et al.* (1978)
	Streptomyces griseus	(D-Glucose/candicidin)	Vieth and Venkatasubramanian (1979)
Crude egg white + glutaraldehyde	*Caldariella acidophila*	(Lactose/D-glucose + D-galactose)	De Rosa *et al.* (1981)
Gelatin + glutaraldehyde	*Arthrobacter* X4		Tramper *et al.* (1979)
	S175N (Human skin fibroblast)	Cell growth	Nilsson and Mosbach (1981)
	DMH W1073 (Rat colon carcinoma)	Cell growth	Nilsson and Mosbach (1981)
Polycondensation of epoxides	*E. coli*	(Penicillin G/6-aminopenicillanic acid)	Klein *et al.* (1978); Klein and Wagner (1980)
Silica hydrogel	*S. cerevisiae*	Evidence of metabolic activity	Rouxhet *et al.* (1981)

as polymer solvents and coagulants, which may in some cases cause damage to the cells. The polymer most commonly used in this procedure is cellulose acetate (Table XI), because of its low costs, good biological resistance, and chemical resistance toward weak acids and solvents.

A similar process was developed by Linko and co-workers (1977), who suspended dried *Actinoplanes missouriensis* cells in α-cellulose dissolved in *N*-ethylpyridinium chloride and formamide, and precipitated them as beads or fibers by contacting the suspension with water. Cell leakage was prevented by glutaraldehyde cross-linking.

3. Microencapsulation Method of Immobilization

The microencapsulation method for enzymes has been extensively used by Chang and co-workers since the first report in the mid-1960s (Chang, 1964) of immobilization of enzymes by entrapping the molecules within microcapsules. However, the droplets are too fragile to be used in heavier, large-scale industrial processes, and therefore this technique appears to be limited to medical and analytical applications.

Mohan and Li (1974, 1975) describe the preparation of liquid surfactant membrane-encapsulated whole cells of *Micrococcus denitrificans*. Unlike other methods of enzyme microencapsulation, where there is a water-insoluble semipermeable membrane, encapsulation of cells was obtained by means of a liquid membrane. Encapsulation of cells was achieved by emulsifying a suspension of viable cells with a mixture of a hydrocarbon solvent (86%), surfactant (2%), polyamine (10%)—which acts as a membrane-strengthening additive—and an anion transport. The emulsion drops are 20–40 μm in diameter and each droplet contains approximately 500–600 cells. Batch and continuous studies with this system have demonstrated the sequential reduction of nitrate and/or nitrite by encapsulated cells. These authors (Mohan and Li, 1975) have suggested that the inclusion of nutrients with cells may enhance the stability of the system even further.

The major advantages of this procedure are its nonchemical nature and its reversibility. However, possible leakage of cells and the fact that diffusion of substrates and products through the membrane is solubility dependent, yet independent of pore size of membrane as this is a liquid membrane, can be disadvantageous.

D. Immobilized-Free-Cell Method

All the methods of immobilization of whole cells described thus far involve the modification of the cell or its microenvironment with alteration of its kinetics and metabolism, and so with a reduced activity

TABLE XI
PRINCIPAL EXAMPLES OF FIBER-ENTRAPPED WHOLE CELLS

Material	Cells	Reaction (substrate/product)	Reference
α-Cellulose	*Actinoplanes missouriensis*	(D-Glucose/D-fructose)	Linko *et al.* (1977, 1978, 1979)
	Bacillus coagulans	(D-Glucose/D-fructose)	Linko *et al.* (1978, 1979)
	Kluyveromyces fragilis	(Lactose/D-glucose + D-galactose)	Linko *et al.* (1978, 1979)
	Kluyveromyces lactis	(Lactose/D-glucose + D-galactose)	Linko *et al.* (1978, 1979)
	Saccharomyces cerevisiae	(Sucrose/D-glucose + D-fructose)	Linko *et al.* (1978, 1979)
	Serratia sp.	Degradation of urea	Linko *et al.* (1978, 1979)
	Streptomyces albus	(D-Glucose/D-fructose)	Linko *et al.* (1978, 1979)
Cellulose triacetate	*Actinomycetes* sp.	Biodegradation of steroids	Germain *et al.* (1981)
	Erwinia herbicola	Production of tyrosine	Yamada *et al.* (1976)
	Escherichia coli	(Penicillin G/6-amino-penicillanic acid)	Marconi *et al.* (1975)
	Sarcina ureae	Degradation of urea	Ghose and Kannan (1979)
	Streptomyces phaeochromogenes	(D-Glucose/D-fructose)	Kolarik *et al.* (1979)
Ethylcellulose	*S. cerevisiae*	(CMP + choline chloride/CDP-choline)	Samejima *et al.* (1978)
Nitrate cellulose	*Zymomonas mobilis*	Production of ethanol	Margaritis and Rowe (1981)

relative to the corresponding free cell. In order to utilize a cell in its native state continuously over a long period of time, several techniques have been in existence for several years, even when they were not at the time of their inception identified with immobilized-cell technology. These techniques include pelletization, dialysis, filtration, and sedimentation of whole cells, which can be recycled to or maintained within the immobilized-cell reactor.

Principal examples of living cells that have been immobilized by this method are shown on Table XII.

1. Pelletization and Flocculation

The multiple reuse of mycelial pellets is one of the simplest forms of cell immobilization and one that has found industrial application. In a recent review Metz and Kossen (1977) describe the factors that influence mycelial-pellet formation, the types of pellets, and the growth of molds in pellets.

Pellet formation has been attributed to several mechanisms of agglomeration, namely of hyphae (Foster, 1949), spores and hyphae (Burkholder and Sinnot, 1945), and solid particles and hyphae (Foster, 1949; Machlis, 1957); and the influence of pellet–pellet and pellet–mold interactions was considered to be important for pellet formation (Pirt and Callow, 1959).

The main factors that influence the pellet formation are agitation, nutrient medium, pH, oxygen concentration, and additives. Agitation is essential, initially, for the dispersion of spore agglomerates but it also influences the structure and survival of pellets once they have been formed. The general tendency is for more agitation to give smaller and more compact pellets. However, strong agitation can cause breaking of pellets (Clark and Lentz, 1963). Mechanical agitation causes development of mycelium around the impeller, and aeration causes filamentous growth.

In the growth nutrient medium, it has been shown (Darby and Mandels, 1954) that nitrogen compounds, either by their concentration or by their chemical type, have a main role in the pellet structure. Other elements and compounds have an effect; for example, manganese at concentrations higher than 2 ppb causes filamentous growth (Metz and Kossen, 1977), and the absence of cyanide prevents the formation of pellets (Rehm, 1971).

The pH plays an important role in the coagulation of spores (Galbraith and Smith, 1969). At pH 5, coagulation was very pronounced, whereas at pH 2 no coagulation took place.

Oxygen concentration in the nutrient medium is important, as oxy-

TABLE XII
PRINCIPAL EXAMPLES OF IMMOBILIZED FREE LIVING CELLS

System	Cells	Reaction (substrate/product)	Reference
Pelletization	*Streptomyces* sp.	Production of lytic enzymes	Antezak *et al.* (1981)
Flocculation	Denitrifying mixed bacteria	Biological denitrification	Ngian *et al.* (1977)
	Saccharomyces carlbergensis	Production of ethanol	Engelbert and Dellweg (1976)
	Saccharomyces uvarum	Growth of cells	Amri *et al.* (1979a,b)
Sedimentation and cell recycle	*Saccharomyces cerevisiae*	Production of ethanol	Cysewski and Wilke (1977); Walsh and Bungay (1979)
	Zymomonas mobilis	Production of ethanol	Lee *et al.* (1980)
Retention within acetyl-cellulose membranes	*Brevibacterium lactofermentus*	Sugars sensor	Hikuma *et al.* (1980)
	Trichosporon brassicae	Alcohol sensor	Hikuma *et al.* (1979a)
	Trichosporon cutaneum	BOD sensor	Hikuma *et al.* (1979b)
Hollow-fiber devices	*Pseudomonas fluorescens*	(Urocanic acid/L-histidine)	Webster *et al.* (1979); Kan and Shuler (1978)
Ultrafiltration devices	*Clostridium histolyticum*	Production of protease	Wang *et al.* (1970)
Dialysis	*Bacillus polymyxa*	Production of proteases and amylases	Hardt *et al.* (1971)
	Clostridium acetobutylicum	Production of proteases and amylases	Hardt *et al.* (1971)
	Escherichia coli	(Sorbitol/L-threonine)	Abbot and Gerhardt (1970a)
	Fleischman's yeast	Diacetyl degradation	Tolls *et al.* (1970)
	Hansenula anomala	BOD sensor	Kulys and Kadziavskieni (1980)
	Lactobacillus bulgaricus	(Whey/ammonium lactate)	Coulman *et al.* (1977); Stieber *et al.* (1977)
	Lactobacillus delbrueckii	(D-Glucose/lactic acid)	Friedman and Gaden (1970)
	S. carlbergensis	Diacetyl degradation	Tolls *et al.* (1970)
	Serratia marcescens	Production of proteases and amylases	Hardt *et al.* (1971)
	Streptococcus thermophilus	Production of proteases and amylases	Hardt *et al.* (1971)
	Streptomyces griseus	(D-Glucose/cycloheximide)	Kominek (1975a,b)
Filtration	*S. cerevisiae*	Production of ethanol	Margaritis and Wilke (1978)

gen depletion in the center of the pellets can occur very easily and cause autolysis of the cells with an eventual formation of a hollow center (Steel *et al.*, 1962).

Some additives, such as polymers (carboxymethyl cellulose, dextrans, etc.), have been shown to increase the pellet size (Kobayashi and Suzuki, 1972) and the metabolic activity (Moo-Young *et al.*, 1969). However, with nonionic surface-active agents such as "Spans," a decrease of coagulation of spores and hyphae has been found (Takahashi *et al.*, 1960).

Flocculation of yeasts has been demonstrated by several authors (Patel and Ingledew, 1975; Amri *et al.*, 1979a). Numerous hypotheses about flocculation have been presented, but the subject is still controversial. Several genetic, physiological, and physicochemical approaches to an explanation have been proposed. Patel and Ingledew (1975) attribute the flocculation to an effect of an intracellular accumulation of acid-soluble glycogen. According to Masschelein and collaborators (1963) and Barker and Kinsop (1972), a reduction in the amount of cell wall D-mannan proteins is involved in the process. The physicochemical approach considers flocculation to depend on the presence, in the cell wall, of free anionic groups (Taylor and Orton, 1973; Stewart *et al.*, 1975), phosphodiester linkages in wall phospho-D-mannans (Lyons and Hough, 1971), or carboxyl groups in wall proteins (Stewart *et al.*, 1975), which are responsible for the intracellular bond formation with Ca^{2+}. In order to form aggregates, potentially flocculent cells must be in a peculiar state, called "state of competence," which appears at the end of the exponential phase of growth (Amri *et al.*, 1979a).

In the form of pellets or flocs, several cells can be maintained within reactors and can be continuously used—that is, they are immobilized (recycled or reused) by incorporation of mechanical-separation devices into fermentation vessels or reactors.

2. Dialysis, Filtration, and Sedimentation

Cells, like enzymes, can be immobilized by confining them with semipermeable membranes, hollow-bore films, or ultrafiltration membranes, where the membrane is impermeable to the biocatalyst, but permeable to products and in some cases to substrates.

Although in the past (Schultz and Gerhardt, 1969) it was not identified as an immobilized-cell method, dialysis culture constitutes a form of cell immobilization, according to the definition of immobilized cells. This method of immobilization offers several advantages relative to other immobilization methods. Thus this method allows the study of free cells, namely their operational stabilities and their application in continuous reactors.

These methods are specially suited for conversion of high molecular weight water-soluble or water-insoluble substrates, as it allows the intimate contact of the free cell with substrate, achieving an efficient conversion of these types of substrates, unlike the immobilized modified cells, which usually have lower catalytic efficiencies toward the same substrates.

Advantages of this method include the simplicity of methodology required to immobilize the cell, because this immobilization consists of placing the cellular suspension on one side of a semipermeable membrane. Other significant advantages are selectivity; control of substrates and products through membrane selectivity; large ratio of surface area to volume (hollow fibers); absence of cell leakage, when properly constructed membranes are chosen; and the favorable ease with which the membrane reactors can be loaded with cells, operated, cleaned, sterilized, and regenerated compared with other methods of immobilization.

However, some disadvantages, inherent to the methods, also exist: the possible reduction of reaction velocity as a result of the permeability resistance of the membrane; the difficulty in certain instances of working with very low substrate concentration due to substrate adsorption to membranes; and, inter alia, the need for a careful control of the residence time of low molecular weight substrates in order to achieve high conversions.

This method can, of course, be combined with other immobilization methods; for example, adsorption of the cells on the walls of the membranes can occur.

Various processes come under the scope of this section as follows. Filter fermenters, which allow the continuous removal of a cell-free effluent, have been described (Sortland and Wilke, 1969; Margaritis, 1974). Other cell-recycle devices based on centrifugation and sedimentation have also been used, for ethanol and beer production. Of particular interest is the high-productivity ethanol manufacture using cell recycle with vacuum fermentation. Cell recycle was accomplished with a settling device, which maintains a high cell population within the fermenter (Cysewski and Wilke, 1977). Fermentation under vacuum is very important as the inhibitory product (ethanol) evaporates and boils away at temperatures compatible with yeast metabolism.

E. Miscellaneous Methods of Whole-Cell Immobilization

Various methods for immobilization of whole cells other than those that fit within the foregoing classification have also been reported. Biospecific adsorption of whole cells to a variety of specific natural materials, such as antibodies or lectins—(phyto) hemagglutinins—has great promise. Lectin is first bound to a carrier and thereby used to immobilize the whole cells. The lectin concanavalin A bound to magnetite has been used (Horisberger, 1976) to immobilize *Candida utilis, Lactobacillus plantarum, Streptococcus faecalis,* and *Bacillus subtilis* cells, all by selective and specific interaction of the lectin with the D-mannan in the cell wall of the microbes. In this particular instance the type of support chosen enables magnetic removal of the cells.

Immobilization of enzymes onto the surface of cells has also been reported. By using the metal-chelation method (Section II,B,2), glucoamylase (EC 3.2.1.3), α-amylase (EC 3.2.1.1), and a protease have been coupled to titanium(IV)-treated *Saccharomyces cerevisiae* cells. Aminoacylase (EC 3.5.1.14) and D-glucose oxidase (EC 1.1.3.4) have been bound on to mycelial pellets of *Aspergillus ochraceus* by the co-cross-linking method (Section II,C,1,*c*), using glutaraldehyde and albumin (Hirano *et al.*, 1977; Karube *et al.*, 1977c).

Some examples of immobilized cells obtained by miscellaneous methods of cell immobilization are shown in Table XIII.

F. Comparison of Different Immobilization Techniques

Although a number of immobilization techniques have been applied to whole cells, it is recognized that no one particular process can be classified as an ideal general method for cell immobilization. Each method of immobilization has specific disadvantages, and for a particular application it is necessary to find an immobilization procedure that would be simple and inexpensive, and would yield the immobilized-cell product such that it has a good retention of activity and a proper operational stability, a prerequisite to both of which is the viability of the whole cell.

However, one can make a general comparison of the different cell-immobilization processes, based on the main characteristics of these methods already described and on the support matrix. Table XIV summarizes some of the relative advantages and disadvantages of the different processes of living-cell immobilization.

Immobilization of cells by chemical methods (i.e., intercellular chemical cross-linking and covalent bonding) involves chemical mod-

TABLE XIII
MISCELLANEOUS METHODS OF IMMOBILIZING LIVING CELLS

Method	Cells	Reaction (substrate/product)	Reference
Biospecific adsorption (polyacrylamide + agglutinin)	L 1210 (Mouse leukemia cells)		Zabriskie *et al.* (1973)
Biospecific adsorption	HeLa Cells		Kinzel *et al.* (1976)
Agarose + lectin	Red blood cells		Mattiasson and Borrebaeck (1978)
Biospecific adsorption	Lymphoma		Rutishauser and Sachs (1975)
Nylon + concanavalin A	Myeloid leukemia fibroblast cells		
Raschig rings coated with gelatin + glutaraldehyde	*Saccharomyces cerevisiae*	Production of ethanol	Gaddy and Sitton (1978)
Berl saddles coated with gelatin + glutaraldehyde	Lactobacilli yeasts	(Whey/lactic acid) Production of ethanol	Compere and Griffith (1976); Griffith and Compere (1976)
Polyacrylamide entrapment of *S. cerevisiae* in concanavalin A + D-glucose oxidase	*S. cerevisiae* + D-glucose oxidase	(Sucrose/D-fructose + D-gluconic acid	D'Souza and Nadkanni (1980)

TABLE XIV
COMPARISON OF THE ATTRIBUTES OF DIFFERENT CLASSES OF IMMOBILIZATION TECHNIQUES

Characteristic	Cross-linking	Adsorption	Chelation	Covalent binding	Entrapping
Preparation	Intermediate	Simple	Simple	Difficult	Intermediate
Binding force	Strong	Weak	Intermediate	Strong	Intermediate
Retention of activity	Low	High	Intermediate	Low	Intermediate
Regeneration of carrier	Impossible	Possible	Possible	Rare	Impossible
Cost of immobilization	Intermediate	Low	Low	High	Low
Stability	High	Low	Intermediate	High	High
General applicability	No	Yes	Yes	No	Yes
Protection from microbial attack	No	Yes	Yes	No	Yes
Viability	No	Yes	Yes	No	Yes

ification of the whole cell; this may cause damage to the cell because of the toxic character of the chemicals used. To reduce this main disadvantage, these processes must be carried out under conditions as mild as possible. However, because of the strength of the bonds between cell and cell or cell and carrier—bonds not easily destroyed by substrate or salt—the operational stability of the immobilized-cell preparation may be high. A problem of growing cells arises from these techniques, in that the new cells may not become coupled to the carrier or may become cross-linked. However, very recently, Jirkú *et al.* (1980) observed multiplication of the immobilized cells of a *Saccharomyces cerevisiae* population without any significant leakage of progeny into the medium. The cell division proceeds, without separation of the daughter cells, but with rod-shaped forms because the newly produced cells are immobilized during their longitudinal growth.

Relative to their applications, cross-linking techniques are generally unsuitable due to lack of good mechanical properties of the cross-linked cell preparations; cross-linking, however, has the major advantage of being a high-cell-density method. Covalently bonded cell preparations, where organic matrices have been used, are either difficult or, in most cases, even impossible to regenerate.

Immobilized cells can be produced by adsorption and chelation simply under mild conditions, although with physical adsorption the forces between the cell and the carrier are generally weak, and leakage of the cell from the matrix can easily occur during operation—mainly with substrates of high molecular weight or with changes in ionic strength of pH of the medium. These characteristics lead to preparations with low operational stabilities. One advantage of these preparations is the possibility of regeneration.

With the entrapping method, in theory no binding between cell and carrier should occur, and the preparations should present a high retention of activity. However, limitations of activity do occur and this method is limited to low molecular weight substrate and product molecules, as there are steric resistances to diffusion of macromolecules.

With respect to the viability of the cells, the nonchemical methods are preferred especially when nontoxic materials are used, such as κ-carrageenan, calcium alginate, and adhesion materials.

III. EFFECTS OF IMMOBILIZATION ON THE KINETICS AND PROPERTIES OF LIVING CELLS

Although many advantages of cell immobilization are recognized and accepted, as already stated, due to immobilization the kinetics and properties of whole cells change, with decrease of free-cell specific activity. The decrease of catalytic activity of immobilized whole cells may be ascribed to several factors, such as toxicity of the materials used in a specific immobilization method and others that have been implicated in the modification of immobilized-enzyme kinetics (Goldstein, 1976):

1. Conformational and steric effects are present when conformational change of the enzyme molecules occurs by binding to a carrier, or the interaction of the substrate with the enzyme is affected by steric hindrance, respectively.
2. Partitioning effects, related to the chemical nature of the support material, may arise from electrostatic or hydrophobic interactions between the matrix and low molecular weight species present in the solution, leading to a modified microenvironment.
3. Mass-transfer diffusional effects, arising from diffusional resistances to the translocation of substrate from the bulk solution to the catalytic sites and the diffusion of products of the reaction back to bulk solution, may operate. These diffusional resistances can be classified as

(*a*) internal or intraparticular mass-transfer effects, when the biocatalyst—cell or enzyme—is located in a porous medium; and (*b*) external or interparticular mass-transfer effects between the bulk solution and the outer surface of biocatalyst particles.

Besides these factors, common to immobilized-enzyme kinetics, the kinetics of immobilized cells is usually more complex, because two additional effects are impinging on them. One is the presence of an additional diffusion barrier created by the presence of an osmotically intact cell wall and cytoplasmic membrane, and the other is the possibility of cell division of immobilized living cells, which allows an increasing cell density or catalytic activity, which in turn causes a nonsteady-state condition within reactors, and consequently a (potential) increase of production rate.

As the enzymes of the immobilized cells do not interact directly with the support material, microenvironment, conformational, and steric effects on the enzyme per se are unlikely. The partitioning effects and the diffusional effects that may occur by virtue of the cell structure are the most important modifiers of the immobilized-cell kinetics.

In view of the effects just outlined, when the kinetic behavior of the immobilized cell can be controlled by one or more of these effects, it is useful to distinguish among (*a*) intrinsic parameters (i.e., the kinetic parameters determined for the free cell); (*b*) inherent rate parameters (i.e., the kinetic parameters that are observed in the absence of any diffusional effects); and (*c*) effective rate parameters (i.e., the kinetic parameters determined when mass-transfer effects are present and operate in the presence or the absence of partition effects).

A. Partition Effects

In the carrier-binding method, when the support matrix is charged, the kinetic behavior of the immobilized biocatalyst may differ from that of the free biocatalyst, even in the absence of mass-transfer effects. This difference is commonly attributed to partitioning effects that cause concentrations of charged species, substrates, products, hydrogen ions, hydroxyl ions, and so forth, in the domain of the biocatlyst, to be different from those in the bulk solution, as a result of electrostatic interactions with fixed charges on the support.

These differences in the equilibrium concentrations in the cases of charged soluble species or compounds may be described by the partition coefficient *P*, given by

$$P = C_i/C_o,$$

where C_i and C_0 are the local and bulk concentrations, respectively. The main consequence of these effects is a shift in the optimum pH, with a displacement of the pH–activity profile of the immobilized cell, toward more alkaline or acidic pH values for negatively or positively charged carriers, respectively. Hattori (1973) has studied the biochemical activities of *Escherichia coli* and *Azotobacter agilis* adsorbed on anion-exchange resin. In both cases the maximum of activity of the adsorbed cells was at a pH higher than that for the cells in suspension.

Goldstein and co-workers (1969) expressed the qualitative considerations on the displacement of pH–activity profiles of immobilized enzymes mathematically. Assuming the Boltzmann distribution, the partitioning of hydrogen ions between the local activity (a_i^{H+}) and the bulk activity (a_0^{H+}) is given by

$$P_H = a_i^{H+}/a_0^{H+} = \exp(-e\psi/kT),$$

or by definition of pH

$$\text{pH} = \text{pH}_i - \text{pH}_0 = 0.43(e\psi/kT),$$

where e is the electrostatic charge, ψ is the electrostatic potential, k is the Boltzmann constant, T is the absolute temperature, and pH_i and pH_0 are the local and the bulk pH values, respectively. This equation shows that the local pH is higher if the support is negatively charged.

By similar considerations, the partitioning of charged compounds, substrate or product, between a charged cell particle and the bulk solution can be represented in the following form:

$$S_i = S_0 \exp(-Ze\psi/kT)$$

where Ze is the substrate charge, Z is an integer, S_i and S_0 are the local and bulk substrate concentrations, respectively.

Thus, for positively charged substrate, when using a negatively charged cell particle, a higher concentration of substrate is obtained in the local environment or microenvironment than in the bulk solution, and thus a higher value of relative activity is obtained than with a neutrally charged matrix. However, when effects other than partitioning ones are also present, it is possible to obtain nil shift of the pH optimum with charged supports by a counterbalance process.

B. Internal and External Mass-Transfer Effects

When a cell is immobilized on or within a solid matrix, mass-transfer effects may be in existence because the substrate must diffuse from the bulk solution to the immobilized cell. If the cell is attached to nonporous carriers, there are only external mass-transfer effects. The effects are due to the catalytically active outer surface, in the reaction solution, being surrounded by a stagnant film, a Nernst layer, across which substrate and product transports are effected by diffusion; the corresponding driving force is the concentration difference between surface and bulk.

For instance, the rate of flow of substrate V_{dif}, from the bulk solution to the catalytic surface is given by

$$V_{dif} = k_L a(S_B - S_S),$$

where k_L is the mass-transfer coefficient, a is the particle surface area per unit of volume, and S_B and S_S are the bulk and surface concentrations of substrate, respectively.

In a surface reaction, the flow of substrate to the catalytic surface and the transformation reaction of substrate take place consecutively. At steady state, the rate of external mass transfer of substrate V_{dif} will be equal to its internal removal by reaction. Hence for a reaction that obeys Michaelis–Menten kinetics, the overall rate of reaction V_{obs} will be:

$$V_{obs} = k_L a(S_B - S_S) = V_{max} S_S/(K_m + S_S)$$

This equation may be solved for S_S, if $K_L a$ and the kinetic constants are known, or S_S may be obtained graphically using the following

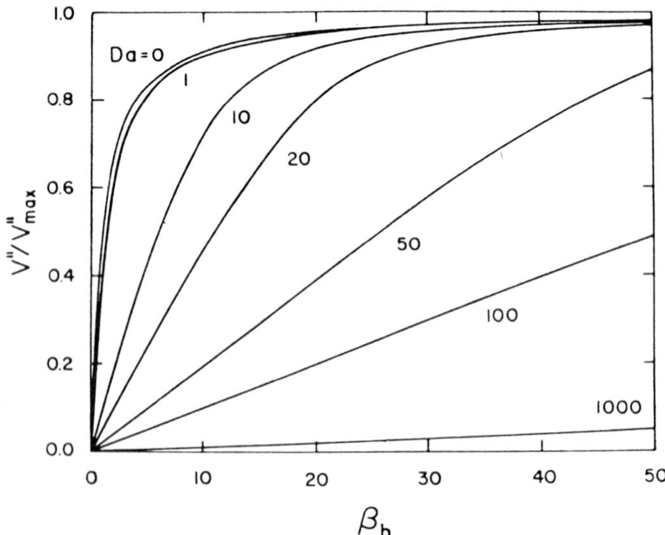

Fig. 12. V_{obs}/V_{max} versus the dimensionless bulk concentration β, for different values of the substrate modulus μ, for external diffusion. (From Horvath and Engasser, 1974.)

dimensionless equation:

$$\frac{V_{obs}}{V_{max}} = \frac{\beta_B - \beta_S}{\mu} = \frac{\beta_S}{1 + \beta_S}$$

where β is the dimensionless substrate concentration ($\beta = S/K_m$), S is the substrate concentration, β_B and β_S are dimensionless substrate concentrations in the bulk solution and at the surface of the immobilized enzyme, and μ is the dimensionless substrate modulus, known as Damkohler number ($\mu = V_{max}/k_L a K_m$). The dependence of V_{obs}/V_{max} on β for different values of μ is shown in Fig. 12.

The external mass-transfer effects on the activity of an immobilized biocatalyst can be quantitatively expressed by the effectiveness factor η, defined as the ratio of the observed reaction rate V_{obs}, to the kinetic rate V_{kin}

$$\eta = V_{obs}/V_{kin}$$

In Fig. 13 the dependence of the effectiveness factor η on β and μ is shown.

The rate flow of substrate V_{dif}, or the rate of reaction V_{kin}, may play

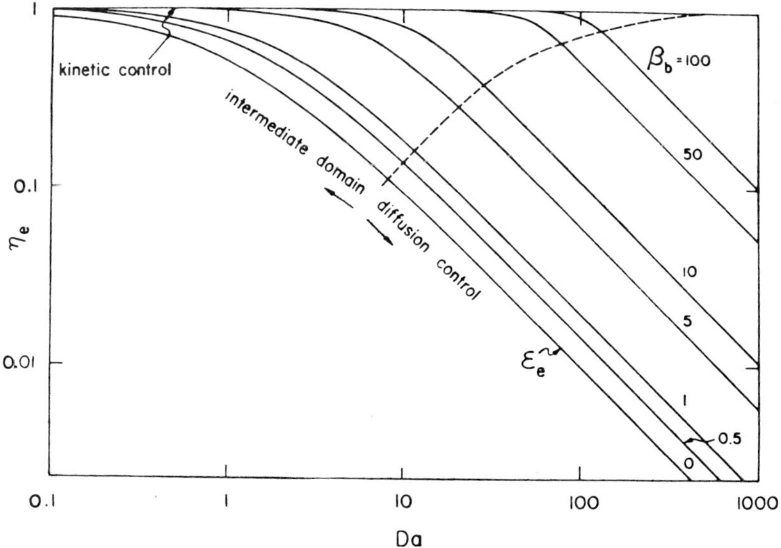

Fig. 13. Plots of the external effectiveness factor η, as a function of the substrate modulus μ, for different values of the bulk substrate concentration β; ε is the limiting first-order effectiveness factor, attained at sufficiently low concentrations. (From Horvath and Engasser, 1974.)

predominant roles, depending on their relative magnitudes, as the lower rate step will be the controller step.

As can be seen in Fig. 14, when the reaction is zero order, V_{obs} will always be equal to V_{max} and the reaction will be kinetically controlled. For first-order reactions, the reaction can be kinetically or diffusionally controlled, depending on $k_L a \gg V_{max}/K_m$ (i.e., mass transport is much faster than the enzyme reaction), or $k_L a \ll V_{max}/K_m$ (i.e., the enzyme reaction is much faster than the diffusion of substrate), respectively.

Decrease of the resistance to external mass transfer is achieved with an increase of linear velocity of the fluid, as this velocity reduces the resistance to a point at which S_B and S_S may be considered equal, and the reaction rate V_{kin} will be the controlling step.

When a cell is immobilized within a porous support, besides possible external mass-transfer effects, there could also exist resistances to internal (intraparticlar) diffusion of substrate, as this must diffuse through the pores in order to reach the cell, and of product, as it must diffuse back to the bulk solution. Consequently, a substrate-concentration gradient is established within the pores, resulting in concentration decreasing with increased distance (in depth) from the surface of the immobilized cell. A corresponding product concentration gradient is obtained in the opposite direction.

Unlike external diffusion, internal mass transfer proceeds in parallel with the reaction and takes into account the depletion of substrate,

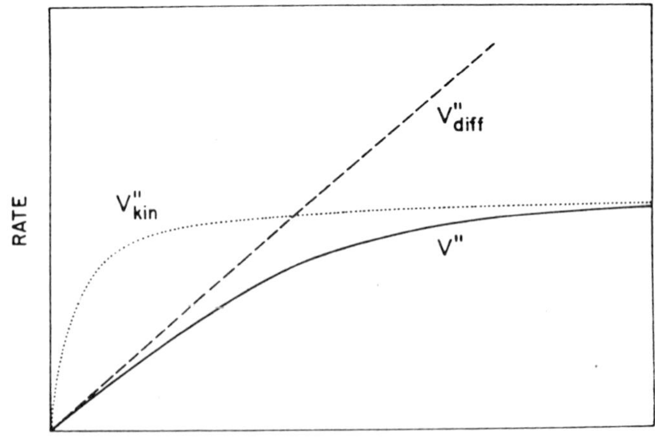

Fig. 14. Schematic plot of the overall rate of reaction *V*, catalyzed by a surface-bound enzyme versus the bulk substrate concentration. (From Horvath and Engasser, 1974.)

within the pores, with increasing distance from the surface of the cell support. The rate of reaction will also decrease for the same reason. As can be easily deduced, the overall reaction is dependent both on the substrate concentration and on the distance from the outside support surface.

The usual way to study this problem is by way of considering that there is a coupled reaction–diffusion process that can be solved, at the steady state, when the rates of internal diffusion and enzyme reaction are equal, using appropriate differential equations for the various geometries considered and isothermal conditions:

$$\frac{d^2S}{dx^2} + \frac{(p+1)}{x}\frac{dS}{dx} = \frac{V_{kin}}{D_{eff}} = \frac{V_{max}\,S}{D_{eff}\,(K_m + S)}$$

where S is the substrate concentration; x the distance from the outer surface; p a geometrical factor with the values of +1 (for spherical pellets), 0 (for cylindrical pellets), and −1 (for rectangular membranes); and D_{eff} is the effective diffusivity of the substrate inside the support and is given by

$$D_{eff} = D\varepsilon/\tau,$$

where D is the substrate diffusivity, ε is the void fraction in the porous matrix, and τ is a tortuosity factor that takes into account the pore geometry and by definition is larger than unity.

The analytical solutions of these equations are easily obtained for first- or constant-order reactions, but numerical solutions are required for Michaelis–Menten type reactions. The equations just given, in these cases, are usually rewritten in terms of dimensionless variables. For a spherical pellet this equation is:

$$\frac{d^2\beta}{dZ^2} + \frac{2}{Z}\frac{d\beta}{dZ} = 9\phi^2\,\frac{\beta}{1+\beta}$$

with the boundary conditions

$$\beta = \beta_S \qquad \text{for } Z = 1$$

and

$$d\beta/dZ = 0 \qquad \text{for } Z = 0.$$

In this equation β is the dimensionless substrate concentration, Z is the dimensionless position in the porous support given by $Z = x/R$, R is the radius of the spherical pellet, and ϕ is the substrate modulus (a modified Thiele modulus) defined by

$$\phi = R/3(V_{max}/K_m D_{eff})^{1/2}$$

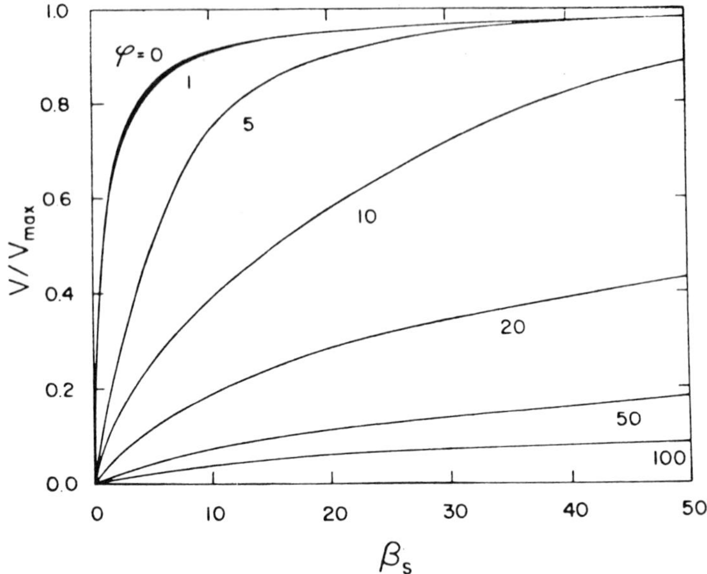

Fig. 15. Normalized overall rate as a function of the dimensionless bulk concentration of substrate β, for different values of ϕ, the substrate modulus for internal diffusion in an enzyme membrane. (From Horvath and Engasser, 1974.)

Numerical integration yields the effective rate of reaction V_{obs}, as a function of the concentration with the modulus ϕ (Fig. 15). The same results can also be represented in the form of graphics of effectiveness factor η against the modified Thiele modulus ϕ (Fig. 16).

The internal mass-transfer effects can, however, be reduced by decreasing the particle dimensions of the porous support containing the cell. Particle diameter decrease results in a reduction of the distance from the outer support surface that the substrate must cross, and consequently also results in a decrease of the substrate concentration gradient.

In the carrier-binding methods, as cells are much larger than enzymes, the matrices with very much larger pore sizes must be used for immobilization. This allows faster diffusion of compounds, so intraparticlar diffusional restrictions are minimized. The large pores also enable the use of much larger substrate molecules.

When external and internal diffusion resistances affect the rate of the reaction simultaneously, the relative contributions of each effect must be estimated separately and quantified by the corresponding effectiveness factors. Hence, the overall reaction rate is given by

$$V_{obs} = \eta_{ext}\, \eta_{int}\, V_{kin}$$

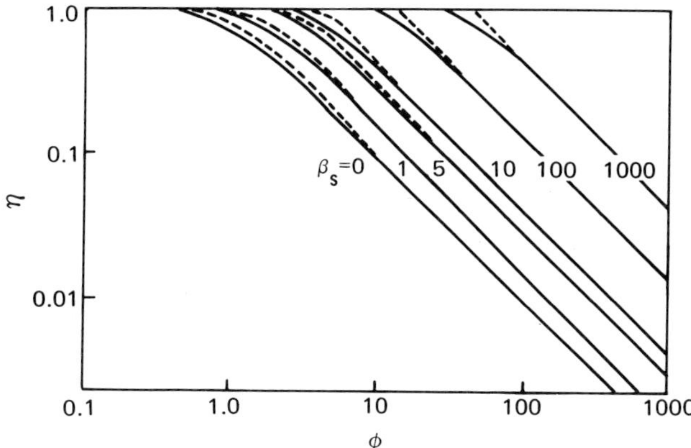

Fig. 16. Graph illustrating the effectiveness factor η, as defined by the equation on page 249, as a function of the modulus ϕ, as defined by the equation on page 251, with the dimensionless surface concentration as the parameter. The effectiveness factors for spherical particles and membranes are represented by solid and broken lines, respectively. (From Engasser and Horvath, 1973.)

In general, effectiveness factors in excess of unity can be obtained, in single enzyme reactions by permeabilization of the walls of the immobilized cell, and in multienzyme reactions by increasing the immobilized-cell concentration by *in situ* cell division.

C. Effect of Immobilization on Stability of Cells

Apart from the influence of the aforementioned factors in partition effects, and external and internal mass-transfer effects (Section III,A and III,B, respectively) on the activity of the cell, other properties of the whole cell, namely the operational stability of the immobilized cell, can change. From the viewpoint of practical utilization, the operational stability of a biocatalyst is one of the most important properties. Its enhancement is advantageous in the continuous use of the immobilized cell, particularly for industrialization of these systems.

For single-enzyme systems, the operational stability of an immobilized-cell preparation is usually expressed as a half-life. However, even for multienzyme reactions, when activity is dependent on the preservation of the integrity of the cell membrane, this activity declines just like that of a single enzyme, presumably because a single enzyme limits the overall stability of the metabolic pathway.

The stability of immobilized-cell preparations is affected by several factors, such as microbial contamination, leakage of cells, cell lysis followed by leakage of intracellular contents (particularly divalent ions), and the effects of endogenous proteases. The stabilities of immobilized cells always appear to be greater than the corresponding cells in suspension (Durand and Navarro, 1978).

IV. IMMOBILIZED-CELL REACTORS

Among the applications of immobilized cells, industrial utilization is perhaps the most important field and consequently the most active one reported. Thus the use of immobilized cells in industrial processes is carried out in basic chemical reactors. However, biochemical reactions differ from chemical reactors, chiefly because biocatalysts operate at normal pressure and comparatively little heat is generated or consumed during the reaction.

Several classifications of enzyme reactors have been proposed (Lilly and Dunnill, 1972), on the basis of mode of operation and flow characteristics of substrate and product. Immobilized-cell reactors can also be classified according to these parameters. A classification is presented in Table XV.

A. Batch Reactors

Batch reactors or fermenters are the most commonly used type of reactor when free cells are used as catalyst. In a batch immobilized-cell reactor, a separation process is required to recover the cell preparation, to permit its reuse. During the operation of this recovery process, appreciable loss of catalyst material may occur as well as loss of activity. So, for industrial processes the use of immobilized cells in a batch operation will be limited to the production of rather small amounts of fine chemicals.

TABLE XV
CLASSIFICATION OF CELL REACTORS

Mode of operation	Flow pattern	Type of reaction
Batch	Well mixed	Batch-stirred tank reactor (BSTR)
	Plug flow	Total-recycle reactor
Continuous	Well mixed	Continuous-feed stirred-tank reactor (CSTR)
		CSTR with continuous-ultrafiltration membrane
	Plug flow	Packed-bed reactor (PBR)
		Fluidized-bed reactor (FBR)
		Tubular reactor
		Hollow-fiber reactor

Traditionally, the stirred-tank reactor has been chosen for batchwise work, because it is the simplest type of reactor, composed of a tank, an agitator, and/or a system of air distribution. Agitation allows good mixing of the medium and dispersion of the incoming air, as well as relative ease of temperature and pH control. However, some matrices, such as inorganic supports, are broken up by attrition in such vessels; alternative designs have therefore been attempted. A possible laboratory alternative is the "basket" reactor, in which the catalyst is retained within a basket, forming either the impeller "blades" or the baffles of the tank reactor. Another alternative is obtained by changing the flow pattern by using a plug-flow type reactor—the total-recycle reactor, or batch-recirculation reactor—which may be a packed- or fluidized-bed reactor, or even a cell-coated tubular reactor. This type of reactor may be useful in cases where a single pass gives inadequate conversions. However, it has found greatest application in the laboratory for the acquisition of kinetic data, when the recycle rate is adjusted so that the conversion in the reactor is low and the reactor can be considered as a differential reactor. One advantage of this type of reactor is that the external mass-transfer effects can be reduced by the high operational fluid velocities.

B. Continuous Reactors

The major field of application of immobilized cells is in continuous operation. This mode of operation generally has some advantages when compared with batch processes, such as ease of automatic control, of operation, and of quality control of products. With respect to the flow pattern, the continuous reactors can be divided into two basic and ideal types: the continuous-feed stirred-tank reactor (CFSTR) and the plug-flow reactor (PFR).

In the ideal CFSTR the conversion degree is independent of the position of reactants in the vessel, as a complete mixing is obtained with stirring and the conditions within the CFSTR are the same as the outlet stream, that is low substrate and high product concentrations. With the ideal PFR the conversion degree is dependent on the length of the reactor as no mixing device at all exists and the conditions within the reactor are never uniform.

Whereas a nearly ideal CFSTR is readily obtained in practice, it only being necessary to have good stirring to obtain complete mixing, an ideal PFR is very difficult to obtain because several adverse factors can occur, such as temperature and velocity gradient normal to the flow direction, and axial dispersion of substrate.

Several factors influence the type of continuous reactor to be chosen for a particular application. One of the most important criteria is based on kinetic considerations. For Michaelis–Menten kinetics, the PFR is preferable to the CFSTR as with this latter reactor more catalyst is required to obtain the same degree of conversion than with a PFR. If product inhibition occurs, this disadvantage of the CFSTR is more pronounced, as a relative high product concentration is always in direct contact with all of the catalyst. There is only one situation in which a CFSTR is more favorable kinetically than a PFR; this is when substrate inhibition occurs.

The nature of support material, the type of cell used, the method of immobilization, and the chemical and physical properties of the substrate and product also influence the choice of reactor type; moreover, operational requirements are still another factor to be taken into account. Thus, for example, when pH control is necessary, the CFSTR is more suitable than the PFR: Due to possible disintegration of support through mechanical shearing, only durable preparations of immobilized cells should be used in CFSTR. With very small immobilized-cell particles, problems such as great pressure drop and plugging arise from the utilization of this catalyst in packed-bed reactors (the most used type of PFR). To overcome these problems, a fluidized-bed reactor, which provides a degree of mixing intermediate to the CFSTR and the idea PFR, can be used with minimal pressure drop. Fluidized-bed and stirred reactors are also especially advantageous, particularly when gas mixing, or efficient pH or temperature control, or use of high flow rates is required.

Reactant characteristics can also influence the choice of reactor. Insoluble substrates and highly viscous fluids are better processed in fluidized-bed reactors or CFSTRs, where no plugging of the reactor is likely to occur, as would be the case in a packed-bed reactor.

As can be deduced from this outline, there are no simple rules for choosing the reactor type and the different factors mentioned must be analyzed individually for each specific case.

V. APPLICATIONS OF IMMOBILIZED LIVING CELLS

Several reviews (Klein and Wagner, 1979; Durand and Navarro, 1978; Cheetam, 1980) describe uses for immobilized cells (dead or alive) in industrial and analytical fields. In this section these same fields are presented in a series of tables (Tables XVI–XIX), which illustrate how far work on the application of a wide variety of living cells has gone.

TABLE XVI
INDUSTRIAL USES OF IMMOBILIZED LIVING CELLS

Use	Cell	Immobilization method	Reference
Production of organic acids			
Acetic acid (vinegar)	*Acetobacter* sp.	Chelation	Kennedy (1979); Kennedy *et al.* (1980)
	Acetobacter aceti	Entrapment (κ-carrageenan)	Mori *et al.* (1981)
	Bacterium scheutzenbachii	Adsorption	Fetzer (1930); Prescott and Dunn (1959)
Citric acid	*Aspergillus foetoidus*	Adsorption	Perez (1976); Atkinson *et al.* (1979)
	Aspergillus niger	Entrapment (collagen)	Vieth and Venkatasubramanian (1978, 1979)
	Candida lipolytica	Entrapment (polyacrylamide)	Stottmeister (1980)
	Saccharomyces lipolytica	Adsorption	Briffaud and Engasser (1978, 1979)
Erythorbic acid	*Penicillium cyaneofulvum*	Entrapment (polyacrylamide)	Kato (1974)
α-Keto acids	*Trigonopsis variabilis*	Entrapment (calcium alginate)	Brodelius *et al.* (1980)
2-Ketogluconic acid	*Serratia marcescens*	Entrapment (collagen)	Venkatasubramanian *et al.* (1978)
2-Keto-L-gulonic acid	*Gluconobacter melanogenes* + *Pseudomonas* sp.	Entrapment (polyacrylamide)	Martin and Pedman (1976)
Lactic acid	*Arthrobacter oxydans*	Entrapment (polyacrylamide)	Yagi *et al.* (1976)
	Lactobacilli	Adsorption to berl saddles coated with gelatin and activated with glutaraldehyde	Compere and Griffith (1976)
	Lactobacillus bulgaricus	Dialysis membrane	Coulman *et al.* (1977); Stieber *et al.* (1977)
	Lactobacillus delbrueckii	Dialysis membrane	Friedman and Gaden (1970)
	L. delbrueckii	Entrapment (polyacrylamide)	Divies (1977)

(*Continued*)

TABLE XVI (*Continued*)

Use	Cell	Immobilization method	Reference
Malic acid	*Brevibacterium ammoniagenes*	Entrapment (κ-carregeenan)	Takata *et al.* (1979)
	Brevibacterium flavum	Entrapment (κ-carrageenan)	Chibata *et al.* (1978); Tosa *et al.* (1979); Takata *et al.* (1979, 1980)
	Brevibacterium helvolum	Entrapment (κ-carrageenan)	Takata *et al.* (1979)
	Proteus vulgaris	Entrapment (κ-carrageenan)	Takata *et al.* (1979)
	Pseudomonas fluorescens	Entrapment (κ-carrageenan)	Takata *et al.* (1979)
	Sarcina aurantiaca	Entrapment (κ-carrageenan)	Takata *et al.* (1979)
	Sarcina variabilis	Entrapment (κ-carrageenan)	Takata *et al.* (1979)
	Sarcina ureae	Entrapment (κ-carrageenan)	Takata *et al.* (1979)
Pantothenic acid	*Escherichia coli*	Entrapment (agar)	Kawabata and Demain (1979, 1980)
Salicylic acid	*Ps. fluorescens*	Dialysis membrane	Abbot and Gerhardt (1970b,c)
Production of amino acids			
L-Alanine	*Pseudomonas dacunhae*	Entrapment (κ-carrageenan)	Yamamoto *et al.* (1980)
L-Aspartic acid	*E. coli*	Entrapment (κ-carrageenan)	Chibata *et al.* (1978); Tosa *et al.* (1979); Wada *et al.* (1979)
L-Glutamic acid	*Corynebacterium lilium*	Entrapment (collagen)	Venkatasubramanian *et al.* (1978)
L-Glutamine	*Corynebacterium glutamicum*	Entrapment (polyacrylamide)	Slowinski and Charm (1973)
L-Histidine	*Ps. fluorescens*	Hollow-fiber devices	Webster *et al.* (1979); Kan and Shuler (1978)
L-Isoleucine	*Serratia marcescens*	Entrapment (κ-carrageenan)	Wada *et al.* (1979, 1980); Chibata (1980)
L-Threonine	*E. coli*	Dialysis	Abbot and Gerhardt (1970a)
L-Tryptophane	*E. coli*	Entrapment (polyacrylamide)	Bang *et al.* (1978); Decottignies-Le Maréchal *et al.* (1979); Azerad *et al.* (1980)

L-Tyrosine	*Erwinia herbicola*	Fiber entrapment (cellulose triacetate)	Yamada *et al.* (1976)
Production of ethanol			
	Kluyveromyces marxianus	Entrapment (calcium alginate)	Kierstan and Bucke (1977)
	Saccharomyces cerevisiae	Adsorption (wood)	Moo-Young *et al.* (1980, 1981); Lamptey *et al.* (1981)
	S. cerevisiae	Adsorption (PVC)	Ghose and Bandyopadhyay (1980)
	S. cerevisiae	Adsorption (porous brick)	Ghose and Bandyopadhyay (1980)
	S. cerevisiae	Adsorption (diatomite)	Moo-Young *et al.* (1980); Lamptey *et al.* (1981)
	S. cerevisiae	Adsorption (ceramic raschig rings)	Sitton and Gaddy (1980); Ghose and Bandyopadhyay (1980)
	S. cerevisiae	Adsorption to raschig rings coated with gelatine and glutaraldehyde	Gaddy and Sitton (1978); Sitton and Gaddy (1980); Sitton *et al.* (1981a,b)
	S. cerevisiae	Covalent binding (alkylation)	Gainer *et al.* (1981)
	S. cerevisiae	Entrapment (polyacrylamide)	Mavituna and Sinclair (1978); Divies (1977)
	S. cerevisiae	Entrapment (calcium alginate)	Kierstan and Bucke (1977); Larsson and Mosbach (1979); Linko and Linko (1981); Linko (1981); Birnbaum *et al.* (1981)
	S. cerevisiae	Entrapment (agar)	Margalith and Holceberg (1981)
	S. cerevisiae	Entrapment (κ-carrageenan)	Chibata (1980); Wada *et al.* (1979, 1981)
	S. cerevisiae	Filtration	Margaritis and Wilke (1978)

(*Continued*)

TABLE XVI (*Continued*)

Use	Cell	Immobilization method	Reference
	S. cerevisiae	Sedimentation and cell recycle	Cysewski and Wilke (1977); Walsh and Bungay (1979)
	Saccharomyces carlbergensis	Adsorption (diatomaceous earth)	Grindbergs *et al.* (1977)
		Covalent binding (Schiff's base formation)	Navarro and Durand (1977)
	S. carlbergensis	Flocculation	Engelbert and Dellweg (1976)
	Saccharomyces uvarum	Entrapment (calcium alginate)	Cheetham (1979)
	Zymomonas mobilis	Entrapment (calcium alginate or κ-carrageenan)	Grote *et al.* (1980)
	Z. mobilis	Fiber entrapment (nitrate cellulose)	Margaritis and Rowe (1981)
	Z. mobilis	Sedimentation and cell recycle	Lee *et al.* (1980)
Production of beer	*S. carlbergensis*	Adsorption (wood)	Navarro (1975)
		Adsorption [poly(vinyl chloride)]	Navarro (1975); Corrieu *et al.* (1976)
		Adsorption (porous brick)	Navarro (1975); Corrieu *et al.* (1976)
Production of wine	*S. cerevisiae*	Entrapment (polyacrylamide)	Divies (1977)
Production of hydrogen			
	Alcaligenes eutrophus	Entrapment (κ-carrageenan)	Klibanov and Puglisi (1980)
	Anabaena cylindrica	Adsorption (glass)	Lambert *et al.* (1979)
	Clostridium butylicum	Entrapment (polyacrylamide)	Karube *et al.* (1976, 1977a,b,d); Suzuki *et al.* (1979); Matsunaga *et al.* (1980)

	Cl. butylicum	Entrapment (agar)	Suzuki *et al.* (1979); Matsunaga *et al.* (1980)
	Cl. butylicum	Entrapment (collagen)	Karube *et al.* (1977b); Matsunaga *et al.* (1980)
	Rhodospirillum rubrum	Entrapment (agar)	Weetall and Bennett (1976)
Production of methane	Methanogenic bacteria	Adsorption (asbestos)	Romanov-Skaya *et al.* (1981)
		Adsorption (silochrome)	Romanov-Skaya *et al.* (1981)
		Entrapment (polyacrylamide)	Karube *et al.* (1980)
		Entrapment (agar)	Karube *et al.* (1980)
Degradation reactions			
Degradation of phenol	*Candida tropicalis*	Entrapment (polyacrylamide)	Hackel *et al.* (1975)
	C. tropicalis	Entrapment (polymethacrylamide)	Klein *et al.* (1978b)
		Entrapment (polystyrene)	Hackel *et al.* (1975)
		Entrapment (aluminum alginate)	Hackel *et al.* (1975); Klein *et al.* (1979)
		Entrapment (styrene-maleic acid copolymer + Al^{3+})	Klein *et al.* (1978)
	Pseudomonas sp.	Adsorption (anthracite)	Scott and Hancher (1976)
	Waste-treatment bacteria	Adsorption (coal)	Holladay *et al.* (1978)
Denitrification			
	Denitrifying mixed bacteria	Adsorption (glass)	Mulcahy and La Motta (1978)
	Denitrifying mixed bacteria	Adsorption (sand)	Jeris and Owens (1975)
	Denitrifying mixed bacteria	Flocculation	Stephenson (1978); Ngian *et al.* (1977)
	Hyphomicrobium sp.	Entrapment (polyacrylamide)	Divies (1977)
	Pseudomonas aeruginosa	Adsorption (polypropylene)	Holló *et al.* (1979)
	Ps. aeruginosa	Adsorption [poly(vinyl chloride)]	Holló *et al.* (1979)
	Ps. aeruginosa	Entrapment (polyacrylamide)	Divies (1977)
	Pseudomonas denitrificans	Entrapment (polyacrylamide)	Nilsson *et al.* (1981)

(*Continued*)

TABLE XVI (*Continued*)

Use	Cell	Immobilization method	Reference
	Ps. denitrificans	Entrapment (calcium alginate)	Nilsson *et al.* (1981); Nilsson and Molin (1981)
	Ps. denitrificans	Entrapment (κ-carrageenan)	Nilsson *et al.* (1981)
	Waste-treatment bacteria	Adsorption (coal)	Scott and Hancher (1976); Hancher *et al.* (1979); Taylor (1978); Ngian and Martin (1980)
	Waste-treatment bacteria	Adsorption [poly(vinyl chloride)]	Holló *et al.* (1980)
Diacetyl degradation	Fleischman's yeast	Dialysis	Tolls *et al.* (1970)
	S. carlbergensis	Dialysis	Tolls *et al.* (1970)
Malic acid degradation	*Lactobacillus casei*	Entrapment (polyacrylamide)	Divies and Siess (1976); Divies (1977)
Urea degradation	*S. ureae*	Fiber entrapment (cellulose acetate)	Ghose and Kannan (1979)
	Serratia sp.	Fiber entrapment (α-cellulose)	Linko *et al.* (1978a, 1979)
Production of antibiotics			
6-Aminopenicillanic acid	*E. coli*	Entrapment (polymethacrylamide)	Klein and Wagner (1978, 1980)
	E. coli	Fiber entrapment (cellulose triacetate)	Marconi *et al.* (1975)
	E. coli	Entrapment (calcium alginate)	Klein and Wagner (1980)
Bacitracin	*Bacillus* sp.	Entrapment (polyacrylamide)	Morikawa *et al.* (1980); Suzuki and Karube (1979)
Penicillin	*E. coli*	Fiber entrapment (cellulose triacetate)	SNAM Progetti (1972)
	Penicillium chrysogenum	Entrapment (polyacrylamide)	Morikawa *et al.* (1979); Suzuki and Karube (1979)
	P. chrysogenum	Entrapment (calcium alginate)	Morikawa *et al.* (1979); Suzuki and Karube (1979)

	P. chrysogenum	Entrapment (collagen)	Morikawa *et al.* (1979); Suzuki and Karube (1979)
Streptomycin	*Streptomyces griseus*	Adsorption (stainless steel)	Atkinson *et al.* (1979)
Transformation of steroids			
Bioconversion of steroids	*Nocardia rhodocrous*	Entrapment (polyurethane)	Fukui *et al.* (1980); Omata *et al.* (1979a)
	N. rhodocrous	Entrapment (maleic polybutadiene)	Fukui *et al.* (1980); Omata *et al.* (1979a)
	N. rhodocrous	Entrapment [poly(ethylene glycol)]	Fukui *et al.* (1980); Omata *et al.* (1979a); Yamane *et al.* (1979)
		Entrapment [poly(propylene glycol)]	Fukui *et al.* (1980); Omata *et al.* (1979a,b); Yamane *et al.* (1979)
Biodegradation of steroids	*Actinomycetes* sp.	Entrapment (polyacrylamide or calcium alginate)	Germain *et al.* (1981)
Cortisol → prednisolone	*Arthrobacter simplex*	Entrapment (polyacrylamide)	Ohlson *et al.* (1978); Larsson *et al.* (1978, 1979)
	A. simplex	Entrapment (calcium alginate)	Ohlson *et al.* (1979); Larsson *et al.* (1979)
Hydrocortisone → prednisolone	*A. simplex*	Entrapment (maleic polybutadiene)	Omata *et al.* (1979b)
		Entrapment [Poly(ethylene glycol) or poly(propylene glycol)]	Fukui *et al.* (1978); Sonomoto *et al.* (1979)
	Corynebacterium simplex	Entrapment (collagen)	Constantinides (1980); Venkatasubramanian *et al.* (1978)
Reichstein compound S → cortisol	*Curvularia lunata*	Entrapment (calcium alginate)	Ohlson *et al.* (1981)
Reichstein compound S → 11-epicortisol	*Rhizopus nigricans*	Entrapment (calcium alginate)	Ohlson *et al.* (1981)

(Continued)

TABLE XVI (*Continued*)

Use	Cell	Immobilization method	Reference
Nitrogen fixation			
	Azotobacter chroococcum	Entrapment (agar)	Karube *et al.* (1981)
	Azotobacter vinelandii	Adsorption (ECTEOLA-cellulose)	Seyhan and Kirwan (1979); Gainer *et al.* (1981); DeNicola and Kirwan (1980)
	A. vinelandii	Adsorption (Bio Rex or AG); covalent binding (alkylation)	Gainer *et al.* (1981)
	Klebsiella pneumoniae	Entrapment (collagen)	Venkatasubramanian and Toda (1981)
Production of enzymes			
α-Amylase	*Bacillus subtilis*	Entrapment (polyacrylamide)	Kokubu *et al.* (1978); Suzuki and Karube (1979)
	B. subtilis	Adsorption (stainless steel)	Atkinson *et al.* (1979)
Lytic enzymes	*Streptomyces* sp.	Pelletization	Antczak *et al.* (1981)
Protease	*Clostridium histolyticum*	Ultrafiltration devices	Wang *et al.* (1970)
Fine chemicals production			
NADP	*B. ammoniagenes*	Entrapment (polyacrylamide)	Murata *et al.* (1979a)
Synthesis of CoA	*B. ammoniagenes*	Entrapment (polyacrylamide)	Shimizu *et al.* (1975, 1979); Yamada *et al.* (1980); Samejima *et al.* (1978)
		Entrapment [poly(vinyl alcohol)]	Shimizu *et al.* (1979)

Production of glutathione	*S. cerevisiae*	Entrapment (polyacrylamide)	Murata *et al.* (1978, 1981)
Production of glucose 6-phosphate	*Achromobacter butyri*	Entrapment (polyacrylamide)	Murata *et al.* (1979)
AMP → ATP	*S. cerevisiae*	Entrapment (polyacrylamide)	Samejima *et al.* (1978)
CMP → CDP	*S. cerevisiae*	Entrapment (polyacrylamide)	Samejima *et al.* (1978)
CMP + choline chloride → CDP-choline	*S. cerevisiae*	Fiber entrapment (ethylcellulose)	Samejima *et al.* (1978)
Synthesis of pyridoxal 5′-phosphate	*Ps. fluorescens*	Entrapment (polyacrylamide)	Yamada *et al.* (1980)
Synthesis of ajmalicine	*Catharanthus roseus*	Entrapment (calcium alginate)	Brodelius *et al.* (1979, 1981); Brodelius and Nilsson (1981)
	C. roseus	Entrapment (agar, agarose, or κ-carrageenan)	Brodelius and Nilsson (1981)
Production of patulin	*Penicillium urticae*	Entrapment (κ-carrageenan)	Deo *et al.* (1981)
Production of L-sorbose	*Acetobacter oxydans*	Entrapment (κ-carrageenan)	Wada *et al.* (1979)
Glucose → candicidin	*S. griseus*	Entrapment (collagen)	Vieth and Venkatasubramanian (1979)
Concentration of plutonium from waste-waters	*Ps. aeruginosa*	Entrapment (collagen)	Vieth and Venkatasubramanian (1979)
Cholesterol → Δ^4-cholesterone	*Mycobacterium rhodochrocous*	Entrapment (collagen)	Vieth and Venkatasubramanian (1979)
Glucose → cycloheximide	*S. griseus*	Dialysis	Kominek (1975a,b)

(*Continued*)

TABLE XVI (*Continued*)

Use	Cell	Immobilization method	Reference
Bioconversion of cardenolides	*Daucus carota*	Entrapment (calcium alginate)	Jones and Veliky (1981); Veliky and Jones (1981)
Bioconversion of glycosides and monoterpenoids	*Digitalis mentha*	Entrapment (calcium alginate)	Alfermann *et al.* (1981b)
Miscellaneous reactions			
Methane oxidation	Methane-oxidizing bacteria	Covalent binding (peptide binding); entrapment (agar)	Romanov-Skaya *et al.* (1981)
Methanol → formaldelyde	*Hansenula polymorpha*	Entrapment (polyacrylamide)	Couderc and Baratti (1980)
Ferrous iron oxidation	*Thiobacillus ferroxydans*	Entrapment (polyacrylamide)	Kutsal and Çaglar (1981)
Glycerol → dihydroxyacetone	*Gluconobacter oxydans*	Entrapment (polyacrylamide)	Somaer and Çaglar (1981)
Butanol production	*Clostridium acetobutylicum*	Entrapment (calcium alginate)	Häggström (1981); Häggström and Molin (1980)
	Clostridium butylicum	Entrapment (calcium alginate)	Krouwell *et al.* (1980)
De novo synthesis	*Morinda citrifolia*	Entrapment (calcium alginate)	Brodelius *et al.* (1979, 1981)
Digitoxin → digoxin	*Digitalis lanata*	Entrapment (calcium alginate)	Brodelius *et al.* (1979, 1981)
Digitoxin → purpurea glycoside A	*D. lanata*	Entrapment (calcium alginate)	Alfermann *et al.* (1981b)

TABLE XVII
ANALYTICAL APPLICATIONS OF IMMOBILIZED LIVING CELLS

Use	Microbial electrode	Immobilization method	Reference
Antibiotics analysis			
Nystatin	*Saccharomyces cerevisiae* + thermistor	Entrapment (polyacrylamide)	Mattiasson (1979)
Cephalosporins (7-phenylacetyl-ADCA; cephaloridine; cephalothin)	*Citrobacter freundii* + pH electrode	Entrapment (collagen)	Suzuki and Karube (1978a, 1979)
D-Glucose analysis			
	Pseudomonas fluorescens + oxygen electrode	Entrapment (collagen)	Suzuki and Karube (1979); Karube *et al.* (1979)
Assimilable sugars analysis	*Brevibacterium lactofermentis* + oxygen electrode	Retention within acetylcellulose membranes	Suzuki and Karube (1979); Hikuma *et al.* (1980)
Ethanol analysis	*Trichosporon* sp. + oxygen electrode	Retention within acetylcellulose membranes	Suzuki and Karube (1979); Hikuma *et al.* (1979a)
Acetic acid analysis	*Trichosporon brassicae* + oxygen electrode	Entrapment (agar)	Suzuki and Karube (1979); Hikuma *et al.* (1979a)
BOD Determination	*Clostridium butylicum* + fuel cell	Entrapment (collagen)	Matsunaga *et al.* (1980); Karube *et al.* (1977b)
	S. cerevisiae + thermistor	Entrapment (polyacrylamide)	Mattiasson *et al.* (1977)
	Mixed bacteria from activated sludge + oxygen electrode	Entrapment (collagen)	Karube *et al.* (1976); Suzuki and Karube (1978a,b)
	Trichosporon cutaneum + oxygen electrode	Retention within acetylcellulose membranes	Hikuma *et al.* (1979b)
	Hansenula anomala + oxygen electrode	Dialysis	Kulys and Kadziavskieni (1980)
Nicotinic acid analysis	*Lactobacillus arabinous* + pH electrode	Entrapment (collagen)	Suzuki and Karube (1978a)

TABLE XVIII
APPLICATIONS OF IMMOBILIZED LIVING ANIMAL CELLS

Use	Cell	Immobilization method	Reference
Virus production			
Poliomyelitis virus	Primary and finite life-span cell strains	Adsorption (DEAE-Sephadex)	A. L. Van Wezel (1967, 1972, 1973)
Foot-and-mouth disease virus	BHK 21 Cells	Adsorption (DEAE-Sephadex or glass)	Spier and Whiteside (1976)
Sindbis virus	CHO Cells	Adsorption (DEAE-Sephadex)	Levine *et al.* (1979)
	Chicken fibroblast	Adsorption (DEAE-Sephadex)	Levine *et al.* (1979)
Murine leukemia virus	NIH 3T3, Clone 1 (mouse fibroblast)	Adsorption (DEAE-Sephadex)	Levine *et al.* (1979)
	JLSV 9 Cells	Adsorption (DEAE-Sephadex)	Levine *et al.* (1979)
Interferon production	HEL 299 Cells (human fibroblast)	Adsorption (DEAE-Sephadex)	Levine *et al.* (1979); Giard *et al.* (1979)
	FS-4 Cells	Adsorption (DEAE-Sephadex)	Levine *et al.* (1979); Giard *et al.* (1979)
	Cell line 316 (skin cells)	Adsorption (DEAE-Sephadex)	Giard *et al.* (1979)
	Cell line 292 (skin cells)	Adsorption (DEAE-Sephadex)	Giard *et al.* (1979)

TABLE XIX
APPLICATIONS OF IMMOBILIZED LIVING PLANT CELLS

Use	Cell	Immobilization method	Reference
De novo synthesis	*Morinda citrifolia*	Entrapment (calcium alginate)	Brodelius *et al.* (1979, 1981)
Digitoxin → digoxin	*Digitalis lanata*	Entrapment (calcium alginate)	Brodelius *et al.* (1979, 1981)
Digitoxin → purpurea-glycoside A	*D. lanata*	Entrapment (calcium alginate)	Alfermann *et al.* (1981b)
Bioconversion of glycosides and monoterpenoids	*Digitalis mentha*	Entrapment (calcium alginate)	Alfermann *et al.* (1981b)
Bioconversion of cardenolides	*Daucus carota*	Entrapment (calcium alginate)	Jones and Veliky (1981); Veliky and Jones (1981)
Ajmalicine synthesis	*Catharanthus roseus*	Entrapment (calcium alginate)	Brodelius *et al.* (1979, 1981); Brodelius and Nilsson (1981)
	C. roseus	Entrapment (agar, agarose, or κ-carregeenan)	Brodelius and Nilsson (1981)

The main areas of application of immobilized living cells in chemical processes are in synthetic reactions for the production of bioenergy (ethanol, hydrogen, and methane) and the production of organic acids and amino acids, although other applications—such as waste treatment, production of antibiotics, and bioconversion of steroids—are important too. However, in these fields relatively very few applications are of industrial scale.

The first truly industrial immobilized-cell process appears to be the production of prednisolone from Reichstein compound S via cortisol, using *Curvularia lunata* and *Corynebacterium simplex* immobilized in polyacrylamide, which commenced operation in Sweden in 1978 (K. Mosbach, personal communication). In other processes, such as vinegar production and waste treatment, living cells have been used, immobilized by natural adhesion to the present support material.

In the analytical field, several immobilized living cells have been applied to production of electrochemical sensors, and these microbial electrode sensors have been developed for an efficient control of fermentation processes, by the determination of organic materials such as whole cells, nutrients, and products in the fermentation broth.

In the biomedical area, mammalian cells have been attached by adsorption to microcarriers for growth and production of vaccines and interferon.

VI. FUTURE TRENDS IN IMMOBILIZED-LIVING-CELL TECHNOLOGY

Although natural immobilized living cells have been used in fermentation processes for many years and the first artificially immobilized cell was presented in the early 1960s, it was only in the late 1970s that academic and industrial institutions began to make great strides toward the widespread use of immobilized-living cell technology.

To date only three processes based on immobilized cells are operative on an industrial scale:

1. Production of L-aspartic acid using immobilized dead cells of *Escherichia coli* with aspartate ammonia lyase (EC 4.3.1.1) (in Japan since 1973)
2. Production of L-malic acid using immobilized dead cells of *Brevibacterium ammoniagenes* with fumarate hydratase (EC 4.2.1.2) activity (in Japan since 1974)

3. Production of prednisolone using immobilized cells of *Curvularia lunata* and *Corynebacterium simplex* with 11β-hydroxylase and Δ^1-dehydrogenase activities, respectively (in Sweden since 1978).

This limited number of industrial applications, in contrast to the relatively large number of published reports, is due to several factors, mainly economic ones—carrier or reagent costs for the immobilization procedure are high. Other factors are linked with low efficiency of immobilization, poor operational stability, the toxic character of chemicals used in the immobilization of viable cells, relatively complicated equipment of continuous operation, low demand for the products being insufficient to justify large scaling up, and the well-established standard methods of the traditional fermentation industry.

In answer to these problems, the search must continue for new immobilization processes and for matrices that permit facile, secure immobilization with good interaction with substrate, and that conform in shape, size, and other parameters to the use for which they are intended. Opportunities for introduction of immobilized-cell applications into new industrial plants must be seized and the relevant processes incorporated at the planning stages.

The most active area of immobilized-living-cell applications will probably be in the production of materials now obtained by fermentation and in the synthesis of new useful compounds. Due to rising costs and the need for new energy sources, another area of development, where much research remains to be done, will be the application of immobilized living cells for energy production—namely converting sunlight into chemical energy and chemical energy into electrical energy (biochemical fuel cells). Other applications of immobilized living cells might be expected in organic synthesis and degradation.

Workers in this field are now directing their attention to industrial-scale applications of immobilized cells. No reference hitherto has been made to the other types of activities of cells—including recognition, immunological, hormonal, transport, and dictational activities.

Besides their usefulness in enzymology, immobilized cells have potential in the area of clinical chemistry—for example in diagnostic aids including cell-recognition and cell-behavior tests. It is conceivable that surfaces of nontissue inplants to the human body could have their autorejection properties reduced or annulled by prior attachment of appropriate cells. Use of various cell types in immobilized form will serve as models of the phenomena of cell activities and functions in real life. Using parasitic-receptor properties of some tissues, specialist cells might be attached to living creatures as a means of giving additional protection against invasion of disease, whereas production of multicell-type surfaces may be the first stages of development of artificial tissues. Synthetic immobilized cells could be the forerunners of artificial blood vessels. Immobilized cells in the form of stabilized cells may ultimately figure as therapeutic agents.

The immobilization of animal and plant cells is still in its infancy. This is at least partly because such cells are far more labile and sensitive than microbial cells, and most immobilization procedures hitherto used have been too drastic for the animal and plant cells to survive. However, new methods of immobilization are expected to be developed to allow the use of such cells in different processes, such as production of vaccines, interferon, hormones, and alkaloids.

Many other possibilities are potentials only, but will be developed ultimately, with the aid of genetic engineering and development of the stabilization of microbial cells, to give a whole variety of applications and aids to human life.

REFERENCES

Abbott, B. J. (1976). *Adv. Appl. Microbiol.* **20,** 203.

Abbott, B. J. (1977). *Annu. Rep. Ferment. Processes* **1,** 205.

Abbott, B. J. (1979). *Annu. Rep. Ferment. Processes* **2,** 91.

Abbott, B., and Gerhardt, P. (1970a). *Biotechnol. Bioeng.* **12,** 577.

Abbott, B., and Gerhardt, P. (1970b). *Biotechnol. Bioeng.* **12,** 591.

Abbott, B., and Gerhardt, P. (1970c). *Biotechnol. Bioeng.* **12,** 603.

Abramson, H. A., Moyer, L. S., and Goren, M. H. (1942). "Electrophoresis of Proteins and the Chemistry of Cell Surfaces," Van Nostrand-Reinhold, Princeton, New Jersey.

Alfermann, A. W., Schuller, I., and Reinhard, E. (1981a). *Adv. Biotechnol.* [*Proc. Int. Ferment. Symp.*], *6th, 1980* Abstracts, F-12.1.15, p. 123.

Alfermann, A. W., Aviv, D., Dantes, A., Figur, C., Galun, E., Krochmal, E., Reinhard, E., and Schuller, I. (1981b). *Abstr. Commun., Eur. Congr. Biotechnol., 2nd, 1981* p. 195.

Amri, M. A., Bonaly, R., Duteurtre, B., and Moll, M. (1979a). *Eur. J. Appl. Microbiol. Biotechnol.* **7,** 227.

Amri, M. A., Bonaly, R., Duteurtre, B., and Moll, M. (1979b). *Eur. J. Appl. Microbiol. Biotechnol.* **7,** 241.

Antczak, T., Bielecki, S., and Galar, E. (1981). *Abstr. Commun., Eur. Congr. Biotechnol., 2nd, 1981* p. 155.

Atkinson, B., and Fowler, H. W. (1974). *Adv. Biochem. Eng.* **3,** 221.

Atkinson, B., Black, G. M., Lewis, P. J. S., and Pinches, A. (1979). *Biotechnol. Bioeng.* **21,** 193.

Azerad, R., Calderon-Seguin, R., and Decottingnies-Le Maréchal, P. (1980). *Bull. Soc. Chim. Fr.* **1-2,** II-83.

Bang, W. G., Lang, S., Sahm, H., and Wagner, F. (1978). *Prepr., Eur. Congr. Biotechnol., 1st, 1978* Part 1, p. 2/186.

Barker, D., and Kinsop, B. H. (1972). *J. Inst. Brew.* **78,** 454.

Birnbaum, S., Larsson, P. O., and Mosbach, K. (1981). *Abstr. Commun., Eur. Congr. Biotechnol., 2nd, 1981* p. 150.

Briffaud, J., and Engasser, J. M. (1978). *Prepr., Eur. Congr. Biotechnol., 1st, 1978* Part 2, p. 2/122.

Briffaud, J., and Engasser, J. M. (1979). *Biotechnol. Bioeng.* **21,** 2093.

Brodelius, P., and Nilsson, K. (1981). *Abstr. Commun., Eur. Congr. Biotechnol., 2nd, 1981* p. 193.

Brodelius, P., Deus, B., Mosbach, K., and Zenk, M. H. (1979). *FEBS Lett.* **103,** 93.

Brodelius, P., Hagerdal, B., and Mosbach, K. (1980). *Enzyme Eng.* **5,** 383.

Brodelius, P., Nilsson, K., and Mosbach, K. (1981). *Adv. Biotechnol.* [*Proc. Int. Ferment. Symp.*], *6th, 1980* Abstracts, F-12.1.19, p. 124.

Broun, G. (1976). *In* "Methods in Enzymology" (K. Mosbach, ed.), Vol. 44, p. 263. Academic Press, New York.

Browstein, A. H., Vieth, W. R., and Constantinides, A. (1974). *Symp. Adv. Immobilized Enzyme Syst., Am. Chem. Soc. Meet.*

Burkholder, P. R., and Sinnot, E. W. (1945). *Am. J. Bot.* **32,** 424.

Chang, T. M. S. (1964). *Science* **146,** 524.

Cheetham, P. S. J. (1979). *Enzyme Microb. Technol.* **1,** 183.

Cheetham, P. S. J. (1980). *Top. Enzyme Ferment. Biotechnol.* **4,** 189.

Chibata, I. (1977). *Adv. Appl. Microbiol.* **22,** 1.

Chibata, I. (1979). *ACS Symp. Ser.* **106,** 187.

Chibata, I. (1980). *Enzyme Eng.* **5,** 393.

Chibata, I., Tosa, T., and Sato, T. (1974). *Appl. Microbiol.* **27,** 878.

Chibata, I., Tosa, T., Sato, T., Yamamoto, K., Takata, I., and Nishida, Y. (1978). *Enzyme Eng.* **4,** 335.

Chipley, J. R. (1974). *Microbios* **10,** 115.

Clark, D., and Lentz, C. (1963). *Biotechnol. Bioeng.* **5,** 193.

Compere, A. L., and Griffith, W. (1976). *Dev. Ind. Microbiol.* **17,** 247.

Constantinides, A. (1980). *Biotechnol. Bioeng.* **22,** 119.

Corrieu, G., Blachere, A., Ramirez, A., Navarro, J. M., Durand, G., Duteurtre, B., and Moll, N. (1976). *Proc. Int. Ferment. Symp., 5th, 1976* p. 294.

Couderc, R., and Baratti, J. (1980). *Biotechnol. Bioeng.* **22,** 1155.

Coulman, G. A., Stieber, R. W., and Gerhardt, P. (1977). *Appl. Environ. Microbiol.* **39,** 725.

Cysewski, G. R., and Wilke, C. R. (1977). *Biotechnol. Bioeng.* **19,** 1125.

Daniels, S. L. (1971). *Dev. Ind. Microbiol.* **13,** 211.

Daniels, S. L., and Kempe, L. L. (1966). *Chem. Eng. Prog., Symp. Ser.* **62,** 142.

Darby, R., and Mandels, G. R. (1954). *Mycologia* **46,** 276.

Decottignies-Le Maréchal, P., Calderón-Seguin, R., Vandecastele, J. P., and Azerad, R. (1979). *Eur. J. Appl. Microbiol. Biotechnol.* **7,** 33.

De Nicola, K., and Kirwan, D. J. (1980). *Biotechnol. Bioeng.* **22,** 1283.

Deo, Y. M., and Gaucher, G. M. (1983). *Adv. Biotechnol.* [*Proc. Int. Ferment. Symp., 6th, 1980* Abstracts, F-12.1.9, p. 122 (1981)].

De Rosa, M., Gambacorta, A., Lama, L., and Nicolaus, B. (1981). *Biotechnol. Lett.* **3,** 183.

Dinelli, D. (1972). *Process Biochem.* **7,** 9.

Divies, C. (1977). French Patent 844,766.

Divies, C., and Siess, M. (1976). *Ann. Microbiol.* (*Paris*) **127B,** 525.

D'Souza, S. F., and Nadkanni, G. B. (1980). *Biotechnol. Bioeng.* **22,** 2179.
Durand, G., and Navarro, J. M. (1978). *Process Biochem.* **13,** 14.
Emery, A. N., Hough, J. S., Novais, J. M., and Lyons, T. P. (1972). *Chem. Eng.* (*London*) **258,** 79.
Engasser, J. M., and Horvath, C. (1973). *J. Theor. Biol.* **42,** 437.
Engelbart, W., and Dellweg, H. (1976). *Proc. Int. Ferment. Symp., 5th, 1976.*
Fetzer, W. R. (1930). *Food Ind.* **2,** 130.
Foster, J. W. (1949). "Chemical Activities of Fungi." Academic Press, New York.
Franks, N. E. (1971). *Biochim. Biophys. Acta* **252,** 246.
Franks, N. E. (1972). *Biotechnol. Bioeng. Symp.* **3,** 327.
Friedman, M. R., and Gaden, E. L. (1970). *Biotechnol. Bioeng.* **12,** 961.
Fukui, S., Tanaka, A., and Gellf, G. (1978). *Enzyme Eng.* **4,** 299.
Fukui, S., Omata, T., Yamane, T., and Tanaka, A. (1980). *Enzyme Eng.* **5,** 347.
Gaddy, J. L., and Sitton, O. C. (1978). *Prepr., Eur. Congr. Biotechnol., 1st, 1978* Part 1, p. 2/202.
Gainer, J. L., Kirwan, D. J., Foster, J. A., and Seylan, E. (1981). *Biotechnol. Bioeng. Symp.* **10,** 35.
Galbraith, J. C., and Smith, J. E. (1969). *Trans. Br. Mycol. Soc.* **52,** 237.
Germain, P., Miclo, R., Glomon, C., Engasser, J. M., and Nguyen, Q. T. (1981). *Abstr. Commun., Eur. Congr. Biotechnol., 2nd, 1981* p. 245.
Gerson, D. F., and Zajic, J. E. (1979). *ACS Symp. Ser.* **106,** 29.
Ghose, T. K., and Bandyopadhyay, K. K. (1980). *Biotechnol. Bioeng.* **22,** 1489.
Ghose, T. K., and Kannan, V. (1979). *Enzyme Microb. Technol.* **1,** 47.
Giard, D. J., Loeb, D. H., Thilly, W. G., Wang, D. I. C., and Levine, D. W. (1979). *Biotechnol. Bioeng.* **21,** 433.
Goldstein, L. (1976). *In* "Methods in Enzymology" (K. Mosbach, ed.), Vol. 44, p. 397. Academic Press, New York.
Goldstein, L., Levin, Y., and Katchalski, E. (1969). *Biochemistry* **3,** 1913.
Griffith, W. L., and Compere, A. L. (1976). *Dev. Ind. Microbiol.* **17,** 241.
Grindbergs, M., Hildebrand, R., and Clarke, B. (1977). *J. Inst. Brew.* **83,** 25.
Grinnell, F. (1978). *Int. Rev. Cytol.* **53,** 65.
Grote, W., Lee, K. J., and Rogers, P. L. (1980). *Biotechnol. Lett.* **2,** 481.
Gulaya, V. E., Turková, J., Jirkú, V., Frydrychova, A., Coupek, J., and Anachenko, S. N. (1979). *Eur. J. Appl. Microbiol. Biotechnol.* **8,** 43.
Hackel, V., Klein, J., Megnet, R., and Wagner, F. (1975). *Eur. J. Appl. Microbiol.* **1,** 291.
Häggström, L. (1981). *Adv. Biotechnol.* [*Proc. Int. Ferment. Symp.*], *6th, 1980* Abstracts, F-9.2.14, p. 80.
Häggström, L., and Molin, M. (1980). *Biotechnol. Lett.* **2,** 241.
Hancher, C. W., Taylor, P. A., and Napier, J. M. (1979). *Biotechnol. Bioeng. Symp.* **8,** 361.
Hardt, F. W., Young, J. C., Clesceri, L. S., and Washington, D. R. (1971). *Environ. Sci. Technol.* **5,** 345.
Hattori, R. (1972). *J. Gen. Appl. Microbiol.* **18,** 319.
Hattori, T. (1973). "Microbiol Life in the Soil." Dekker, New York.
Hattori, T., and Furusaka, C. (1960). *Biochemistry* (Tokyo) **48,** 331.
Hattori, T., and Furusaka, C. (1961). *Biochemistry* (Tokyo) **50,** 312.
Hayashi, T., Tanaka, Y., and Kawashima, K. (1979). *Biotechnol. Bioeng.* **21,** 1019.
Heinrich, M., and Rehm, H. J. (1981a). *Eur. J. Appl. Microbiol. Biotechnol.* **11,** 139.
Heinrich, M., and Rehm, H. J. (1981b). *Abstr. Commun., Eur. Congr. Biotechnol., 2nd, 1981* p. 277.
Heukelekian, A., and Heller, A. (1940). *J. Bacteriol.* **40,** 547.

Heukelekian, H., and Dondero, N. C. (1964). "Principles and Applications in Aquatic Microbiology." Wiley, New York.

Hikuma, M., Kubo, T., Yasuda, T., Karube, I., and Suzuki, S. (1979a). *Anal. Chim. Acta* **109,** 33.

Hikuma, M., Suzuki, H., Yasuda, T., Karube, I., and Suzuki, S. (1979b). *Eur. J. Appl. Microbiol. Biotechnol.* **8,** 289.

Hikuma, M., Obana, H., Yasuda, T., Karube, I., and Suzuki, S. (1980). *Enzyme Microb. Technol.* **2,** 234.

Hirano, K., Karube, I., and Suzuki, S. (1977). *Biotechnol. Bioeng.* **19,** 311.

Holladay, D. W. *et al.* (1978). *Chem. Eng. Prog., Symp. Ser.* **74,** 241.

Holló, J., Tóth, J., Tengerdy, R. P., and Johnson, J. E. (1979). *ACS Symp. Ser.* **106,** 73.

Holló, J., Weinbrenner, Zs., Czakó, L., and Tóth, J. (1980). *Biotechnol. Lett.* **2,** 87.

Holst, O., Enfors, S. O., and Matthiasson, B. (1981). *Abstr. Commun., Eur. Congr. Biotechnol., 2nd, 1981* p. 156.

Horisberger, M. (1976). *Biotechnol. Bioeng.* **18,** 1647.

Hughes, R. C. (1968). *Biochem. J.* **106,** 49.

Hughes, R. C., and Thurman, P. F. (1970). *Biochem. J.* **119,** 925.

Jack, T. R., and Zajic, J. E. (1977a). *Adv. Biochem. Eng.* **5,** 125.

Jack, T. R., and Zajic, J. E. (1977b). *Biotechnol. Bioeng.* **19,** 631.

Jeris, J. S., and Owens, R. W. (1975). *J. Water Pollut. Control Fed.* **47,** 2043.

Jirkú, V., Turková, J., Kuchynková, P., and Krumphanzl, V. (1979). *Eur. J. Appl. Microbiol. Biotechnol.* **6,** 217.

Jirkú, V., Turková, J., and Krumphanzl, V. (1980). *Biotechnol. Lett.* **2,** 509.

Jones, A., and Veliky, I. A. (1981). *Adv. Biotechnol.* [*Proc. Int. Ferment. Symp.*], *6th, 1980* Abstracts, F-12.1.16, p. 123.

Kan, J. K., and Shuler, M. L. (1978). *Biotechnol. Bioeng.* **20,** 217.

Karube, I., Matsunaga, T., Tsuru, S., and Suzuki, S. (1976). *Biochim. Biophys. Acta* **444,** 338.

Karube, I., Matsunaga, T., and Suzuki, S. (1977a), *J. Solid-Phase Biochem.* **2,** 97.

Karube, I., Mitsuda, S., Matsunaga, T., and Suzuki, S. (1977b). *J. Ferment. Technol.* **55,** 243.

Karube, I., Matsunaga, T., Tsuru, S., and Suzuki, S. (1977c). *Biotechnol. Bioeng.* **19,** 1233.

Karube, I., Matsunaga, T., Mitsuda, S., and Suzuki, S. (1977d). *Biotechnol. Bioeng.* **19,** 1535.

Karube, I., Mitsuda, S., and Suzuki, S. (1979). *Eur. J. Appl. Microbiol. Biotechnol.* **7,** 343.

Karube, I., Kuriyama, S., Matsunaga, T., and Suzuki, S. (1980). *Biotechnol. Bioeng.* **22,** 847.

Karube, I., Matsunaga, T., Otomine, Y., and Suzuki, S. (1981). *Enzyme Microb. Technol.* **3,** 309.

Kato, E. (1974). *Meiji Daigaku Kagaku Gijutsu Kenkyushao Nempo* **16,** 26.

Kawabata, Y., and Demain, A. L. (1979). *ACS Symp. Ser.* **106,** 133.

Kawabata, Y., and Demain, A. L. (1980). *Enzyme Eng.* **5,** 389.

Kennedy, J. F. (1979). *ACS Symp. Ser.* **106,** 119.

Kennedy, J. F. (1980). *Enzyme Microb. Technol.* **2,** 164.

Kennedy, J. F., and Cabral, J. M. S. (1983). *In* "Solid Phase Biochemistry: Analytical and Synthetic Aspects" (W. H. Scouten, ed.). Wiley, New York. in press.

Kennedy, J. F., and Kay, I. M. (1976). *J. Chem. Soc., Perkin Trans. 1* p. 329.

Kennedy, J. F., Barker, S. A., and White, C. A. (1977). *Carbohydr. Res.* **54,** 1.

Kennedy, J. F., and Pike, V. W. (1979). *Enzyme Microb. Technol.* **1,** 31.

Kennedy, J. F., and Pike, V. W. (1980). *Enzyme Microb. Technol.* **2,** 28.

Kennedy, J. F., Barker, S. A., and Humphreys, J. D. (1976a). *J. Chem. Soc., Perkin Trans. 1* p. 962.

Kennedy, J. F., Barker, S. A., and Humphreys, J. D. (1976b). *Nature (London)* **261,** 242.

Kennedy, J. F., Humphreys, J. D., Barker, S. A., and Greenshields, R. N. (1980). *Enzyme Microb. Technol.* **2,** 209.

Kennedy, J. F., Humphreys, J. D., and Barker, S. A. (1981). *Enzyme Microb. Technol.* **3,** 129.

Kierstan, M., and Bucke, C. (1977). *Biotechnol. Bioeng.* **19,** 387.

Kinzel, V., Kubler, D., Richards, J., and Stohr, M. (1976). *Science* **192,** 487.

Klein, J., and Wagner, F. (1978). *Pap., Eur. Congr. Biotechnol. Conf., 1st, 1978* p. 142.

Klein, J., and Wagner, F. (1980). *Enzyme Eng.* **5,** 335.

Klein, J., Hackel, U., Schara, P., Washansen, P., Wagner, F., and Martin, C. K. A. (1978). *Enzyme Eng.* **4,** 339.

Klein, J., Hackel, U., and Wagner, F. (1979). *ACS Symp. Ser.* **106,** 101.

Klibanov, A. M., and Puglisi, A. V. (1980). *Biotechnol. Lett.* **2,** 445.

Kobayashi, H., and Suzuki, H. (1972). *J. Ferment. Technol.* **50,** 625.

Kokubu, F., Karube, I., and Suzuki, S. (1978). *Eur. J. Appl. Microbiol. Biotechnol.* **5,** 233.

Kolarik, M. J., Chen, B. J., Emery, A. H., and Lim, H. C. (1974). *In* "Immobilized Enzymes in Food and Microbial Processes" (A. Olson and C. Cooney, eds.), p. 71. Plenum, New York.

Kolot, F. B. (1980). *Process Biochem.* **10,** 2.

Korminek, K. A. (1975a). *Antimicrob. Agents Chemother.* **7,** 856.

Korminek, K. A. (1975b). *Antimicrob. Agents Chemother.* **7,** 861.

Krouwell, P. G., Van der Laan, W. F. M., and Kossan, N. W. F. (1980). *Biotechnol. Lett.* **2,** 253.

Kulys, J., and Kadziavskieni, K. (1980). *Biotechnol. Bioeng.* **22,** 221.

Kumakura, M., Yoshida, M., and Kaetsu, I. (1978). *Eur. J. Appl. Microbiol. Biotechnol.* **6,** 13.

Kumakura, M., Yoshida, M., and Kaetsu, I. (1979). *Biotechnol. Bioeng.* **21,** 679.

Kutsal, T., and Caglar, M. A. (1981). *Adv. Biotechnol.* [*Proc. Int. Ferment. Symp.*], *6th, 1980* Abstracts, F-12.1.11, p. 122.

Lambert, G. R., Daday, A., and Smith, G. D. (1979). *FEBS Lett.* **101,** 125.

Lamptey, J., Moo-Young, M., and Robinson, C. W. (1981). *Adv. Biotechnol.* [*Proc. Int. Ferment Symp.*], *6th, 1980* Abstracts, F-12.2.9, p. 125.

Larreta-Garde, V., Thomasset, B., and Barbotin, J. N. (1981). *Enzyme Microb. Technol.* **3,** 216.

Larsen, D., and Dimmick, R. L. (1964). *J. Bacteriol.* **88,** 1380.

Larsson, P. O., and Mosbach, K. (1979). *Biotechnol. Lett.* **1,** 502.

Larsson, P. O., Ohlson, S., and Mosbach, K. (1978). *Enzyme Eng.* **4,** 317.

Larsson, P. O., Ohlson, S., and Mosbach, K. (1979). *Appl. Biochem. Bioeng.* **2,** 291.

Lartigue, D. J., and Weetall, H. H. (1976). U.S. Patent 3,939,041.

Lee, C. K., and Long, M. E. (1974). U.S. Patent 29,130.

Lee, K. J., Lefebree, H., Tribe, D. E., and Rogers, P. L. (1980). *Biotechnol. Lett.* **2,** 487.

Levine, D. W., Wang, D. I. C., and Thilly, W. G. (1979). *Biotechnol. Bioeng.* **21,** 821.

Lilly, M. D., and Dunnill, P. (1972). *Enzyme Eng.* **1,** 221.

Linko, P. (1981). *Adv. Biotechnol. Proc. Int. Ferment. Symp.*], *6th, 1980* Abstracts, F-12.2.7, p. 128.

Linko, Y.-Y., and Linko, P. (1981). *Abstr. Commun. Eur. Congr. Biotechnol., 2nd, 1980* p. 56.

Linko, Y. Y., Viskari, R., Pohjola, L., and Linko, P. (1977). *J. Solid-Phase Biochem.* **2,** 203.

Linko, Y. Y., Poutannen, K., Viskari, R., Weckström, L., and Linko, P. (1978). *Prepr., Eur. Congr. Biotechnol., 1st, 1978* Part 1, p. 2/194.

Linko, Y. Y., Poutannen, K., Weckström, L., and Linko, P. (1979). *Enzyme Microb. Technol.* **1,** 26.

Long, M. E. (1976). U.S. Patent 3,935,069.

Lyons, T. P., and Hough, J. S. (1971). *J. Inst. Brew.* **77,** 300.

Machlis, L. (1957). *Am. J. Bot.* **44,** 113.

Malouf, E., and Prater, J. (1961). *J. Metals* **13,** 353.

Marcipar, A., Cochet, N., Brackenbridge, L., and Lebeault, J. M. (1978). *Prepr., Eur. Congr. Biotechnol., 1st, 1978* Part 1, p 2/178.

Marcipar, A., Cochet, N., Brackenbridge, L., and Lebeault, J. M. (1979). *Biotechnol. Lett.* **1,** 65.

Marconi, W., Bartoli, F., Cecere, F., Galli, G., and Morisi, F. (1975). *Agric. Biol. Chem.* **39,** 277.

Margalith, P., and Holcberg, I. (1981). *Abstr. Commun., Eur. Congr. Biotechnol., 2nd, 1981* p. 160.

Margaritis, A. (1974). Doctoral Dissertation, University of California, Berkeley.

Margaritis, A., and Rowe, G. E. (1981). *Adv. Biotechnol.* [*Proc. Int. Ferment. Symp.*], *6th, 1980* Abstracts, F-12.1.13, p. 130.

Margaritis, A., and Wilke, C. R. (1978). *Biotechnol. Bioeng.* **20,** 727.

Marshall, K. C. (1976). "Interfaces in Microbiol. Ecology." Harvard Univ. Press, Cambridge, Massachusetts.

Martin, C. K., and Perlman, D. (1976). *Eur. J. Appl. Microbiol. Biotechnol.* **3,** 91.

Martin, P. (1978). M.Sc. Thesis, University of Western Ontario.

Mason, J. R., Pirt, S. J., and Sommerville, H. J. (1978). *Enzyme Eng.* **4,** 343.

Masschelein, C. A., Ramos, C. J., Castian, C., and Devreux, A. (1963). *J. Inst. Brew.* **69,** 332.

Mattiasson, B. (1979). *ACS Symp. Ser.* **106,** 203.

Mattiasson, B., and Borrebaeck, C. (1978). *FEBS Lett.* **85,** 119.

Mattiasson, B., Larsson, P.-O., and Mosbach, K. (1977). *Nature* **268,** 519.

Matsunaga, T., Karube, I., and Suzuki, S. (1980). *Biotechnol. Bioeng.* **22,** 2607.

Mavituna, F., and Sinclair, C. G. (1978). *Prepr., Eur. Congr. Biotechnol., 1st, 1978* Part 1, p 2/182.

Messing, R., ed. (1975). "Immobilized Enzymes for Industrial Reactors." Academic Press, New York.

Messing, R., and Oppermann, R. A. (1979). *Biotechnol. Bioeng.* **21,** 49.

Messing, R., Oppermann, R. A., and Kolot, F. (1979). *ACS Symp. Ser.* **106,** 13.

Metz, B., and Kossen, N. W. F. (1977). *Biotechnol. Bioeng.* **19,** 781.

Mohan, R. R., and Li, N. N. (1974). *Biotechnol. Bioeng.* **16,** 513.

Mohan, R. R., and Li, N. N. (1975). *Biotechnol. Bioeng.* **17,** 1137.

Monsan, P. (1978). *Eur. J. Appl. Microbiol. Biotechnol.* **5,** 1.

Moo-Young, M., Hirose, T., and Geiger, K. H. (1969). *Biotechnol. Bioeng.* **11,** 725.

Moo-Young, M., Lamptey, J., and Robinson, C. W. (1980). *Biotechnol. Lett.* **2,** 541.

Moo-Young, M., Lamptey, J., and Robinson, C. W. (1981). *Adv. Biotechnol.* [*Proc. Int. Ferment. Symp.*], *6th, 1980* Abstracts, F-9.2.5., p. 78.

Mori, A., Suzue, H., Osuga, J., and Wada, M. (1981). *Adv. Biotechnol.* [*Proc. Int. Ferment. Symp.*], *6th, 1980* Abstracts, F-12.1.10, p. 122.

Morikawa, Y., Karube, I., and Suzuki, S. (1979). *Biotechnol. Bioeng.* **21,** 261.

Morikawa, Y., Karube, I., and Suzuki, S. (1980). *Biotechnol. Bioeng.* **22,** 1015.

Mosbach, K. (1981). *Abstr. Commun., Eur. Congr. Biotechnol., 2nd, 1981* p. 48.

Mosbach, K., and Mosbach, R. (1966). *Acta Chem. Scand.* **20,** 2807.

Muir, L., and Berry, V. (1976). *Hydrometallurgy* **2,** 11.
Mulcahy, L., and La Motta, E. J. (1978). Report No. Env. E. 59-78-2. Department of Civil Engineering, University of Massachusetts, Amherst & Boston.
Munton, T., and Russell, A. (1973). *Appl. Microbiol.* **26,** 508.
Murata, K., Tani, K., Kato, J., and Chibata, I. (1978). *Eur. J. Appl. Microbiol. Biotechnol.* **6,** 23.
Murata, K., Kato, J., and Chibata, I. (1979a). *Biotechnol. Bioeng.* **21,** 887.
Murata, K., Uchida, T., Tani, I., Kato, J., and Chibata, I. (1979b). *Eur. J. Appl. Microbiol. Biotechnol.* **7,** 45.
Murata, K., Tani, K., Kato, J., and Chibata, I. (1981). *Eur. J. Appl. Microbiol. Biotechnol.* **11,** 72.
Navarro, J. M. (1975). Thesis Doct. Ing., University of Toulouse, France.
Navarro, J. M. (1978). S M I Collogue, Toulouse, France.
Navarro, J. M., and Durand, G. (1977). *Eur. J. Appl. Microbiol. Biotechnol.* **4,** 243.
Nelson, R. P. (1976). U.S. Patent 3,957,580.
Ngian, K. F., and Martin, W. B. (1980). *Biotechnol. Bioeng.* **22,** 1007.
Ngian, K. F., Lin, S. H., and Martin, W. B. (1977). *Biotechnol. Bioeng.* **19,** 1773.
Nilsson, I., and Molin, N. (1981). *Abstr. Commun., Eur. Congr. Biotechnol., 2nd, 1981* p. 159.
Nilsson, I., Häggström, L., and Molin, N. (1981). *Adv. Biotechnol.* [*Proc. Int. Ferment. Symp.*], *6th, 1980* Abstracts, F-12.1.18, p. 124.
Nilsson, K., and Mosbach, K. (1981). *Abstr. Commun., Eur. Congr. Biotechnol., 2nd, 1981* p. 199.
Nishimaru, H., Izumi, C., Narita, S., and Yamata, K. (1975). Japanese Patent 14068.
Novais, J. M. (1971). Ph.D. Thesis, University of Birmingham.
Ohlson, S., Larsson, P. O., and Mosbach, K. (1978). *Biotechnol. Bioeng.* **20,** 1267.
Ohlson, S., Larsson, P. O., and Mosbach, K. (1979). *Eur. J. Appl. Microbiol. Biotechnol.* **7,** 103.
Ohlson, S., Larsson, P. O., and Mosbach, K. (1981). *Adv. Biotechnol.* [*Proc. Int. Ferment. Symp.*], *6th, 1980* Abstracts, F-12.1.5., p. 121.
Omata, T., Tanaka, A., Yamane, T., and Fukui, S. (1979a). *Eur. J. Appl. Microbiol. Biotechnol.* **6,** 207.
Omata, T., Lida, T., Tanaka, A., and Fukui, S. (1979b). *Eur. J. Appl. Microbiol. Biotechnol.* **8,** 143.
Pasteur, L. (1869). "Memoires sur la fermentation acétique." Paris.
Patel, G. B., and Ingledew, W. H. (1975). *Can. J. Microbiol.* **21,** 1608.
Perez, F. (1976). M.Sc. Thesis, University of Manchester Institute of Science and Technology.
Petre, D., Noel, C., and Thomas, D. (1978). *Biotechnol. Bioeng.* **20,** 127.
Pirt, S. J., and Callow, D. (1959). *Nature (London)* **184,** 307.
Poulsen, P., and Zittan, L. (1976). *In* "Methods in Enzymology" (K. Mosbach, ed.), Vol. 44, p. 809. Academic Press, New York.
Prescott, S. C., and Dunn, C. G. (1959). "Industrial Microbiology." McGraw-Hill, New York.
Rehm, H. J. (1971). "Einführung in die Industrielle Mikrobiologie." Springer-Verlag, Berlin and New York.
Richards, F. M., and Knowles, J. R. (1968). *J. Mol. Biol.* **37,** 231.
Romanov-Skaya, V. A., Karpenko, V. I., Pantskhava, E. S., Grinberg, T. A., and Malashenko, Y. R. (1981). *Adv. Biotechnol.* [*Proc. Int. Ferment. Symp.*], *6th, 1980* Abstracts, F-12.1.14, p. 123.
Rouxhet, P. G., Van Haecht, J. L., Didelez, J., Gerard, P., and Biquet, M. (1981). *Enzyme Microb. Technol.* **3,** 49.

Russell, H. L. (1891). *Zeitschr. F. Hyg. Infektionekranh.* **11,** 65.

Rutishauser, V., and Sachs, L. (1975). *J. Cell Biol.* **65,** 247.

Samejima, H., Kimura, K., Ado, Y., Suzuki, Y., and Tadokoro, T. (1978). *Enzyme Eng.* **4,** 237.

Scheurich, P., Schnabl, H., Zimmermann, U., and Klein, J. (1980). *Biochim. Biophys. Acta* **598,** 645.

Schultz, J. S., and Gerhardt, P. (1969). *Bacteriol. Rev.* **33,** 1.

Scott, C. D., and Hancher, C. W. (1976). *Biotechnol. Bioeng.* **18,** 1393.

Seyhan, E., and Kirwan, D. J. (1979). *Biotechnol. Bioeng.* **21,** 271.

Shimizu, S., Morioka, H., Tani, Y., and Ogata, K. (1975). *J. Ferment. Technol.* **53,** 77.

Shimizu, S., Tani, Y., and Yamada, H. (1979). *ACS Symp. Ser.* **106,** 87.

Sitton, O. C., and Gaddy, J. L. (1980). *Biotechnol. Bioeng.* **22,** 1735.

Sitton, O. C., Magruder, G. C., and Gaddy, J. L. (1981a). *Adv. Biotechnol.* [*Proc. Int. Ferment. Symp.*], *6th, 1980* Abstracts, F-12.2.4, p. 127.

Sitton, O. C., Magruder, G. C., Book, N. L., and Gaddy, J. L. (1981b). *Biotechnol. Bioeng. Symp.* **10,** 213.

Slowinski, W., and Charm, S. (1973). *Biotechnol. Bioeng.* **15,** 973.

SNAM Progetti (1972). Belgian Patent 782,646.

Sommerville, H. J., Mason, J. R., and Ruffell, R. N. (1977). *Eur. J. Appl. Microbiol. Biotechnol.* **4,** 75.

Sonaer, A., and Caglar, M. A. (1981). *Abstr. Commun., Eur. Congr. Biotechnol., 2nd, 1981* p. 157.

Sonomoto, K., Tanaka, A., Omata, T., Yamane, T., and Fukui, S. (1979). *Eur. J. Appl. Microbiol. Biotechnol.* **6,** 325.

Sortland, L. D., and Wilke, C. R. (1969). *Biotechnol. Bioeng.* **11,** 305.

Spier, R. E., and Whiteside, J. P. (1976). *Biotechnol. Bioeng.* **18,** 659.

Steel, R., Martin, S. M., and Leutz, C. P. (1954). *Can. J. Microbiol.* **1,** 150.

Stephenson, J. P. (1978). M.E. Project, McMaster University, Westdale, Ontario.

Stewart, G. G., Russel, I., and Garrisson, I. F. (1975). *J. Inst. Brew.* **81,** 248.

Stieber, R. W., Coulman, G. A., and Gerhardt, P. (1977). *Appl. Environ. Microbiol.* **34,** 733.

Stottmeister, V. Z. (1980). *Allg. Mikrobiol.* **10,** 763.

Suzuki, S., and Karube, I. (1978a). *Prepr., Eur. Congr. Biotechnol., 1st, 1978* p. 31.

Suzuki, S., and Karube, I. (1978b). *Enzyme Eng.* **4,** 329.

Suzuki, S., and Karube, I. (1979). *ACS Symp. Ser.* **106,** 59.

Suzuki, S., Karube, I., and Matsunaga, T. (1979). *Biotechnol. Bioeng. Symp.* **8,** 501.

Takahashi, J., Abekawa, Y., and Yamada, K. (1960). *J. Agric. Chem. Soc. Jpn.* **34,** 1043.

Takata, I., Tosa, T., and Chibata, I. (1978). *J. Solid-Phase Biochem.* **2,** 225.

Takata, I., Yamamoto, K., Tosa, T., and Chibata, I. (1979). *Eur. J. Appl. Microbiol. Biotechnol.* **7,** 161.

Takata, I., Yamamoto, K., Tosa, T., and Chibata, I. (1980). *Enzyme Microb. Technol.* **2,** 30.

Tanaka, A., Jin., T. N., Kawamoto, S., and Fukui, S. (1979). *Eur. J. Appl. Microbiol. Biotechnol.* **7,** 351.

Taylor, N. W., and Orton, W. L. (1973). *J. Inst. Brew.* **79,** 249.

Taylor, P. A. (1978). Report Contract W-7405. U.S. Department of Energy, U.S. Government, Washington, D.C.

Thiele, H. (1954). *Kolloid-Z.* **136,** 80.

Thiele, H., and Awad, A. (1969). *J. Biomed. Mater. Res.* **3,** 431.

Toda, K. (1975). *Biotechnol. Bioeng.* **17,** 1729.

Toda, K., and Shoda, H. (1975). *Biotechnol. Bioeng.* **17,** 481.

Tolls, T., Shovers, J., Sandine, W., and Elliker, P. (1970). *Appl. Microbiol.* **19,** 649.

Tosa, T., Sato, T., Mosi, T., Yamamoto, K., Takata, I., Nishida, Y., and Chibata, I. (1979). *Biotechnol. Bioeng.* **21,** 1697.

Tramper, J., Van der Plaas, H. E., Van der Kaaden, A., Müller, F., and Middlehoven, W. J. (1979). *Biotechnol. Lett.* **1,** 397.

Uchida, T., Watanabe, T., Kato, J., and Chibata, I. (1978). *Biotechnol. Bioeng.* **20,** 255.

Updike, S. M., Harris, D. R., and Shrago, E. (1969). *Nature (London)* **224,** 1122.

Vandamme, E. J. (1976). *Chem. Ind. (London)* p. 1070.

Van Hemert, D., Kilbunn, D. G., and Van Wezel, A. L. (1969). *Biotechnol. Bioeng.* **11,** 875.

Van Oss, C. J., Gillman, G. F., and Newman, A. W. (1975). "Phagocytic Engulfment and Cell Adhesiveness." Dekker, New York.

Van Wezel, A. L. (1967). *Nature (London)* **216,** 64.

Van Wezel, A. L. (1972). *Prog. Immunobiol. Stand.* **5,** 187.

Van Wezel, A. L. (1973). *In* "Tissue Culture Methods and Applications" (M. K. Patterson and P. Kruuse, Jr., eds.), p. 372. Academic Press, New York.

Veliky, J. A., and Jones, A. (1981). *Adv. Biotechnol.* [*Proc. Int. Ferment. Symp.*], *6th, 1980* Abstracts, F-12.1.17, p. 124.

Venkatasubramanian, K., ed. (1979). *ACS Symp. Ser.* **106.**

Venkatasubramanian, K., and Toda, Y. (1981). *Biotechnol. Bioeng. Symp.* **10,** 237.

Venkatasubramanian, K., Constantinides, A., and Vieth, W. R. (1978). *Enzyme Eng.* **3,** 29.

Vieth, W. R., and Venkatasubramanian, K. (1978). *Enzyme Eng.* **4,** 307.

Vieth, W. R., and Venkatasubramanian, K. (1979). *ACS Symp. Ser.* **106,** 1.

Vorlop, K., and Klein, J. (1981). *Biotechnol. Lett.* **3,** 9.

Wada, M., Kato, J., and Chibata, I. (1979). *Eur. J. Appl. Microbiol. Biotechnol.* **8,** 241.

Wada, M., Uchida, T., Kato, J., and Chibata, I. (1980). *Biotechnol. Bioeng.* **22,** 1175.

Wada, M., Kato, J., and Chibata, I. (1981). *Eur. J. Appl. Microbiol. Biotechnol.* **11,** 67.

Walsh, T. J., and Bungay, H. R. (1979). *Biotechnol. Bioeng.* **21,** 1081.

Wang, D., Sinsky, A., and Butterworth, T. (1970). *In* "Membrane Science and Technology" (J. Flynn, ed.). Plenum, New York.

Warth, A., and Strominger, J. (1968). *Bacteriol. Proc.* p. 64.

Webster, I. A., Shuler, M. L., and Rony, P. R. (1979). *Biotechnol. Bioeng.* **21,** 1725.

Weetall, H. H. (1976). *In* Methods in Enzymology" (K. Mosbach, ed.), Vol. 44, p. 134. Academic Press, New York.

Weetall, H. H., and Bennett, M. A. (1976). *Proc. Int. Ferment. Symp., 5th, 1976* p. 299.

Whipple, G. C. (1901). *Tech. Q.* **14,** 21.

Yagi, S., Toda, Y., and Minoda, T. (1976). *Annu. Meet. Agric. Chem. Soc. Jpn.* p. 414.

Yamada, H., Okamura, S., Kojima, H., Okamoto, Y., and Ito, Y. (1976). Japanese Patent 76/144779.

Yamada, H., Shimizu, S., Shimada, H., Tani, Y., Takahashi, S., and Ohashi, T. (1980). *Biochimie* **62,** 395.

Yamamoto, K., Tosa, T., Yamashita, K., and Chibata, I. (1977). *Biotechnol. Bioeng.* **19,** 1101.

Yamamoto, K., Tosa, T., and Chibata, I. (1980). *Biotechnol. Bioeng.* **22,** 2045.

Yamane, T., Nakatani, H., Sada, E., Omata, T., Tanaka, A., and Fukui, S. (1979). *Biotechnol. Bioeng.* **21,** 2133.

Yang, H. S., and Studebaker, J. F. (1978). *Biotechnol. Bioeng.* **20,** 17.

Zabriskie, D., Ollis, D. F., and Burger, M. M. (1973). *Biotechnol. Bioeng.* **15,** 981.

ZoBell, C. E. (1943). *J. Bacteriol.* **46,** 39.

Zvyagintsev, D. G. (1971). *Mikrobiologiya* **31,** 339.

Energy Production with Immobilized Cells

Shuichi Suzuki and Isao Karube

Research Laboratory of Resources Utilization
Tokyo Institute of Technology
Yokohama, Japan

I. INTRODUCTION

Hydrogen is now attracting attention as a clean fuel source. Because various bacteria and algae produce hydrogen under anaerobic conditions, they may be suitable sources of commercial hydrogen. The biophotolysis of water by microorganisms also is a hydrogen-producing system. It is well known that photosynthetic bacteria, green algae, and blue-green algae evolve hydrogen in the presence of light. In the case of photosynthetic bacteria, an organic acid such as malate is required as an electron donor for hydrogen production. Therefore, the

APPLIED BIOCHEMISTRY AND BIOENGINEERING
Volume 4

ISBN 0-12-041104-0

photosynthetic bacteria do not appear to be suitable for commercial hydrogen production. However, the blue-green algae are attracting attention as hydrogen-photoproduction systems. By coupling the photosynthetic system in plant chloroplasts with a hydrogenase, one could accomplish the light-driven splitting of water into hydrogen and oxygen (Benemann *et al.*, 1973).

Methane is also an attractive energy source, and it can be produced by various microorganisms from waste biomass. However, the enzymes involved in the biogas-producing systems just described, such as hydrogenase and nitrogenase, are very unstable, and therefore it is difficult to use whole cells for continuous biogas production. Biochemical energy-conversion systems in principle are attractive as new energy-production methods; various approaches are now under investigation. The high-energy, electron-rich substances, such as carbohydrates, lipids, and proteins, are not usually electroactive in a fuel cell, but the intermediates produced during biological oxidation may often be active at an electrode. The following are some of the suggested bioanode reactions and bacteria being considered (Lewis, 1966): carbohydrate → ethyl alcohol (*Saccharomyces sp.*, *Pseudomonas lindneri*); carbohydrate → hydrogen (*Clostridium butyricum*); sulfate → hydrogen sulfide (*Desulfovibrio desulfuricans*). Hydrogen exhibits excellent reactivity in these electroactive materials.

In this chapter, the immobilization of hydrogen- and methane-producing microorganisms in synthetic or natural polymers is described, and gas production with these immobilized cells is discussed. Subsequent application of the hydrogen to the hydrogen–oxygen fuel cell system is also reviewed.

II. BIOGAS PRODUCTION BY IMMOBILIZED CELLS

A. Hydrogen Production from Biomass by Immobilized *Clostridium butyricum*

It is well known that hydrogen is produced from the fermentation of glucose by *Clostridia*. This conversion of carbohydrate to hydrogen is achieved by a multienzyme system. In bacteria the route is believed to involve glucose conversion to 2 mol of pyruvate, with 2 mol of ATP and 2 mol of NADH formed by the Embden–Meyerhof pathway (Thauer *et al.*, 1972; Raeburn and Rabinowitz, 1971a,b). The pyruvate may be oxidized through a pyruvate–ferredoxin oxidoreductase or through a pyruvate–formate lyase. The products of the ferredoxin oxidoreductase reaction are acetyl-CoA, CO_2, and reduced ferredoxin.

Pyruvate formate lyase decomposes pyruvate to acetyl-CoA and formate, with the formate then oxidized to CO_2. Furthermore, NADH–ferredoxin oxidoreductase also oxidizes NADH and reduces ferredoxin. The reduced ferredoxin is reoxidized to form hydrogen by the hydrogenase. As a result, 4 mol of hydrogen are produced from 1 mol of glucose under ideal conditions. However, as described already, the hydrogenase system in bacteria is very unstable. The immobilization of hydrogen-producing bacteria has great value because this stabilizes the hydrogenase system. With the immobilized bacteria the multienzyme system and cofactors such as NAD, and ATP can be used for production of hydrogen (Karube *et al.*, 1976).

1. Hydrogen Evolution by Various Bacteria

Table I shows the hydrogen-evolving activity of several types of whole cells. *Cl. butyricum* IFO 3847 evolved the largest amount of hydrogen and was used in the subsequent studies.

2. Immobilization of *Cl. butyricum* in Polyacrylamide Gel

The optimum concentration of acrylamide for hydrogen evolution from glucose by whole cells was examined. Preliminary experiments

TABLE I
HYDROGEN-EVOLVING ACTIVITY OF WHOLE CELLS[a]

Bacteria	H_2 Evolved (μmol)
Escherichia coli (IFO 12173)	46
Clostridium butyricum	
(IFO 3847)	63
(IAM 19002)	52
(IAM 19003)	60
Clostridium acetobutylicum	
(IAM 190011)	30
(IAM 190012)	18
Clostridium perfringens	
(A)	4
(B)	5
(C)	18

[a] Medium, 2 ml (0.05 *M* glucose); cells, 0.1 g wet weight per ml, 37°C, 18 h, under anaerobic conditions.

showed that 10% *N,N'*-methylenebisacrylamide was a suitable concentration of cross-linking agent for hydrogen evolution with the immobilized whole cells. The optimum concentration of acrylamide was approximately 5%. However, this gel preparation was too soft to use. Thus, a 10% gel (90% acrylamide, 10% *N,N'*-methylenebisacrylamide) was employed in the experiments.

3. Optimum Conditions for Hydrogen Production

Whole cells, whether free or immobilized, also produce organic acids. The acids produced by immobilized whole cells (glucose fermentation) were identified as formic, acetic, and butyric by gas chromatography. Lactic acid was analyzed using lactate dehydrogenase. No significant difference in the organic acids formed was observed between immobilized cells and native cells. The pH of the reaction mixture fell gradually during incubation of the cells. The effect of the initial pH on hydrogen evolution by whole cells was examined. No difference in the optimum pH for hydrogen evolution was observed between immobilized whole cells and native cells. The native cells did not evolve hydrogen below pH 5, whereas this limitation was not observed with the immobilized cells. The optimum temperature for hydrogen evolution with both the native and immobilized cells was 37°C.

4. Continuous Hydrogen Production from Glucose

Figure 1 shows the effect of oxygen on the hydrogen evolution of *Cl. butyricum*. Native cells and immobilized cells were incubated in air for 24 h at 37°C, and the hydrogen evolved was measured. The native cells were then centrifuged, and both native cells and immobilized cells were resuspended in fresh media and incubated again for 24 h at 37°C. As shown in Fig. 1, no hydrogen was observed with the native cells after the first 24-h incubation, whereas the immobilized cells continuously evolved hydrogen under aerobic conditions (saturated dissolved oxygen at 37°C). Furthermore, the amount of hydrogen evolved by the immobilized cells under aerobic conditions was almost the same as the amount under anaerobic conditions. The hydrogen-producing system of the immobilized whole cells was thus protected from the deleterious effects of oxygen that were observed with the native cells, although the rate of hydrogen evolution of the immobilized cells decreased with extended incubation periods. However, the rate of hydrogen evolution by the immobilized cells increased again after the cells were resuspended in fresh medium. The organic acids produced by glucose fermentation may inhibit hydrogen evolution by the immobilized whole cells, but no inactivation occurred

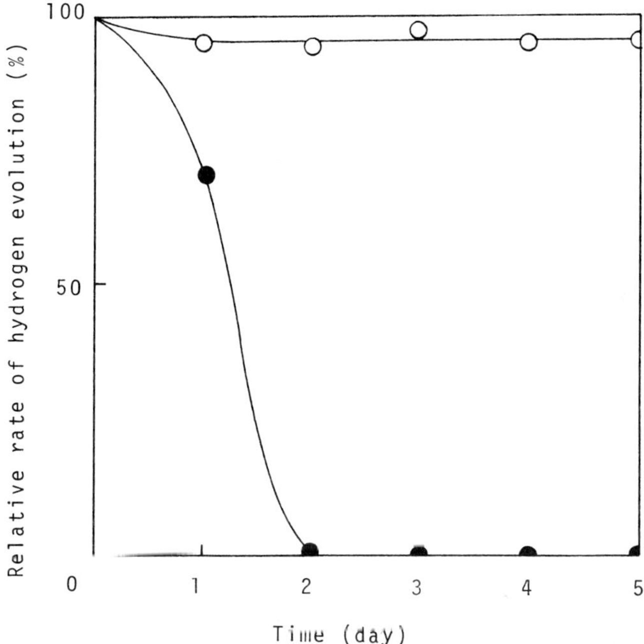

Fig. 1. Effect of oxygen on hydrogen evolution of *Cl. butyricum*. Native cells (0.2 g wet weight) and immobilized whole cells (0.2 g wet weight) in 0.1 *M* phosphate buffer (pH 7.7) containing 0.25 *M* glucose were incubated in air at 37°C for 24 h. Native cells (●——●); immobilized cells (○——○).

during the incubation. The results showed that the polyacrylamide network also stabilized the hydrogen-producing system.

Continuous hydrogen production by immobilized whole cells was carried out in a batch system (Fig. 2). Native whole cells and immobilized whole cells were incubated under anaerobic conditions for 24 h at 37°C. The hydrogen evolved was determined by gas chromatography. The native whole cells were centrifuged under anaerobic conditions, and both the native cells and immobilized cells were resuspended in similar media and incubated again for 24 h. Only traces of hydrogen were evolved by the native cells after the first or second 24-h incubation, whereas the immobilized whole cells continued to evolve hydrogen over a 20-day period.

5. Hydrogen Production by Cells Entrapped in Acetylcellulose Filters

Cl. butyricum cells were immobilized in acetylcellulose filters with agar (Karube *et al.*, 1981a,b). Agar (0.8 g) was dissolved in physiological saline (36 ml) at 100°C and cooled to 50°C, then *Cl. butyricum* (2 g

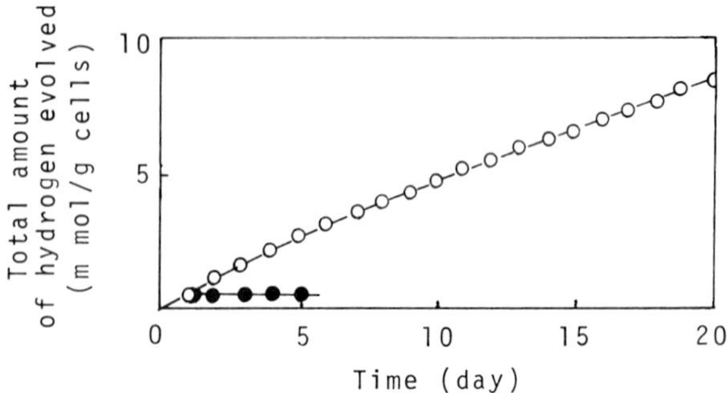

Fig. 2. Continuous hydrogen production from wastewater by immobilized whole cells. Native (●——●) and immobilized (○——○) whole cells were incubated under anaerobic conditions. Cells (0.2 g wet weight) were incubated in phosphate buffer (0.1 *M*, pH 7.7) containing 0.25 *M* glucose at 37°C.

wet cells in 1 ml) and acetylcellulose filters (5 g) were added and the mixture cooled to 30°C. The acetylcellulose filter-entrapped whole cells were washed with physiological saline and used for hydrogen production.

The effects of the cell and agar concentrations on the hydrogen-producing activity of the immobilized whole cells were examined. The optimum cell concentration in the agar solution was 20%. Further increase or decrease of the cell concentration resulted in diminished activity. The agar solution containing 20% whole cells was therefore employed for the subsequent work. The optimum agar concentration for hydrogen production was between 1.5 and 2.5%. A 2.0% agar solution exhibited good mechanical rigidity and fairly good hydrogen-evolution activity, and was thus employed in most of the experiments.

The optimum pH range for hydrogen evolution was 6–7. No difference in the optimum pH was observed between immobilized whole cells and native cells. As mentioned earlier, only the native cells did not evolve hydrogen below pH 5. The cells were still intact after being entrapped in the agar–acetylcellulose filters.

6. Continuous Hydrogen Production from Wastewaters

Continuous hydrogen production by immobilized whole cells was carried out in a batch system. A 5-liter jar fermenter was employed for continuous hydrogen production. About 2 kg of immobilized whole cells (wet) and 3 liters of wastewater from an alcohol factory (BOD 70,000 ppm) were employed for these experiments. The wastewater was continuously replaced with water of BOD 70,000 ppm at a flow rate of 10 ml/min. Hydrogen was evolved over a 30-day period at a rate of 40 ml/min. Initially, the amount of hydrogen produced increased with increasing incubation time. Because molasses was used as a raw material for fermentation at the alcohol factory, the wastewater contained carbohydrates and other nutrients. Therefore, activation of the hydrogen-production system may have been caused by the growth of the bacteria in the media. The hydrogen-producing bacteria in agar–acetylcellulose filters were living and maintained their hydrogen-producing activity for a long time. As mentioned before, under ideal conditions 4 mol of hydrogen is produced from 1 mol of glucose using *Clostridia*. Of the total glucose consumed, about 30% was converted to hydrogen under optimum conditions. Mutational improvement of the hydrogen-producing bacteria could be considered for the future.

B. Photoproduction of Hydrogen by Immobilized Blue-Green Algae

A heterocystous blue green algae, *Anabaena* spp., can evolve hydrogen from water when irradiated by light. Sustained hydrogen production for a week using nitrogen-starved cultures of *Anabaena cylindrica* has been reported (Benemann and Weare, 1974); however, the evolution of hydrogen was strongly inhibited by gaseous nitrogen. This inhibition was due to hydrogen production and nitrogen fixation competing for the same processes in the algae. This system suffered further from problems of filament breakage and structural degeneration of the algae. Recently, Asada *et al.* (1979) isolated an *Anabaena* sp. that was capable of evolving hydrogen in air and where the production of hydrogen was not inhibited so much by gaseous nitrogen. Therefore, these algae were attractive for use in light-induced hydrogen production. When the *Anabaena* sp. was immobilized in agar gel, it evolved hydrogen in the presence of air (Kayano *et al.*, 1981a).

1. Immobilization of *Anabaena* spp.

Whole cells of blue-green algae were immobilized in agar. Table II shows the rate of hydrogen production by immobilized and free algae under various conditions. The hydrogen productivity of the immobilized algae was three times that of the free algae. Therefore, the immobilization of *Anabaena* spp. was effective for hydrogen production.

The maximum activity for hydrogen production was observed at an agar concentration of 2% (w/v), and increasing or decreasing the agar concentration lowered the activity. For further experiments, a 2% gel

TABLE II
HYDROGEN PRODUCTION BY IMMOBILIZED ALGAE AND FREE ALGAE UNDER VARIOUS CONDITIONS[a]

Variable reaction conditions	Relative rate of H_2 production (%)
Free cells	
shaken	100
stirred	40
static	137
Immobilized cells	
stirred	382

[a] Constant reaction conditions: pH 8, 30°C, 3000 lux.

was used to immobilize the algae. The optimum algae content (dry cells in the gel) was 3.3 mg/g. The activity decreased at higher or lower microbial concentrations.

2. Optimum Conditions for Hydrogen Production by Immobilized Blue-Green Algae

Hydrogen production by immobilized cells was performed in various media. The rate of hydrogen production by immobilized algae in the modified BG-11 medium (pH 8) without a nitrogen source was higher than in either 0.1 *M* phosphate buffer (pH 8) or distilled water. Carbon dioxide or sodium carbonate was tested as the carbon source. A mixed gas of 5% carbon dioxide and 95% argon was suitable for hydrogen production. However, it was difficult to keep the carbon dioxide concentration in the reactor constant. Therefore, a liquid or solid carbon source was also tested. The maximum hydrogen productivity was observed in the medium containing 10 m*M* sodium carbonate. Therefore, 10 m*M* sodium carbonate was employed in the remaining experiments. The hydrogen-production rate also increased with increasing light intensity up to 10,000 lux.

3. Continuous Hydrogen Production by Immobilized *Anabaena* spp.

The time course of hydrogen production is shown in Fig. 3. The gel was incubated in BG-11 medium at pH 8 and 10 m*M* sodium carbonate. The immobilized algae were preincubated for 40 h before hydrogen production. Hydrogen was continuously evolved at a rate of

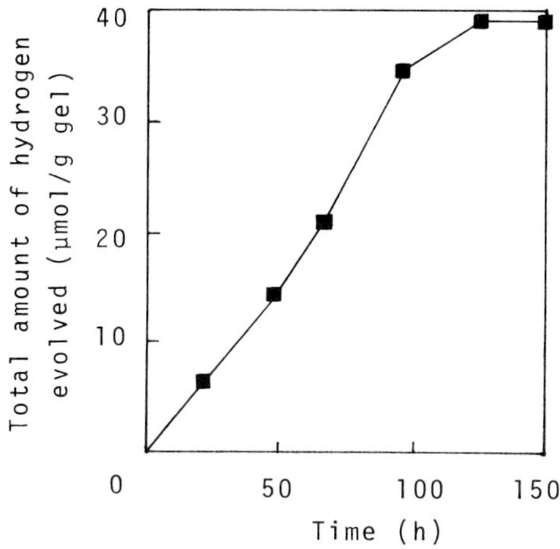

Fig. 3. Continuous hydrogen production by immobilized *Anabaena* sp. Immobilized whole cells (1 g wet agar gel) of *Anabaena* sp. N-7363 (3.3 mg dry cells, 50 μg of chlorophyll *a*) were incubated in 9 ml of incubation medium without nitrogen sources. The medium (pH 8) contained 10 m*M* $NaHCO_3$. Reaction was carried out at 30°C under illumination of 3000 lux.

0.16–0.52 μmol/h/g for 7 days. The immobilized *Anabaena* sp. also produced oxygen, which was removed by a reactor containing *Bacillus subtilis*. The immobilized algae did not settle out or form clumps. Moreover, the algae were protected from filament breakage and structural degeneration, which sometimes occurs as a result of agitation. Consequently, the continuous production of hydrogen was possible using these immobilized algae.

C. Photoproduction of Hydrogen by Immobilized Green Algae–*Clostridium butyricum* System

Green algae are known to reduce NADP under anaerobic conditions, and immobilized *Cl. butyricum* cells evolve hydrogen from NADPH. As described already, the hydrogenase in the immobilized living cells was protected from the deleterious effects of oxygen and was not inactivated. Therefore, a hydrogen-evolution system, using the immobilized *Chlorella vulgaris, Cl. butyricum,* and NADP (Fig. 4), was possible. NADP was reduced by photosystems I and II in algae under light irradiation, and hydrogen was evolved through oxidation of NADPH.

In this section, the photolysis of water is described using the immobilized *Ch. vulgaris*–NADP–*Cl. butyricum* system.

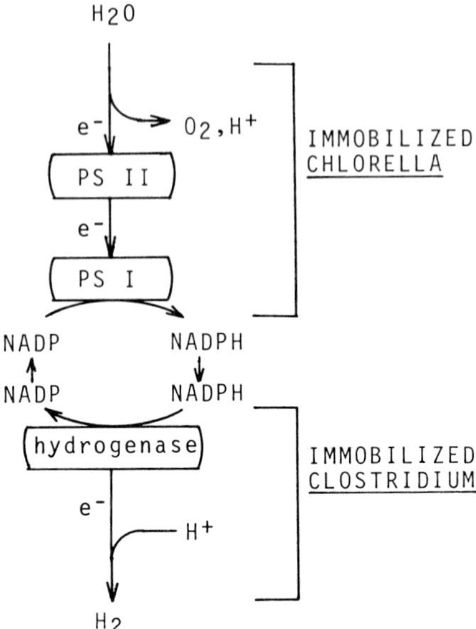

Fig. 4. Principle of a hydrogen-evolution system using immobilized *Chlorella vulgaris*, *Clostridium butyricum*, and NADP. PS, Photosystem; NADP, nicotinamide adenine dinucleotide phosphate.

1. NADP Reduction by Immobilized *Ch. vulgaris*

The rate of NADP reduction by immobilized *Ch. vulgaris*, under illumination when preincubated in the dark, was 4.3 μmol/h/mg Chl. The immobilized *Ch. vulgaris* reduced NADP under light irradiation, and the immobilized *Cl. butyricum* evolved hydrogen from NADPH. Consequently, a light-driven splitting of water into hydrogen was possible by coupling the immobilized *Ch. vulgaris* and *Cl. butyricum*.

As described already, an electron carrier was needed for hydrogen evolution using immobilized *Ch. vulgaris* and *Cl. butyricum*. NADP, NAD, and FAD were employed as electron carriers. The largest amount of hydrogen was evolved when NADP was used. Preliminary experiments showed that immobilized *Cl. butyricum* could evolve hydrogen from NADH and NADPH. However, the rate of NAD reduction by immobilized *Ch. vulgaris* was slower than that of NADP reduction. Therefore, the rate of hydrogen evolution was slow when NAD was used as the electron carrier. In contrast, immobilized *Ch. vulgaris* could not reduce FAD. Therefore, the immobilized *Ch. vulgaris*–FAD–*Cl. butyricum* system evolved only trace amounts of hydrogen, similar to the system without electron carriers. The rate of hydrogen evolution increased with increasing NADP concentration.

2. Optimum Reaction Conditions

The optimum temperature for hydrogen evolution from NADPH by immobilized *Cl. butyricum* was 37°C, and that for NADP reduction by immobilized *Ch. vulgaris* was 30°C. The rate of hydrogen evolution increased with increasing temperature up to 37°C. The rate of hydrogen evolution decreased below pH 7 and above pH 8. Therefore, further experiments were performed at 37°C and pH 7. The rate of hydrogen evolution increased at higher $NaHCO_3$ concentrations, with 10 m*M* as the optimum. A high concentration of $NaHCO_3$ accelerated the NADP reduction by immobilized *Ch. vulgaris*. The rate of hydrogen evolution increased at higher cell concentrations of *Cl. butyricum*, with 200 mg wet cells/g wet gel of *Cl. butyricum* used for immobilization in agar gel. The rate of hydrogen evolution also increased at higher cell concentrations of *Ch. vulgaris* below 125 mg/g wet gel and became constant above 250 mg/g wet gel.

3. Continuous Hydrogen Evolution by the System

Figure 5 shows the time course of hydrogen evolution by the immobilized *Ch. vulgaris* and immobilized *Cl. butyricum* systems with and without NADP. The complete system with NADP continuously evolved hydrogen at 0.29–1.34 μmol/h/mg Chl for 6 days. In contrast, the system without NADP evolved only a trace amount of hydrogen. The immobilized *Ch. vulgaris* also evolved only a trace amount of hydrogen, and no hydrogen was produced by the immobilized *Cl. butyricum* because the reaction mixture did not contain any organic compounds. Most of the hydrogen was produced by splitting water with light irradiation. The rate of water-splitting hydrogen evolution increased with increasing NADP concentration, and a high concentration of NADP was needed for the system. Therefore, the hydrogen-evolution rate may be limited by the diffusion of NADPH to the hydrogenase system in *Cl. butyricum*.

D. Methane Production by Immobilized Methanogenic Bacteria

The biological formation of methane is the result of a specific type of bacterial energy-yielding metabolism. Bacterial methanogenesis is a ubiquitous process in most anaerobic environments. The association of this event with anaerobic decomposition of organic matter in microbial habitats, such as sewage sludge digesters and the rumen and intestinal tracts of animals, has been recognized and documented for more than a century (Zeikus, 1977).

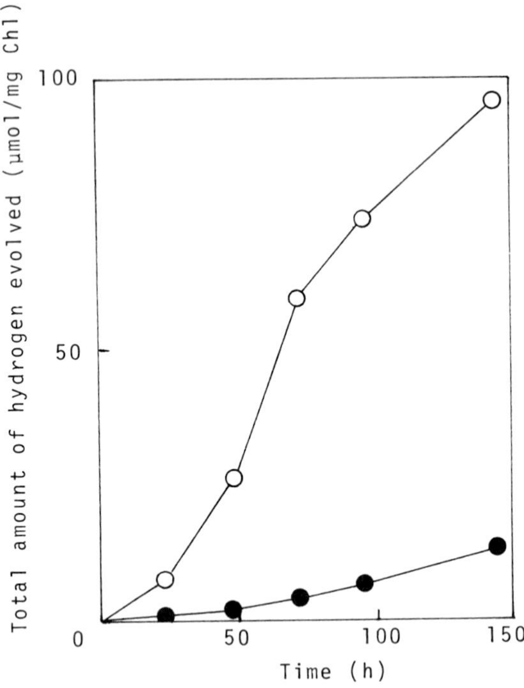

Fig. 5. Continuous hydrogen evolution by the immobilized *Chlorella vulgaris–Clostridium butyricum*. Coimmobilized *Ch. vulgaris* whole cells (250 μg Chl) and *Cl. butyricum* whole cells (10 mg wet cells) were incubated anaerobically in buffer solution with 800 μM of NADP (○——○) or without NADP (●——●) at pH 8 and 37°C under the illumination (3000 lux).

The methanogenic microbial population seems to consist of methanogenic bacteria, hydrogen-producing bacteria, and other microorganisms. Moreover, the methanogenic bacteria in the microbial population produce methane from acetic acid, formic acid, and hydrogen, or carbon dioxide and hydrogen. Consequently, methane was most likely produced according to the scheme described in Equations (1)–(4) by the microbial population employed in the present study. Initially, carbohydrates, lipids, and proteins in the wastewater are hydrolyzed by cellulolytic, lipolytic, and proteolytic bacteria. The products of these hydrolyses, such as sugars, fatty acids, and amino acids, are converted to hydrogen, carbon dioxide, formic acid, and acetic acid by hydrogen-producing bacteria. Finally, anaerobic methanogenic bacteria evolve methane from acetic acid, formic acid, and hydrogen, or carbon dioxide and hydrogen. However, the exact mechanism by which any of these substrates is converted into methane needs to be elucidated. The reactions are summarized as follows:

$$CO_2 + 4H_2 \rightarrow CH_4 + 2H_2O \quad (1)$$

$$4HCOOH \rightarrow CH_4 + 3CO_2 + 2H_2O \quad (2)$$

$$CH_3COOH \rightarrow CH_4 + CO_2 \quad (3)$$

$$CH_3COOH + 4H_2 \rightarrow 2CH_4 + 2H_2O \quad (4)$$

The authors have attempted to develop a continuous methane-production system from wastewater by a methanogenic microbial population. However, the methane-producing system in methanogenic bacteria is unstable under aerobic conditions. Still, methanogenic microbial cells were immobilized in agar gel, and continuous methane production from wastewaters using the immobilized microbial population was carried out in a batch system (Karube *et al.*, 1980a).

1. Immobilization of Methanogenic Bacteria

The methanogenic microbial population was immobilized in agar gel, polyacrylamide gel, and collagen membranes, with the agar gel preparation giving the highest activity. The methane-producing activity of the polyacrylamide gel- and collagen membrane-immobilized microbial populations were very small at the initial time. The maximum activity was observed at an agar concentration between 1.5 and 3% (w/v). At concentrations greater than 3%, changes in the agar gel concentration decreased the activity. The optimum microbial content was 20 mg wet cells/g gel. The activity decreased at higher and lower microbial contents.

2. Optimum Conditions for Methane Production

The rates of methane production by native and immobilized bacteria were examined using both washed and collected native and immobilized bacteria. The maximum methane production by native cells occurred during the first 2 h, whereas a significant but lower rate of methane production by the immobilized microorganism population continued for up to 10 h. This difference may be caused by substrate limitation and methane diffusion through the agar gel matrices. However, the methane productivity of the immobilized microbial population increased linearly with time and attained a higher total output than that of native cells. This phenomenon was also observed in the case of hydrogen production by immobilized whole cells. The optimum pH of the native cells ranged from 6.5 to 7.0. The immobilized

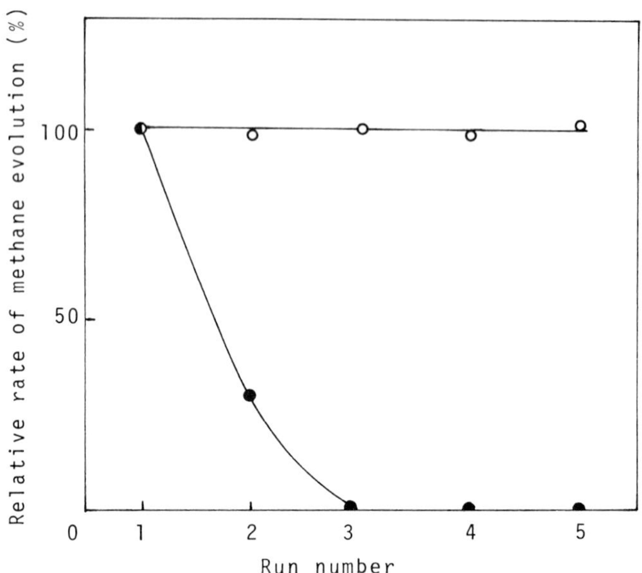

Fig. 6. Effect of oxygen on methanogenic activity of immobilized bacteria. Immobilized whole cells (0.02 g wet weight) in 1 g wet gel (○——○) or free (0.02 g) native cells (●——●) were incubated in an alcohol factory wastewater (5 ml, 1000 ppm) at 37°C in air for 6 h in each run. Runs carried out in sequence of 1 through 5.

bacteria were stable in alkaline conditions, but the native and immobilized whole cells did not produce methane below pH 5.

The optimum temperature of the native cells was 45°C, whereas the immobilized cells showed high methanogenic activities between 37 and 45°C. Methane was not produced below 20°C or above 55°C.

The effect of oxygen on the methanogenic activity is shown in Fig. 6. Native and immobilized bacteria were incubated in air for 6 h at 37°C. Then the methane-production rate was determined by gas chromatography. After determination, fresh medium was added to a Shrenk flask, and both native and immobilized cells were reincubated under the same conditions. The methane productivity of the native cells decreased gradually, and no methane production was observed after 12 h of incubation. By contrast, the immobilized cells produced methane continuously under aerobic conditions.

3. Continuous Methane Production with a Batch System

Continuous methane production by immobilized bacteria was performed in a batch system (Fig. 7) under aerobic conditions for 24 h at 37°C. Methane production was determined by gas chromatography,

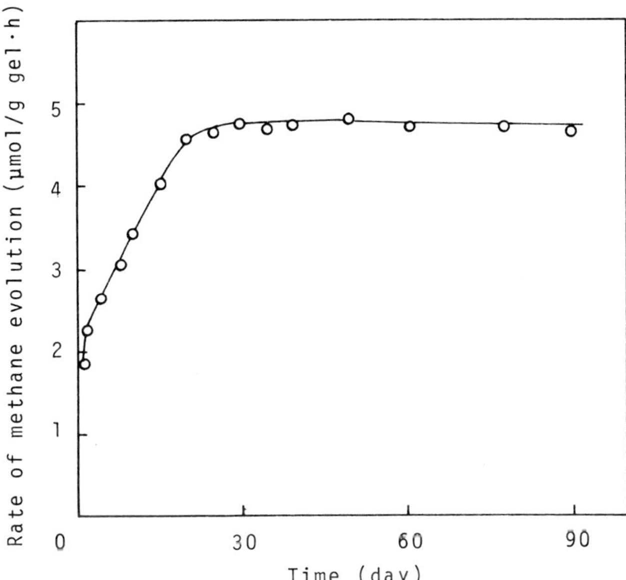

Fig. 7. Continuous methane production by immobilized bacteria. Immobilized whole cells were incubated under the same conditions as in Fig. 6 in air for 24 h.

and then the medium was replaced. The immobilized bacteria were reincubated for 24 h. The rate of methane production was gradually increased and reached a steady state (4.5 μmol/g gel/h) after 25 days. Immobilized cells continuously evolved methane over a 90-day period.

As previously reported, the growth of cells in agar gel matrices during incubation provides an obvious explanation for the increase in the activity of methane production with successive utilization of the immobilized cells (Fig. 7). A steady-state methane-production rate was obtained after 25 days, which means that the interestitial space was filled with active bacteria at that time.

Table III shows the organic compounds in the wastewater before and after reaction with the immobilized bacteria. The experiments were performed at pH 7 and 37°C for 18 h, and the wastewater was sterilized. Glucose and sucrose were measured by enzymatic methods. As shown in Table III, carbohydrates, such as glucose and sucrose, were completely decomposed to organic acids by the immobilized bacteria. However, acetic, propionic, and butyric acids remained in the wastewater. When the wastewater was incubated without immobilized bacteria, no decrease in the organic compounds was found.

TABLE III
ORGANIC COMPOUNDS IN THE WASTEWATERS BEFORE AND AFTER REACTION[a]

Substrates	Contents (mg/liter)	
	Initial	After reaction
Glucose	360	0
Sucrose	540	0
Formic acid	20	0
Acetic acid	60	70
Propionic acid	20	20
Butyric acid	0	20

[a] Reaction conditions: pH 7, 37°C, 18 h.

The limitation on the organic acids remaining in the wastewater may be due to the diffusion gradient required for the organic acid to reach the immobilized bacteria. Further developmental studies in this laboratory are being directed toward using the immobilized bacteria for the continuous production of methane from wastewater.

III. MICROBIAL FUEL CELLS USING IMMOBILIZED CELLS

As described already, various bacteria and algae produce hydrogen under anaerobic conditions. Hydrogen exhibits excellent reactivity in electroactive materials. One of the first biochemical fuel cell systems designed to use hydrogen-producing bacteria was developed by Rohrback *et al.* (1962), who reported the production of hydrogen from the fermentation of glucose by *Cl. butyricum.* However, because the hydrogen-evolution system, especially hydrogenase in bacteria, is unstable, it is difficult to use whole cells for continuous hydrogen production.

Recently, immobilization techniques for enzymes and bacteria have been developed for industrial and clinical applications of these biocatalysts. Hydrogen-producing bacteria, *Cl. butyricum,* were immobilized, and the immobilized whole cells continuously evolved hydrogen from glucose and wastewater under aerobic conditions. Immobilized hydrogen-producing bacteria were applied to a biochemical fuel cell. Solar energy is also very attractive for energy production, and Berk and Canfield (1964) reported on one of the first photochemical fuel cells utilizing microorganisms. In the presence of light, *Rhodo-*

spirillum rubrum produced hydrogen, which was oxidized on the surface of the anode. However, in this system malate was employed as an electron donor for hydrogen production.

As described already, the use of immobilized blue-green algae coupled with immobilized chloroplasts and *Cl. butyricum* made possible water-splitting hydrogen evolution. Therefore, construction of a photochemical fuel cell system using immobilized blue-green algae or immobilized chloroplasts–*Cl. butyricum* is possible. In this section, microbial fuel cells using immobilized whole cells are described.

A. Hydrogen–Oxygen Fuel Cell Using Immobilized *Clostridium butyricum*

As described already, *Cl. butyricum* was immobilized in polyacrylamide gel, and the immobilized whole cells continously evolved hydrogen from glucose under anaerobic conditions. Immobilized cells of *Cl. butyricum* have been applied to a biochemical fuel cell (Karube *et al.*, 1977) that was operated for 15 days and from which a continuous current of 1.2–1.1 mA was obtained. Glucose was used as a nutrient. However, because glucose was expensive, wastewater was tested as the nutrient for the immobilized whole cells. *Cl. butyricum* was immobilized in agar gel, and the immobilized whole cells were employed for the production of hydrogen from industrial wastewater (Suzuki *et al.*, 1978).

1. Wet-Type Hydrogen–Oxygen Fuel Cell

A schematic diagram of the system is presented in Fig. 8. The reactor for hydrogen production was made of acrylic plastic with 90-ml capacity (diameter 2.2 × 24 cm). About 40 g of immobilized whole cells of *Cl. butyricum* were packed into the reactor. The fuel cell consisted of an anode chamber (10 × 10 × 3 cm) and a cathode chamber (10 × 10 × 0.5 cm), separated by an anion-exchange membrane (Selemion type AMV, Asahi Glass Co.). The anode was a platinum black electrode (10 × 20 cm), and the anolyte was 150 ml wastewater from the alcohol factory. The cathode was a carbon electrode (7.5 × 8.0 × 3.0 cm), with the catholyte 50 ml of 0.1 *M* phosphate buffer (pH 7). The reactor for the wastewater treatment was a glass vessel (diameter 9.2 × 45 cm) containing about 300 g immobilized whole cells. The wastewater was saturated with dissolved oxygen and stirred magnetically at 37 ± 1°C. The current, the anode potential, and the cell voltage were measured by a millivolt-milliammeter (Kikusui Electronics, Model 114) and displayed on a recorder (TOA, Model EPR-100A). The anode potential was determined with reference to a saturated calomel electrode (SCE).

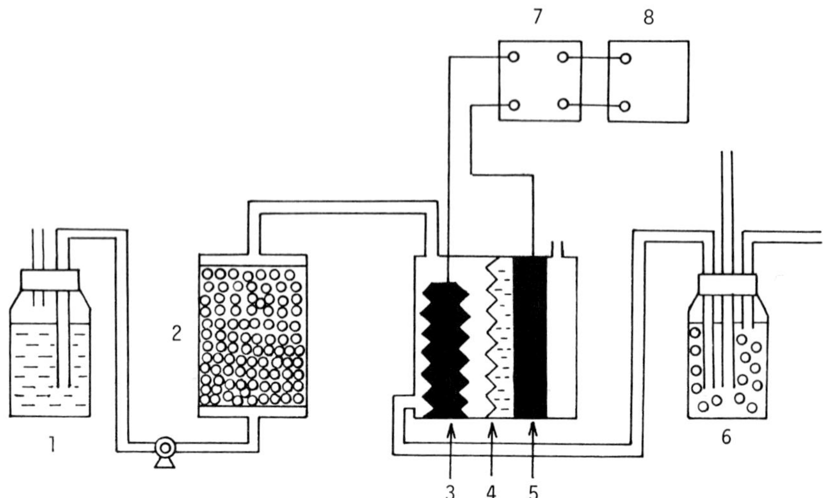

Fig. 8. Schematic diagram of wet-type hydrogen–oxygen fuel cell using immobilized *Clostridium butyricum*. (1) Reservoir for the wastewater; (2) packed-bed reactor for immobilized *Cl. butyricum;* (3) platinized platinum anode; (4) anion-exchange membrane; (5) porous carbon cathode; (6) continuously stirred-tank reactor for immobilized aerobic microorganisms; (7) millivolt-milliammeter; (8) recorder.

The system was composed of three devices: the reactor for the hydrogen production, the fuel cell, and the reactor for wastewater treatment. The chemical reactions involved in the three devices are as follows:

Hydrogen reactor:

$$\text{Wastewater} \xrightarrow{\textit{Cl. butyricum}} H_2 + \text{formic acid} + \text{other compounds}$$

Fuel cell:

$$H_2, \text{formic acid} \rightarrow \text{current}$$

Wastewater reactor:

$$\text{Other compounds} \xrightarrow{\text{microorganisms}} CO_2 + H_2O + \text{unused compounds}$$

Formic acid produced by immobilized *Cl. butyricum* also contributed to the current generation (Karube *et al.*, 1977).

The wastewater from the packed-bed reactor was transferred to a fuel cell at a flow rate of 5 ml/min. The anode potential became more negative because the anode was saturated with the hydrogen produced by the immobilized whole cells. The anode potential was −0.5 V (vs. the SCE), and the cell voltage of the fuel cell was 0.63 V at pH 7.

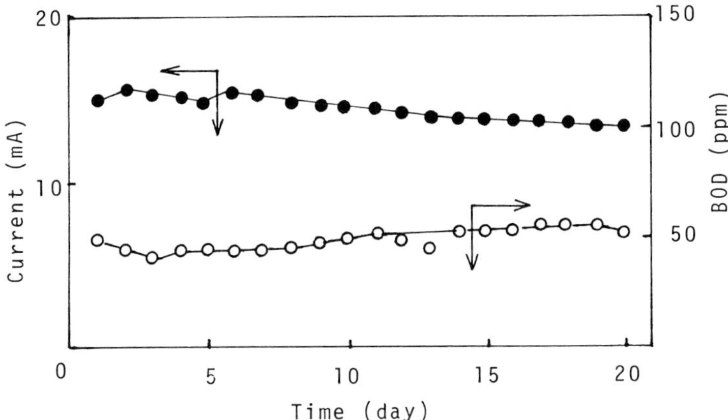

Fig. 9. Continuous operation of the fuel cell system. Wastewater (pH 7, BOD 660 ppm) from an alcohol factory was employed and transferred to the system at a flow rate of 5 ml/min. Current, ●——●; BOD, ○——○.

The limiting current density changed from 10^{-3} to 10^{-1} mA/cm^2 as the resistance between the electrodes changed from 5 to 20,000 Ω. The maximum current density of 1 mA/cm^2 and anode potential of −0.63 V were obtained using the condensed wastewater (BOD 3300 ppm) and a platinum black electrode (2 × 4 cm).

As the flow rate was increased from 1.5 to 12 ml/min, the current increased until a maximum was reached at 10 ml/min. The current also increased with increasing BOD of the wastewater.

The maximum current was about 40 mA when the BOD of the concentrated wastewater was 3300 ppm. However, the utilization ratio of organic compounds in the wastewater decreased with increasing flow rate and BOD. Therefore, a flow rate of 5 ml/min and a BOD of 660 ppm were employed for the fuel cell.

The biochemical fuel cell system was operated at the optimum conditions described already. The fuel cell was operated for 20 days, to produce a continuous current from 15 to 13 mA. This continuous operation indicated that the *Cl. butyricum* in the agar gel was living and maintained hydrogen-evolution activity for a long time. At the same time, the BOD of the wastewater could be maintained below 50 ppm (Fig. 9).

2. Gas-Type Hydrogen–Oxygen Fuel Cell

As described already, the current obtained from the biochemical fuel cell system was still low. In this case, wastewater containing hydrogen was transferred directly to the anode chamber, and diffusion of dissolved hydrogen to the electrode surface was the rate-determining step in the current generation. Another system consisted of a continuous stirred reactor for hydrogen production by the immobilized cells of *Cl. butyricum,* and two gas-type hydrogen–oxygen (air) fuel cells. The alcohol factory wastewater was used, and the system was operated continuously (Suzuki *et al.,* 1980).

A schematic diagram of the system is presented in Fig. 10. A reactor (jar fermenter, Model MD 300, Marubishi Rika Co., Tokyo) with a 5-liter capacity was loaded with about 1 kg of immobilized *Cl. butyricum.* The fuel cell consisted of an anode chamber (diameter 10.4 cm, thickness 0.4 cm) and a cathode chamber (diameter 10.4 cm, thickness 0.4 cm), separated by a nylon–glass filter containing potassium hydroxide solution (8 *N*). A platinum black–nickel mesh anode (diameter 10.4 cm, 100 mesh) and a palladium black–nickel mesh cathode (diameter 10.4 cm, 250 mesh) were used. The reactor was maintained at 37 ± 1°C and the fuel cells at 25 ± 0.5°C. The current and cell voltage were measured by an ammeter (Yokokawa, E-11) and an electrometer (Hokuto Denko Ltd., HE 101A), and displayed on a recorder (TOA, Model EPR-200A).

The effect of the ratio of air to hydrogen-flow rate on the cell voltage was examined. The hydrogen-flow rate was fixed at 6 ml/min, and the fuel cells were operated at 1-Ω discharge conditions. The cell voltage of each fuel cell became constant at an air : hydrogen-flow rate ratio above 3.

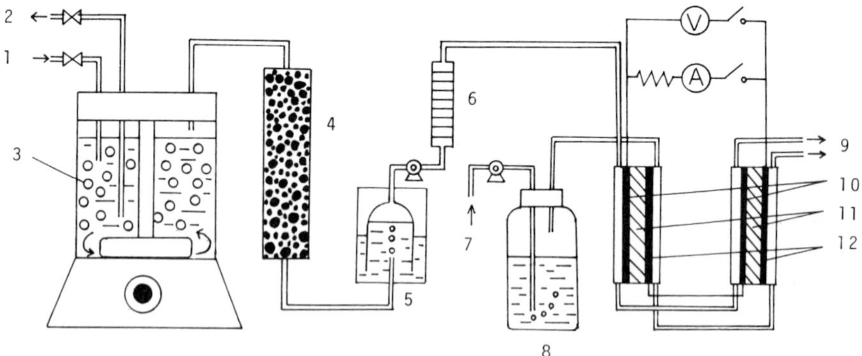

Fig. 10. Schematic diagram of gas-type hydrogen–oxygen fuel cell using immobilized *Clostridium butyricum.* (1) Wastewater from alcohol factory; (2) outlet; (3) immobilized whole cell; (4) soda lime; (5) hydrogen reservoir; (6) flow meter; (7) air; (8) KOH solution; (9) exhaust; (10) platinum black–nickel mesh anode (diameter 10.4 cm); (11) palladium black–nickel mesh cathode (diameter 10.4 cm); (12) nylon–glass filter containing 8 *N* KOH as an electrolyte.

The effect of the hydrogen-flow rate on the cell voltage of the fuel cell was also studied. As flow rates were varied from 2.5 to 30 ml/min, the cell voltage increased and became approximately constant above 6 ml/min. At a fixed flow-rate ratio of air to hydrogen of 3, the maximum cell voltage was 0.55–0.66 V when the flow rate of hydrogen was 10 ml/min. However, the cell voltage became almost constant above 6 ml/min. The reactivity of hydrogen at the anode surface might be the rate-determining step for current generation.

Hydrogen from the reactor was transferred to a fuel cell at a flow rate of 6 ml/min. The anode potential became more negative because the anode was saturated with hydrogen produced by the immobilized whole cells. The cell voltage of each fuel cell was 0.95 V. The limiting current density changed from 0.4 to 40 mA/cm^2 as the resistance between the electrodes changed from 1 to 100 Ω. The maximum current of 1.1 A was obtained at a hydrogen-flow rate of 35 ml/min.

The biochemical fuel cell system was operated under the optimum conditions described already. Figure 11 shows the current–time relationship of the biochemical fuel cell system. The fuel cell was operated for 7 days, and a current of 550–500 mA was obtained continuously over this period. This continuous operation indicated that the *Cl. butyricum* in the agar gel was living and had maintained its hydrogen-evolution activity for a long time.

3. Improved Gas-Type Hydrogen–Oxygen Fuel Cell

A 5-liter reactor (jar fermenter, Model MD 300, Marubishi Rika Co., Tokyo) was charged with about 2 kg of immobilized *Cl. butyricum*.

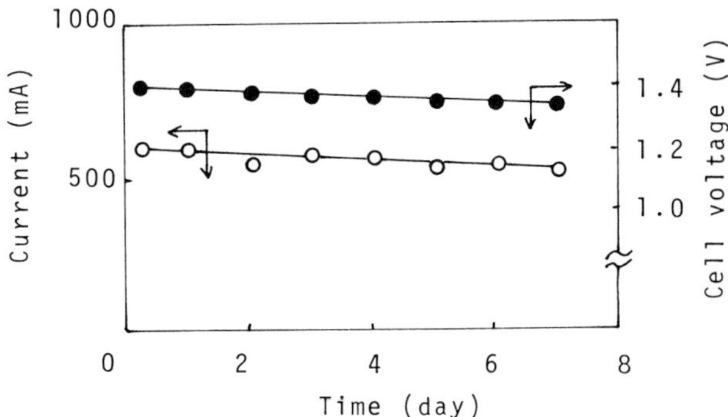

Fig. 11. Current–time relationship of the fuel cell system. The system in Fig. 10 was operated. *Clostridium butyricum* whole cells (100 g wet weight) in 1 kg wet gel were incubated in 4 liters of wastewater (pH 7). Wastewater in the reactor was completely changed after 24 h operation. Cell voltage, ●——●; current, ○——○.

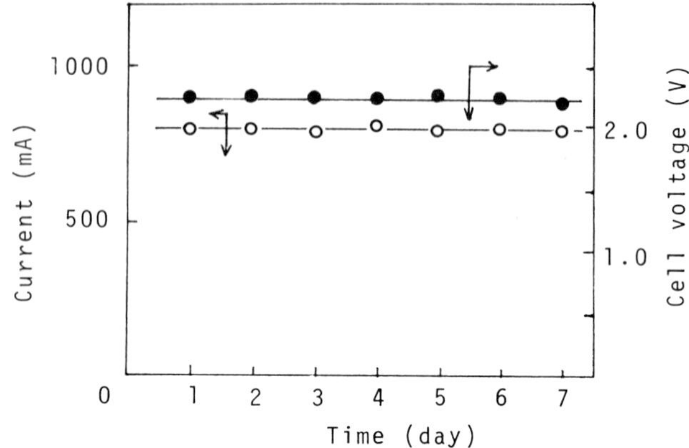

Fig. 12. Current–time relationship of the improved fuel cell system. Wastewater (BOD 80,000 ppm, pH 7) was transferred to the reactor for immobilized hydrogen-producing bacteria of the average flow rate of 10 ml/min. Fuel cells were operated at 2 Ω resistance. Cell voltage, ●——●; current, ○——○.

The immobilized whole cells continuously produced hydrogen at 20 ml/min/kg wet gel. Five fuel cells were connected in series and run using the generated hydrogen at the optimum condition and with wastewater of BOD 80,000 ppm. Figure 12 shows the current–time relationship of the fuel cell system over 7 days of operation. The current of 0.8 A and cell voltage of 2.2 V were obtained continuously over this period (Karube *et al.*, 1981a). The current obtained was about 1.5 times and the cell voltage 4 times higher than that reported previously. A current of 0.8 A was obtained for a long time. However, no attempt was made to optimize the fuel cell system.

B. Photochemical Fuel Cell System Using Immobilized Blue-Green Algae

Hydrogen produced by immobilized blue-green algae, *Anabaena* N-7363, was used in a wet-type hydrogen–oxygen fuel cell system. Figure 13 is a schematic diagram of the photochemical fuel cell system, which consisted of an immobilized *Anabaena* reactor, a reactor for removing the evolved oxygen, a $CaCl_2$ column, and a hydrogen–oxygen fuel cell.

After washing the gels with distilled water, the gels (2 kg) were placed in a transparent acrylate reactor containing 11 liters of modified BG-11 medium without KNO_3 (pH 8). The medium was stirred slowly, and the reactor was illuminated with a fluorescent lamp (Toshiba Co.)

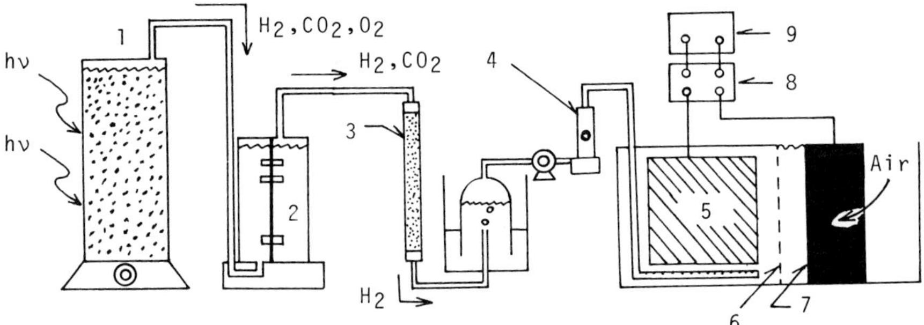

Fig. 13. Schematic diagram of the photochemical fuel cell system using immobilized *Anabaena* sp. (1) Immobilized blue-green algae (*Anabaena* sp.); (2) aerobic bacteria (*Bacillus subtilis*) and cultivation medium; (3) soda lime; (4) flow meter; (5) platinized platinum anode; (6) anion-exchange membrane; (7) porous carbon cathode; (8) millivolt-milliammeter; (9) recorder.

at 10,000 lux. Precultured *Bacillus subtilis* was suspended in 2.6 liters of medium and held in the reactor for removing oxygen at 30°C.

The hydrogen–oxygen fuel cell consisted of a platinized platinum anode (10 × 50 cm), a porous active-carbon cathode (7.5 × 8.0 × 2.5 cm), and the electrolyte (0.1 *M* phosphate buffer solution, pH 8). The anode and cathode were separated by an anion-exchange membrane (Selemion type AMV, Asahi Glass Co.). The fuel cell was operated at room temperature. Hydrogen and oxygen gases evolved in the immobilized *Anabaena* reactor were passed through the *B. subtilis* oxygen-removing reactor, and then through calcium chloride columns to remove evolved carbon dioxide. The pure hydrogen gas thus obtained was stored and fed to the anode chamber of the fuel cell at 0.1 ml/min. Air was supplied to the cathode chamber. The photochemical fuel cell system was operated at the optimum conditions described already. Figure 14 shows the time–current relationship of the illuminated immobilized *Anabaena* sp. A photocurrent of 15–20 mA was obtained for 7 days. This result indicated that hydrogen produced by the immobilized *Anabaena* sp. reacted at the anode to give a current. The conversion ratio from hydrogen to current was 80–100%.

Many reports on photochemical cells using biological materials have been published (Berk and Canfield, 1964; Fong and Winograd, 1976; Takahashi and Kikuchi, 1976; Aizawa *et al.*, 1977; Ochiai *et al.*, 1980). However, the current obtained with the photochemical fuel cell system using *Anabaena* N-7363 was dramatically higher than those reported previously. This improvement is the result of using im-

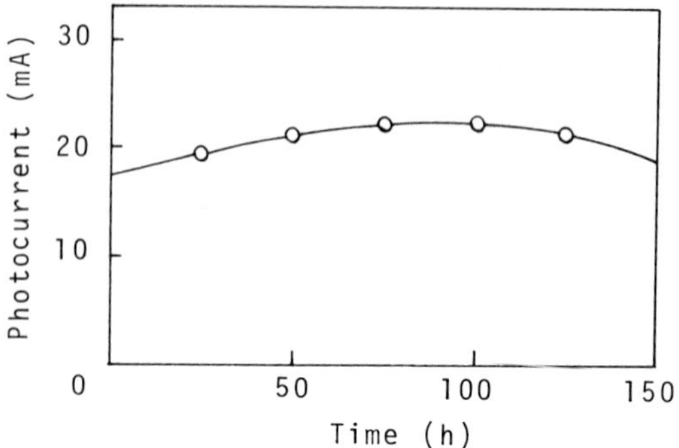

Fig. 14. Time–current relationship of the photochemical fuel cell using immobilized *Anabaena* sp. The system illustrated in Fig. 13 was employed for the operation. The hydrogen evolution by immobilized *Anabaena* sp. (2 kg wet gel, 100 mg chlorophyll *a*) in 11 liters medium was carried out same as that in Fig. 3. Aerobic bacterial cells precultured were suspended in 2.6 liters medium.

mobilized blue-green algae. However, the rate of hydrogen production by the immobilized blue-green algae was still low, and the immobilized algae were inactivated after a long incubation. Further studies are directed toward improving the hydrogen productivity and stability of *Anabaena* sp.

C. Photochemical Fuel Cell Using Immobilized Chloroplasts–*Clostridium butyricum*

The coupling of the photosynthetic system in plant chloroplasts with a hydrogenase could result in the light-driven splitting of water into hydrogen and oxygen (Benemenn *et al.*, 1973). However, the lifetime of the isolated chloroplasts is very short, and hydrogenase is also very unstable. In an attempt to stabilize the systems, the chloroplasts were immobilized in polyacrylamide or agar gel and used for carbon dioxide fixation (Karube *et al.*, 1979) and NADP reduction (Karube *et al.*, 1980b). The lifetime of the immobilized chloroplasts was longer than that of the isolated intact materials. In addition, hydrogenase was stabilized and protected from the deleterious effects of oxygen by immobilization of living whole cells. Spinach chloroplasts and *Clostridium butyricum* were immobilized in agar gel, used for light-induced hydrogen evolution (Karube *et al.*, 1982). Figure 15 shows the principle of the photo-induced hydrogen-evolution system. Ferredoxin is reduced by photosystems I and II in chloroplasts under light irradiation. Hydrogen is produced through oxidation of this reduced ferredoxin by the hydrogenase in *Cl. butyricum*. The hydrogen

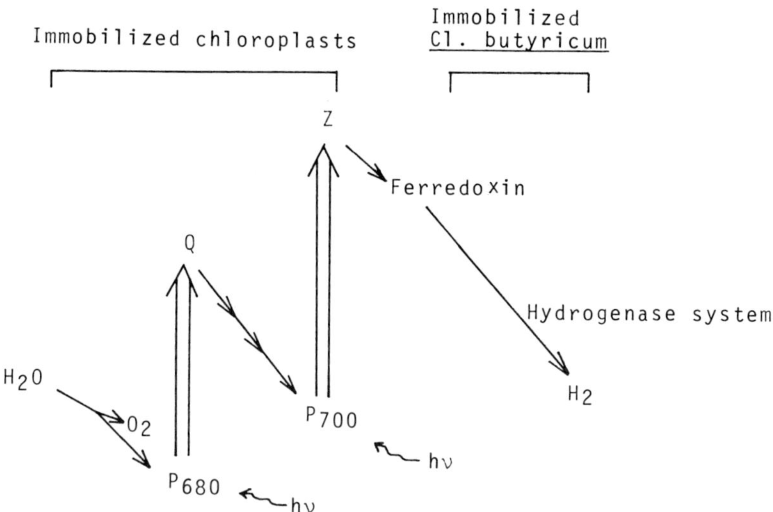

Fig. 15. Principle of the photoinduced hydrogen-evolution system.

then is used to operate a fuel cell (Kayano *et al.*, 1981b; Karube *et al.*, 1981b).

1. Apparatus

Isolation of spinach chloroplasts and crude ferredoxin were performed as described previously (San Pietro and Lang, 1958). Culture of *Clostridium butyricum* IFO 3847 was carried out as described previously (Karube *et al.*, 1976). Spinach chloroplasts and *Cl. butyricum* were immobilized in 2% agar gel by the method described previously (Suzuki *et al.*, 1978). A 300-ml transparent glass reactor (3 × 10 × 10 cm) was charged with 120 g of immobilized-chloroplast gel. A 90-ml reactor of acrylic plastic (diameter 2.2 × 24 cm) was filled with 30 g of immobilized *Cl. butyricum.*

The hydrogen–oxygen fuel cell consisted of a platinized platinum anode (10 × 50 cm), porous active-carbon cathode (7.5 × 8.0 × 2.5 cm), and the electrolyte (0.1 *M* phosphate buffer solution, pH 8). The anode and the cathode were separated by an anion-exchange membrane (Selemion type AMV, Asahi Glass Co.). The chloroplast reactor was illuminated with a 500-W reflector lamp (Toshiba Co.) at 20,000 lux. Nitrogen was bubbled through the immobilized-chloroplast reactor. The chloroplast reactor was maintained at 25°C ± 1°C; the *Cl. butyricum* reactor was maintained at 37°C. The fuel cell was operated at room temperature. The phosphate buffer solution (0.1 *M*, pH 8) containing 8 μM crude ferredoxin was circulated through the chloroplast reactor, the *Cl. butyricum* reactor, and the anode chamber of the fuel cell at a flow rate of 10 ml/min. The current, the anode potential, and the cell voltage were measured by a millivolt-milliammeter and displayed in a recorder.

2. Optimum Conditions for Immobilized Chloroplasts

The immobilized chloroplasts lost their activity below pH 5 and above pH 9. The optimum pH was 8. However, the optimum pH of the hydrogenase in the immobilized bacteria was also 8. Therefore, a phosphate buffer solution at pH 8 was employed for the experiments. The activity of the immobilized chloroplasts increased with increasing ferredoxin concentration with maximum activity obtained at 8 μM ferredoxin. The activity of the immobilized chloroplasts under anaerobic (N_2 bubbling) conditions was higher than under aerobic conditions. The oxygen inhibition of the photoreduction activity of chloroplasts may have been caused by oxidation of reduced ferredoxin by oxygen produced from the immobilized chloroplasts. However, the photoreduction activity of the immobilized chloroplasts with oxygen scavengers was 1.4 times higher than without these scavengers, such as glucose oxidase, catalase, glucose, and ethanol. When oxygen was re-

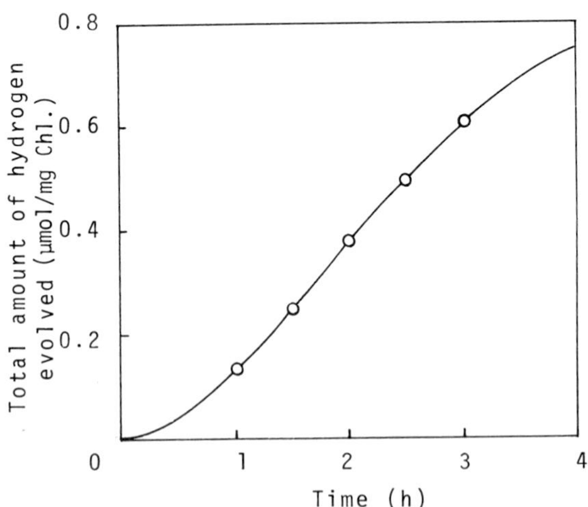

Fig. 16. Hydrogen production by immobilized chloroplasts–*Clostridium butyricum.* Spinach chloroplasts (0.1 mg Chl) and *Cl. butyricum* (0.1 g wet cells) were entrapped in 1 g wet gel. The phosphate buffer solution (0.1 M, pH 8) containing 8 μM crude ferredoxin was employed as an incubation medium. Reaction was carried out at 25°C under illumination (20,000 lux).

moved from the system by N_2 bubbling, the activity of the chloroplasts became 3.7 times higher than that under aerobic conditions.

3. Hydrogen Production by the Immobilized-Chloroplasts Reactor and Immobilized *Cl. butyricum* Reactor

Figure 16 shows the time course of hydrogen production by the immobilized chloroplasts–*Cl. butyricum* system under light irradiation. The rate of hydrogen production increased, and the maximum rate was attained during 1–2 h. Then, the hydrogen-production rate decreased gradually. Hydrogen was produced for 4 h by this system. When ferredoxin was not employed in the system, no hydrogen was evolved. This showed that electrons were not transferred from the immobilized chloroplasts to the immobilized *Cl. butyricum* without ferredoxin.

4. Photoresponse of the System

Figure 17 shows the photoinduced current of the immobilized chloroplasts–immobilized *Cl. butyricum*–hydrogen–oxygen fuel cell system. With light irradiated at 40- to 80-min intervals, the current increased with light irradiation. When the light was cut off, the current

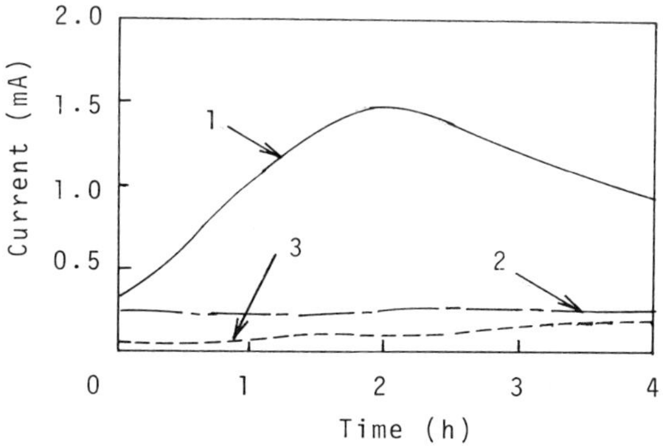

Fig. 17. Photoinduced current of the photochemical fuel cell using immobilized chloroplasts. (1) Complete system consists of immobilized chloroplasts, crude ferredoxin, immobilized *Clostridium butyricum*, and a wet-type hydrogen–oxygen fuel cell; (2) system without immobilized chloroplasts; (3) system without immobilized *Cl. butyricum*. Immobilized chloroplasts (120 g wet gel, 11 mg Chl) were incubated anaerobically in phosphate buffer solution (0.1 *M*, pH 8) containing 8 μM of crude ferredoxin at 25°C under illumination (20,000 lux). The buffer solution containing ferredoxin reduced in the immobilized chloroplasts reactor was carried to the immobilized *Cl. butyricum* reactor (30 g wet gel, 3 g wet cells). The hydrogen produced by hydrogenase in *Cl. butyricum* at 37°C was carried to the fuel cell.

gradually decreased. The photoresponse was repeated twice. No photoresponse was obtained from the individual immobilized chloroplasts–fuel cell system and immobilized *Cl. butyricum*–fuel cell systems. As reduced ferredoxin was not oxidized at the anode, no photocurrent was obtained for the immobilized chloroplast–fuel cell system. Dichlorophenyldimethylurea (DCMU) is known to inhibit electron transport in photosystems. The current decreased significantly when DCMU was added to the system. When ferredoxin was not added to the system, only little current was obtained. These facts also support the schematic diagram in Fig. 15.

5. Continuous Operation of the System

The photochemical fuel cell system was operated at the optimum conditions described already. Table IV shows time–current relationships of the immobilized chloroplasts–immobilized *Cl. butyricum*–hydrogen–oxygen fuel cell system under illumination of 20,000 lux. A photocurrent of 0.4–1.5 mA was obtained for 4 h. This result indicates that hydrogen produced by the immobilized chloroplasts–*Cl. butyricum* system under illumination was reacted at the anode and a current was generated. The current obtained from the system decreased gradually with increasing reaction time. Then, the photoreduction activity of immobilized chloroplasts and the hydrogenase activity of immobilized *Cl. butyricum* were measured after 4 h. The activity of hydrogenase was determined by measuring the hydrogen produced from reduced benzyl viologen by immobilized *Cl. butyricum*. Immobilized *Cl. butyricum* retained the initial hydrogenase activity. However, the activity of immobilized chloroplasts decreased to 30% of the initial activity. Therefore, the decrease in current may have been

TABLE IV
TIME–CURRENT RELATIONSHIP OF THE PHOTOCHEMICAL FUEL CELL USING IMMOBILIZED CHLOROPLASTS[a]

Time (h)	Photocurrent (mA)
0	0.32
1	1.02
2	1.46
3	1.18
4	0.92

[a] Reaction conditions: pH 8, 25°C, 20,000 lux, 8 μM ferredoxin.

caused by the decomposition of chloroplasts. The lifetimes of immobilized chloroplasts became longer than those of the native ones. But, further stabilization of chloroplasts is required for continuous hydrogen production.

IV. CONCLUSIONS

These results suggest that immobilized bacteria, algae, and chloroplasts can be used for continuous production of methane or hydrogen, and can be used in a hydrogen–oxygen fuel cell. Employment of immobilized microorganisms and organelles made possible prolonged gas production. However, the power obtained was weak. Improvement of hydrogen productivity is needed for practical use in a hydrogen–oxygen fuel cell. Therefore, molecular breeding of hydrogen-producing bacteria is important for practical application of a bioenergy conversion system.

REFERENCES

Aizawa, M., Suzuki, N., Takahashi, F., and Suzuki, S. (1977). *J Solid-Phase Biochem.* **2,** 111.

Asada, Y., Tonomura, K., and Nakayama, O. (1979). *J. Ferment. Technol.* **57,** 280.

Benemann, J. R., and Wear, N. M. (1974). *Science* **184,** 174.

Benemann, J. R., Berenson, J. A., Kaplan, N. O., and Kamen, M. D. (1973). *Proc. Natl. Acad. Sci. U.S.A.* **70,** 2317.

Berk, R. S., and Canfield, J. H. (1964). *Appl. Microbiol.* **12,** 10.

Fong, F. K., and Winograd, N. (1976). *J. Am. Chem. Soc.* **98,** 2287.

Karube, I., Matsunaga, T., Tsuru, S., and Suzuki, S. (1976). *Biotechnol. Bioeng.* **20,** 1153.

Karube, I., Matsunaga, T., Tsuru, S., and Suzuki, S. (1977). *Biochim. Biophys. Acta* **444,** 338.

Karube, I., Aizawa, K., Ikeda, S., and Suzuki, S. (1979). *Biotechnol. Bioeng.* **21,** 253.

Karube, I., Kuriyama, S., Matsunaga, T., and Suzuki, S. (1980a). *Biotechnol. Bioeng.* **22,** 847.

Karube, I., Otsuka, T., Kayano, H., Matsunaga, T., and Suzuki, S. (1980b). *Biotechnol. Bioeng.* **22,** 2655.

Karube, I., Kuriyama, S., Matsunaga, T., and Suzuki, S. (1980c). *Energy Dev. Jpn.* **3,** 141.

Karube, I., Suzuki, S., Matsunaga, T., and Kuriyama, S. (1981a). *Ann. N.Y. Acad. Sci.* **369,** 91.

Karube, I., Matsunaga, T., Otsuka, T., Kayano, H., and Suzuki, S. (1981b). *Biochim. Biophys. Acta* **637,** 490.

Karube, I., Suzuki, S., Matsunaga, T., and Kayano, H. (1982). "Advances in Biotechnology," p. 389. Dekker, New York.

Kayano, H., Karube, I., Matsunaga, T., and Suzuki, S. (1981a). *Eur. J. Appl. Microbiol. Biotechnol.* **12,** 1.

Kayano, H., Matsunaga, T., Karube, I., and Suzuki, S. (1981b). *Biotechnol. Bioeng.* **23,** 2283.

Lewis, K. (1966). *Bacteriol. Rev.* **30,** 101.
Ochiai, H., Shibata, H., Sawa, Y., and Kato, T. (1980). *Proc. Natl. Acad. Sci. U.S.A.* **77,** 2442.
Raeburn, S., and Rabinowitz, J. C. (1971a). *Arch. Biochem. Biophys.* **146,** 9.
Raeburn, S., and Rabinowitz, J. C. (1971b). *Arch. Biochem. Biophys.* **146,** 21.
Rohrback, G. H., Scott, W. R., and Canfield, J. H. (1962). *Proc., Annu. Power Sources Conf.* **16,** 18–21.
San Pietro, A., and Lang, H. M. (1958). *J. Biol. Chem.* **231,** 211.
Suzuki, S., Karube, I., and Matsunaga, T. (1978). *Biotechnol. Bioeng. Symp.* **8,** 501.
Suzuki, S., Karube, I., Matsunaga, T., Kuriyama, S., Suzuki, N., Shirogami, F., and Takamura, T. (1980). *Biochimie* **62,** 353.
Takahashi, F., and Kikuchi, R. (1976). *Biochim. Biophys. Acta* **430,** 490.
Thauer, R. K., Kirchiniawy, F. H., and Jungermann, K. A. (1972). *Eur. J. Biochem.* **27,** 282.
Zeikus, J. B. (1977). *Bacteriol. Rev.* **30,** 101.

Chemical Engineering Analysis of Immobilized-Cell Systems

K. Venkatasubramanian

Department of Chemical and Biochemical Engineering
Rutgers University
New Brunswick, New Jersey
and
H. J. Heinz Company
Pittsburgh, Pennsylvania

S. B. Karkare and W. R. Vieth

Department of Chemical and Biochemical Engineering
Rutgers University
New Brunswick, New Jersey

APPLIED BIOCHEMISTRY AND BIOENGINEERING
Volume 4

ISBN 0-12-041104-0

I. INTRODUCTION

Events of the last few years have borne witness to a surging interest in the field of whole-cell immobilization for carrying out a variety of bioconversion processes. Reviews by Chibata and Tosa (1980), Venkatasubramanian and Vieth (1979), and a book on this topic (Venkatasubramanian, 1979) outline the various methods of cell immobilization and their applications. A variety of immobilization techniques ranging from gel entrapment and physical adsorption to covalent attachment are available. Depending on the method and the application, the immobilized cells may be growing, live but resting, or in the dead (intact or autolyzed) state. In some cases, the cells are deliberately killed prior to immobilization, but they retain the activity of some of the enzymes (Venkatasubramanian, 1980). Thus we encounter systems of varying degrees of complexity—starting from simple single-enzyme systems to more complex systems of cells growing in the immobilized state—while performing the necessary bioconversions. Hence, the analysis of an immobilized-whole-cell reactor would depend on the type of immobilization and the specific application.

Much of the literature on the design and performance analysis of immobilized-cell reactors focuses on single-enzyme kinetics. Detailed chemical engineering analyses for such systems are now well documented for a variety of reactor configurations. However, the concept of using live microorganisms in the immobilized state has emerged only recently. It has been convincingly demonstrated that gel-entrapped microorganisms can replicate within the gel matrix (Larretta Garde *et al.*, 1981; Jirkú *et al.*, 1981). This approach holds great potential in terms of production of growth-associated metabolites by immobilized cells as well as from the point of view of rejuvenation of catalytic activity in other cases. Because this concept is still in its infancy, little attention has been paid to the analysis of reactor systems embodying these live biocatalysts. However, some work has been reported in the field of activated-sludge processing for waste treatment (Tanaka *et al.*, 1981).

We begin this chapter by reviewing the available literature on different types of immobilized-cell reactors. We then introduce idealized reactor kinetic equations for the case of immobilized-cell systems and discuss their applications to reactor design in the context of primary and secondary metabolite synthesis. The analysis of such systems presented here is by no means complete. Rather, it suggests a direction for further mathematical analysis that would include the effects of external and internal mass-transfer resistances as well as the physical stability and integrity of the catalyst. Thus, the purpose of this chapter is not to review everything there is on hand, but to outline an approach to analyze these systems and to point out areas where additional work is needed.

II. TYPES OF IMMOBILIZED-CELL REACTORS AND THEIR APPLICATIONS

Several reactor configurations have been used for immobilized-whole-cell processes. Following is a discussion of the more common types and some potentially useful reactor types along with their relative merits and disadvantages. The different reactor configurations discussed here are sketched schematically in Fig. 1.

A. Packed-Bed Reactors

This is by far the most often employed reactor type for immobilized-whole-cell processes. When the biocatalyst is in the form of spheres, chips, disks, sheets, beads, or pellets, it can be packed readily into a column. In a packed-bed reactor, there is a steady movement of the substrate across a bed of immobilized whole cells in a chosen spatial direction. If the fluid-velocity profile is perfectly flat over the cross section, the reactor is said to operate as a plug-flow reactor (PFR). Of course, deviations from this idealized behavior do

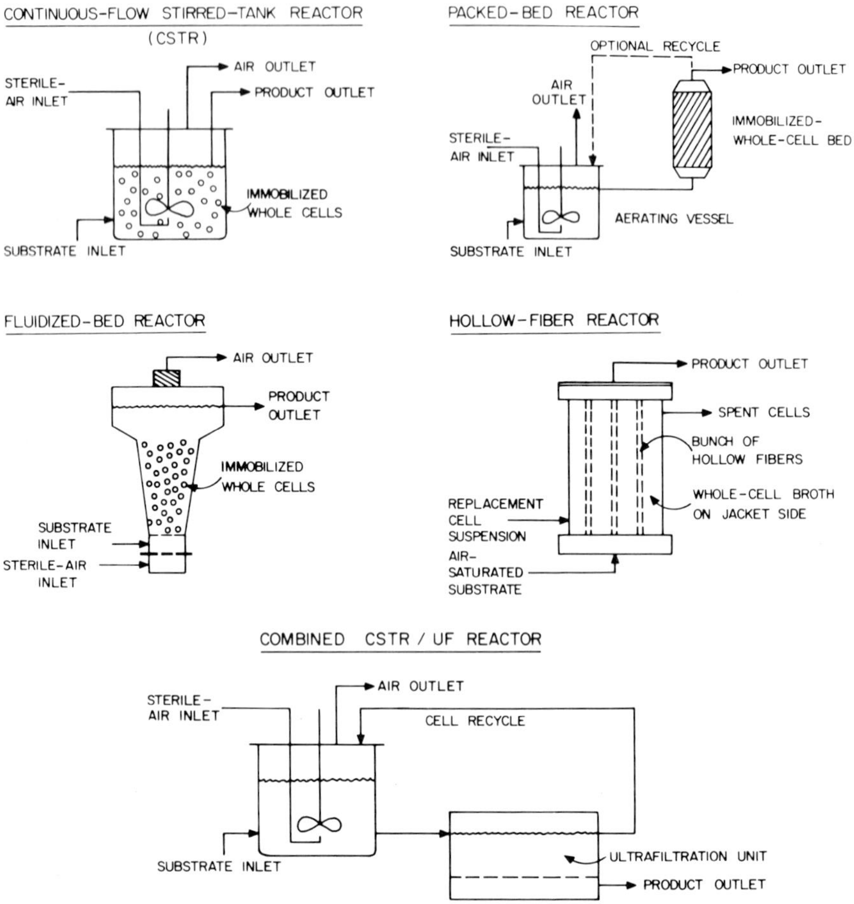

Fig. 1. Types of immobilized-cell reactors.

occur and must be considered when analyzing the performance of these reactors.

Some examples of this type of reactor configuration are (1) collagen-bound whole cells used in a spirally wound packed-bed reactor for glutamic acid production (Constantinides *et al.*, 1981), (2) continuous ethanol production by yeast immobilized on calcium alginate beads (Linko and Linko, 1981), (3) cells immobilized on porous ceramic material for continuous culturing of microorganisms (Messing *et al.*, 1981), and (4) conversion of fumaric acid to aspartic acid by *Escherichia coli* cells entrapped in polyacrylamide gel (Chibata *et al.*, 1974). The first three examples are most recent and indicate the current trend toward immobilization of "living" cells rather than the single-enzyme system described in the fourth example.

Packed-bed reactors have the advantage of simplicity of operation, high mass-transfer rates, and high reaction rates (for non-substrate-inhibited kinetics). In the case of immobilized living cells, oxygenation and carbon dioxide removal are often necessary. Under these conditions, a packed bed is liable to pose problems such as gas removal and inefficient gas–liquid contact. In laboratory scale, the oxygenation problem can be circumvented to a large extent by prior oxygenation of the substrate. However, for high cell densities oxygen depletion along the length of the reactor is very rapid. Oxygen transfer may pose a serious problem in the scale-up of these reactors, unless they are staged or segmented. Another problem in such systems is the periodic fluctuation in the viable cell population due to nutrient depletion along the reactor length.

B. Continuous-Flow Stirred-Tank Reactors

In an ideal CSTR, the contents of the reactor are perfectly mixed. Consequently, all elements of the reactor have essentially the same concentration, which is the same as the concentration of the outflow. Therefore, the reaction rate is determined by the composition of the exit stream from the reactor. Whereas in a PFR, the substrate concentration is maximized with respect to final conversion at every point in the reactor, it is minimized at every point in a CSTR. Thus, in a CSTR, the average reaction rate is lower. Hence, this reactor configuration may be more suitable for substrate-inhibited reaction kinetics. The open construction of the CSTR permits ready replacement of the immobilized-whole-cell catalyst. It also facilitates easy control of temperature and pH. Thus the system may be suitable where substrate costs are not very important and where a stable productivy is essential.

Applications of the CSTR have been restricted mostly to the immobilized-enzyme systems (Smiley, 1971; O'Neill *et al.*, 1971; Weetall and Havewala, 1972; Dutta *et al.*, 1973). It is obvious that in CSTR applications, the catalyst must be strong enough to withstand a rather high degree of shear. This condition restricts the use of live immobilized cells in a CSTR, because such preparations can be sensitive to shear.

C. Fluidized-Bed Reactors

Fluidized-bed reactors can provide a happy mean between the traditional packed-bed and continuous stirred-tank reactors. In a fluidized-bed reactor, the individual catalyst particles are kept in motion by a continuous flow of the substrate. The pressure drop of the fluid flow essentially supports the weight of the bed. The reactor thus provides for free movement of the catalyst particles throughout the bed. The fluidization may be carried out either by liquid or by gas (e.g., air when oxygenation is necessary).

Fluidized-bed reactors offer the advantage of good solid–fluid mixing and minimal pressure drops. The solid–fluid mixing is important in the case of growing immobilized cells because it can help achieve a stable cell population (in contrast to the plug-flow reactor). This is possible because particles with lower cell loading continuously move from substrate-depleted zones to substrate-rich zones, and vice versa, by virtue of their changing densities. This can help achieve a dynamic equilibrium vis-à-vis the viable cell population. Hence, this reactor system holds great potential for growing living immobilized cells where oxygenation is often necessary. The system is also more suitable for gas removal.

Fluidized beds have traditionally been associated with problems of cumbersome and difficult operation. This tendency is somewhat exacerbated by the difficulty of modeling the behavior of fluidized-bed reactors and consequently the inability to predict accurately the performance of scaled-up versions of the reactor.

The effective catalyst-packing density in a fluidized-bed reactor is smaller compared to a PFR. Another problem relates to the density difference between the immobilized-cell particles and the substrate. To achieve good fluidization characteristics, this density difference should be as high as possible. Because some of the most popular immobilization methods are based on entrapping the microbial cells in highly hydrated hydrocolloid gels (e.g., carrageenan, alginate), often the density of the bound cells is not significantly different from that of the liquid substrate. Thus, practical reactor systems can have a range of performance from a mere expanded-bed operation to a truly fluidized-bed system. However, many attempts are now under way to circumvent these problems. Furthermore, the petrochemical industry has pioneered the development of fluid-bed reactors, and established technology is now available that can provide a good basis for adaptation to the immobilized-whole-cell processes.

Applications of fluidized beds in immobilized-whole-cell work have until now been limited to activated-sludge processes. However, a few other immobilized-whole-cell processes have also been carried out in fluidized-bed reactors (e.g., Kennedy *et al.*, 1980; Karkare *et al.*, 1981). It appears from the foregoing discussion that fluidized beds hold great potential for future work in immobilized-cell processes.

D. Hollow-Fiber Reactors

Reactors packed with semipermeable hollow fibers, which permit the passage of only the reactant and the product, but not the whole cells, have also been found to be a useful reactor configuration. They provide a high catalyst surface area in a given reactor volume; reactor selectivity can be achieved by using fibers of dissimilar permeability characteristics. Hollow fibers can be arranged in a reaction system in different ways. For example, Kan and Shuler (1978) entrapped whole cells on the shell side of the fiber bundle and passed the substrate inside the fibers. The substrate diffuses through the fiber walls to the cells and the product diffuses back. Conversely, the catalyst may be trapped inside the fibers and substrate passed over the fiber bundle on the shell side (Rony, 1972; Marconi *et al.*, 1974). Michaels (1980) has described other variations of hollow-fiber reactors and their applications to a variety of bioconversion processes. Among the major advantages of this reactor configuration are economy of immobilization, complete retention of cells, and the ease of cell replacement for maintenance of full activity or the manufacture of a different product. Large-scale ultrafiltration devices are commercially available, hence additional technology development is not necessary in this field. The major disadvantage of this system is the lack of proper control of cell loading to assure uniform catalyst distribution, which is important for the stability of the system. A nonuniform flow distribution may also result, leading to significant resistance to diffusion of substrate and product.

E. Other Reactor Types

Many variations and combinations of the basic types of immobilized-cell reactors discussed here are possible. For example, recycle reactors may find application when reaction rates are too slow or when high bulk mass-transfer coefficient values are necessary. In this type of reactor, a portion of the outflow is recycled and mixed with the inlet stream to the reactor. This permits operation of the reactor at high fluid velocities, which minimizes bulk mass transfer resistance to the transport of substrate to the catalyst surface. Even though high flow rates reduce the contact time of the substrate in the reactor (per pass), the recycling process effectively provides sufficient contact time to achieve desired conversions.

A variation of a typical packed-bed reactor is the trickle-bed filter in which liquid flows down over the packed catalyst in the form of a thin film. Air may be passed either cocurrently or countercurrently to liquid flow to provide efficient oxygenation. For example, Briffaud and Engasser (1979) studied citric acid production by immobilized whole cells in a trickle-bed filter using wood chips as the supporting carrier. It is of interest here that traditional vinegar manufacturing is based essentially on a trickle-bed filter using immobilized cells.

III. CHOICE OF REACTOR TYPE

The choice of reactor type for a particular process would depend on the process requirements and conditions. The following factors (summarized in Table I) would influence prominently the choice of the reactor type.

A. Cell Viability Requirements

For single-enzyme systems where oxygenation is not necessary and when there are no gaseous products, a packed-bed reactor may be the most suitable. Again depending on the catalyst-loading requirement, other types of reactors such as hollow-fiber reactors may be chosen. If viability or growth is essential and oxygenation is important, fluidization with air may be resorted to. If large volumes of CO_2 are to be removed, a slanted packed bed may be of interest.

B. Type of Supporting Carrier

Particulate biocatalysts can usually be used in any of the reactor types. Membranes and fibrous supports have to be used in a packed-bed configuration. Beads of suitable densities may be used in fluidized beds, whereas a process requiring high surface area may be carried out in hollow-fiber/ultrafiltration type of reactors.

TABLE I
FACTORS INFLUENCING CHOICE OF REACTOR TYPE FOR IMMOBILIZED-CELL SYSTEMS

1. Cell viability requirements
2. Type of support matrix and method of immobilization
3. Nature of substrate
4. Kinetics of reactions involved
5. Operational requirements of the process
6. Ease of catalyst replacement and regeneration
7. Hydraulic considerations
8. Ease of design, fabrication, and process scale-up
9. Reactor cost

C. Nature of Substrate

If the substrate is a single-phase, clean liquid without excessive particulate matter, packed beds can be effectively employed. For colloidal or particulate substrates, CSTR or fluidized beds would be more suitable. This is also true when good gas–liquid mixing is desired. The choice between CSTR and fluidized bed would be dictated by the vulnerability of the biocatalyst to shear.

D. Kinetics of the Reactions Involved

Product-inhibited and noninhibited reactions are best carried out in plug-flow reactors in order to compensate partly for the inhibition. Conversely, substrate inhibition may dictate the use of a CSTR. Autocatalytic reactions (such as cell growth-associated products) may require a combination of the two types or a fluidized bed.

E. Operational Requirements of the Process

Some processes may require strict pH control or temperature control for maximum productivity. In such cases a CSTR configuration is most suitable. If a high surface : volume ratio is desired, hollow-fiber reactors can be used. If oxygen transfer is critical, a fluidized-bed reactor or CSTR can be used.

F. Ease of Catalyst Replacement and Regeneration

In a single-enzyme system, if the catalyst life is short, it may be necessary to replace the catalyst often. A CSTR lends itself to easy catalyst replacement, whereas a packed column has to be shut down. However, in the case of live microbial cells, it may be possible to rejuvenate the catalyst *in situ* by passing growth nutrients through the reactor. If so, a packed bed or fluidized bed is equally suitable.

G. Hydraulic Considerations

Catalyst bed compaction due to hydraulic forces can lead to serious increases in pressure drops and reduction in the apparent activity of the whole cells due to reduced permeability. Thus the hydraulic considerations would dictate the maximum permissible bed depth if a packed-bed reactor configuration is selected. Conversely, if the catalyst support forms a compressible bed (e.g., gel-entrapped cells), it may dictate selection of an upflow mode for substrate flow or a fluidized-bed reactor.

H. Ease of Design and Fabrication

Modeling of PFR and CSTR is much easier than that of a fluidized-bed or some other nonconventional reactors. CSTR is also very easy to fabricate because of the relatively simple construction. Moreover, nonconventional reactors may pose problems in scale-up, because not much literature is available on such reactor types.

I. Reactor Cost

The reactor cost becomes important if the catalyst cost is relatively low. Because of their simple construction, CSTRs are the cheapest per unit reactor volume. Other types of reactors have to be designed and built for specific purposes. However, when considering reactor costs, one must also consider the cost of the catalyst. The reactor type may dictate faster or slower catalyst replacement. For example, due to high shear rates in CSTRs, the catalyst half-life is likely to be much shorter than in other types of reactors.

It is clear from the foregoing considerations that there are no simple rules for choosing a reactor type. For an efficient reactor system, one should endeavor to combine the advantageous features of the different types. For example, in cases where high oxygen requirements exist and product inhibition is a problem, one may use several small fluidized-bed units in series. This essentially combines the best features of fluidized beds (good oxygenation) with that of a plug-flow reactor (partial removal of product inhibition). Such a system has been used in our laboratory (Karkare *et al.*, 1981). The same system can also be advantageously employed for substrate-inhibited kinetics by introducing the substrate in between the stages in steps in order to reduce the substrate inhibition.

IV. ANALYSIS OF IMMOBILIZED-CELL REACTORS: BACKGROUND

As mentioned earlier, the analysis of an immobilized-whole-cell reactor would depend primarily on the type of immobilized-cell process to be used. When the cells are nonviable and only a single intracellular enzyme is of interest, it can be treated essentially as a bound-single-enzyme system. The analysis of such a system involves application of enzyme kinetics to the appropriate reactor-performance equations and development of appropriate quantitative indices (e.g., effectiveness factor) to account for external and internal mass-transfer resistances. If, however, cells are viable and/or growing in the immobilized state, the concepts of cell growth kinetics and cell maintenance requirements have to be addressed in developing the reactor performance equations. We thus have two distinct types of immobilized-cell (IMC) processes that require separate treatment.

Because the single-enzyme types of IMC reactors are essentially similar to bound-enzyme reactors, the equations for this system would be the same as those for the bound-enzyme reactors. Vieth *et al.* (1976) have described the various reactor performance equations for this case in great detail. Hence in this chapter, only a summary of these equations will be presented and some of the more recent analyses that have appeared in the literature since then will be discussed.

To our knowledge, no systematic reactor performance equations have yet been published for the case of live, growing cell systems. Hence, in the second category we have attempted to define the various concepts involved and, with the help of certain basic assumptions, developed rudimentary equations to describe immobilized-cell processes for the production of (1) biomass, (2) growth-associated products, and (3) secondary metabolites. Needless to say, the treatment given here needs further refinement in terms of better modeling of mass-transfer resistances and modeling of the cellular metabolism itself. Therefore, this can be considered only as a first step in modeling and analysis of immobilized-live-cell reactors. However, the analysis does shed some light on the exciting possibilities of these systems and quantifies their superiority over submerged fermentation processes in terms of productivities and conversion efficiencies.

The following analysis begins with single-enzyme-type immobilized-cell reactors. After discussing idealized reactor systems first, we move on to include the nonideal behavior of these reactors mainly in terms of mass-transfer effects. This is followed by a similar treatment for the case of immobilized-living-cell reactor systems.

V. SINGLE-ENZYME-TYPE IMC REACTORS

A. Definitions and Assumptions

In developing mathematical models for enzyme reactors, certain general assumptions can be made. These assumptions, which are valid in many cases of practical interest, simplify the mathematical analysis considerably. They are as follows:

1. Because most enzyme-catalyzed reactions take place in the physiological temperature range and exhibit low enthalpies of reaction, isothermal conditions are maintained in an enzyme reactor.

2. The immobilized-enzyme particles are packed in a column or suspended in a stirred vessel in a uniform manner, so that there are no statistically significant variations between two different parts of the reactor.

3. From the arguments set forth by Denbigh (1965) for isothermal packed-bed reactors, the relevant assumptions can be made concerning plug-flow conditions and a negligibly small contribution of longitudinal turbulent dispersion in comparison with the transport due to the bulk flow. Further, for an isothermal reactor with a flat velocity profile and with uniform distribution of catalyst particles over the cross section, there is no concentration gradient in the radial direction. Hence, radial dispersion in the reactor is not a factor in the analysis under the conditions of radial symmetry.

4. The CSTR system may be assumed to be perfectly mixed and the reactor residence-time distribution may be characterized by a single mean residence time.

The reactor spacetime τ is defined as

$$\tau = V_R/Q \tag{1}$$

where V_R is the volume of the reactor and Q is the flow rate through the reactor.

The fractional conversion x is given by

$$x = (S_0 - S)/S_0 \tag{2}$$

where S_0 and S are the inlet and outlet substrate concentrations, respectively. Reactor (volumetric) productivity P_r is defined as

$$P_r = xS_0/\tau \tag{3}$$

This gives an indication of the amount of product produced per unit time per unit reactor volume. It must be mentioned in passing that in conventional submerged fermentations another performance index is often used, that is, specific productivity, expressed as amount of product produced per unit time per unit mass of microbial cells. We believe this parameter is of little value in immobilized-living-cell systems, because the cell density is constantly changing even under apparent steady-state conditions.

The simplest and most commonly used kinetic expression for the enzymatic rate of reaction is given by the Briggs–Haldane monoenzyme, monosubstrate, steady-state model:

$$r = -\frac{dS}{dt} = \frac{k_2 ES}{K_m + S} \tag{4}$$

where r is the reaction rate, S is the substrate concentration, K_m is the Briggs–Haldane (popularly known as the Michaelis–Menten) constant, and k_2E is the maximum reaction rate (V_m) for that system. Sometimes K_m and V_m may be replaced by apparent constants $K_{m'}$ and $V_{m'}$ to account for external influences on intrinsic kinetics. It should be mentioned here that this analysis may also be applicable to several multienzyme systems if it can be validly assumed that a single enzyme is rate controlling.

B. Idealized Reactor Performance Equations

1. Michaelis–Menten Kinetics

The following equations describe the reactor performance of the two main types of reactors when the reactions are kinetically controlled:

$$\text{CSTR:}\quad S_0X + K_{m'}[x/(1-x)] = k_2E\tau \tag{5}$$

$$\text{PFR:}\quad S_0X + K_m \ln(1-x) = k_2E\varepsilon\tau \tag{6}$$

where ε is the void volume of the reactor. It can be deduced from these equations that the amounts of enzyme required for a CSTR as compared to a PFR are related as follows:

$$\frac{E_{\mathrm{CSTR}}}{E_{\mathrm{PFR}}} = \frac{x\varepsilon}{(1-x)\ln(1-x)} \tag{7}$$

In general, the amount of enzyme required for a CSTR is much higher than that for a PFR.

2. Substrate- and Product-Inhibition Kinetics

When the enzymes are subjected to inhibition by substrate and/or product, the reactor performance is altered significantly. Table II [Eqs. (8)–(16)] summarizes the reactor performance equations for both PFR and CSTR for substrate-inhibition and product-inhibition (competitive and noncompetitive) kinetics.

3. Enzyme-Inactivation Kinetics

The ratio of inactivation of an enzyme during its actual use over a period of time can often be described as a pseudo-first-order process.

$$-\frac{dE}{dt} = k_dE \tag{17}$$

where E is the effective enzyme concentration in the reactor at time t, and k_d is the first-order decay constant. Substituting this in the reactor

TABLE II

REACTOR PERFORMANCE EQUATIONS BASED ON KINETIC EXPRESSIONS FOR SUBSTRATE AND PRODUCT INHIBITION, EQS. (8)–(16)

Kinetic form	Kinetic expression	Reactor performance equation[a]: CSTR	Reactor performance equation[a]: PFR
Substrate inhibition	$r = \frac{k_2 E}{1 + (K_{m'}/S) + S/k_s}$ (8)	$S_0 X + K_{m'}\left(\frac{x}{1-x}\right) + \frac{S_0^2}{k_s}(x - x^2) = k_2 E\tau$ (9)	$S_0 X - K_m \ln(1-x) + \frac{S_0^2}{2k_s}(2x - x^2) = k_2 E\varepsilon\tau$ (10)
Product inhibition (competitive)	$r = \frac{k_2 E}{1 + (K_{m'}/S)[1 + (P/K_p)]}$ (11)	$S_0 X + K_{m'}\left(\frac{x}{1-x}\right) + \frac{K_{m'}}{K_p}\frac{S_0^2}{(1-x)} = k_2 E\tau$ (12)	$S_0 x \left(1 - \frac{K_{m'}}{K_p}\right) - K_m \ln(1-x) \left[1 + \frac{S_0}{K_p}\right] = k_2 E\varepsilon\tau$ (13)
Product inhibition (noncompetitive)	$r = \frac{k_2 E}{[1 + (K_{m'}/S)][1 + (P/K_{p'})]}$ (14)	$S_0 X + K_{m'}\left(\frac{x}{1-x}\right) + \frac{K_{m'}}{K_{p'}}\frac{(1-x)}{S_0 x^2} + \frac{S_0^2 x^2}{K_{p'}} = k_2 E\tau$ (15)	$S_0 x \left(1 - \frac{K_{m'}}{K_{p'}}\right) - K_m \ln(1-x)$ (16) $\left[1 + \frac{S_0}{K_{p'}}\right] + \frac{S_0^2 x^2}{2K_{p'}} = k_2 E\varepsilon\tau$

[a] CSTSR, Continuous-flow stirred-tank reactor; PFR, plug-flow reactor.

performance equation, we get

$$\text{CSTR:}\quad \ln\left[\frac{S_0x + K_{m'}[x_0/(1 - x_0)]}{S_0x_t + K_{m'}[x_t/(1 - x_t)]}\right] = k_d t \tag{18}$$

and

$$\text{PFR:}\quad \ln\left[\frac{S_0x_0 - K_{m'}\ln(1 - x_0)}{S_0x_t - K_{m'}\ln(1 - x_0)}\right] = k_d t \tag{19}$$

where x_0 and x_t are conversions at $t = 0$ and when the reactor has been operating for time t.

Substrate-dependent enzyme decay is given by

$$-\left(\frac{dE}{dt}\right) = (k_{d'}/S)E \tag{20}$$

where $k_{d'}$ is a lumped deactivation constant. This yields for a CSTR:

$$S_0(x_t - x_0) + S_0 \ln\left(\frac{x_0}{x_t}\right) + k_{m'} \ln\left[\frac{K_{m'} + S_0(1 - x)}{K_{m'} + S_0(1 - x_t)}\right] = K_{m't} \tag{21}$$

An analytical solution cannot be obtained for PFR in this case and numerical techniques have to be resorted to.

C. Effect of Mass Transfer on the Performance of Immobilized-Cell Reactors

These effects can be broadly classified into external (film) diffusion, diffusive, and electrostatic effects; internal (pore) diffusion; and combined external and internal diffusion effects.

1. External Film Diffusion

The rate of mass transfer of the substrate from the fluid to the catalyst surface is given by

$$r_m = k_L a_m (S_F - S_S) \tag{22}$$

where k_L is the mass-transfer coefficient, a_m is the surface area for mass transfer, and S_F and S_S are the substrate concentrations in the bulk and at the surface, respectively. Several correlations are available to estimate k_L for different particle geometries and operating conditions (Vieth *et al.*, 1976).

For many engineering calculations, first-order kinetics is a reasonable approximation to account for the mass-transfer resistance. For packed beds, this leads to an equation of the type:

$$\tau' = k_f[-\ln(1 - x)] \tag{23}$$

where $k_f = K_{m'}/V_{m'}$, the pseudo-first-order rate constant; and τ' is the reactor spacetime based on reactor fluid volume,

$$\tau' = (V_R\varepsilon)/Q \tag{24}$$

When the external diffusion resistance is significant, the same equation can be used with a modified pseudo-first-order constant defined as

$$k_{f'} = \frac{k_f k_L a_m}{k_f + k_L} \tag{25}$$

2. Diffusive and Electrostatic Effects

The problem of boundary-layer diffusional resistance can often be compounded by partitioning of substrate due to electrostatic forces. Particularly in the case of surface-immobilized whole cells, this effect can be pronounced because of the net negative or positive charge on the cell walls. No analysis of this effect has been presented in the literature yet. Presumably the problem can be solved using the same approach as in the case of immobilized enzymes. Hamilton *et al.* (1973) have presented such an analysis for a wide range of surface potentials using the Gouy–Chapman potential distribution.

3. Internal (Pore) Diffusion

The widely used way to tackle this problem is to express the reaction—with and without internal diffusion resistance—in terms of an effectiveness factor η. The simultaneous mass transfer and reaction of the substrate in the (internal) matrix of a single catalyst element can be described by a second-order differential equation:

$$D_e(\partial^2 S/\partial z^2) - r = 0 \tag{26}$$

The solution of this equation yields the effectiveness factor in terms of the Thiele modulus ϕ, which is defined as

$$\phi = \ell \left[\frac{k_{\text{true}} S_s^{m-1}}{D_e}\right]^{0.5} \tag{27}$$

where z is the distance from the center of the catalyst particle, ℓ is the characteristic dimension of the catalyst particle, D_e is the effective diffusivity, m is the reaction order, and k_{true} is the true kinetic constant without any mass-transfer disguises.

Using this approach, the effectiveness factor for spherical particles in a packed bed is

$$\eta = \left[\frac{1}{\phi}\left(\frac{1}{\tanh 3\phi} - \frac{1}{3\phi}\right)\right] \tag{28}$$

whereas that for a packed bed of rectangular chips is

$$\eta\tau = \left[\frac{\tanh\phi}{\phi}\right] \tag{29}$$

The dependence of η on ϕ is shown in Fig. 2 for the latter case. Once the effectiveness factor is calculated, the actual rate expression is multiplied by this factor and used in the reactor performance equation as before.

Analytical solutions cannot be obtained in case of Michaelis–Menten kinetics. The approach taken in such cases is to separate the concentration-dependent and concentration-independent terms and define a modified Thiele modulus:

$$\phi_m = \ell[V_{m'}/K_{m'}D_e]^{0.5} \quad (30$$

Halwachs (1979) has used this approach recently to generate substrate concentration profiles in pores for various kinetic orders. Because the catalytic activity of an immobilized-whole-cell system depends on the amount of enzyme exposed to the medium, the surface area of the support should be as great as possible. For this reason, porous materials are often used as support for the whole cells. Hence, the performance of a reactor with pore-diffusion resistance is of particular interest. The equations presented here apply to packed-bed reactors. A detailed analysis of such systems in a CSTR was recently done by Lin (1978) and Lin and Wei (1979) with Michaelis–Menten as well

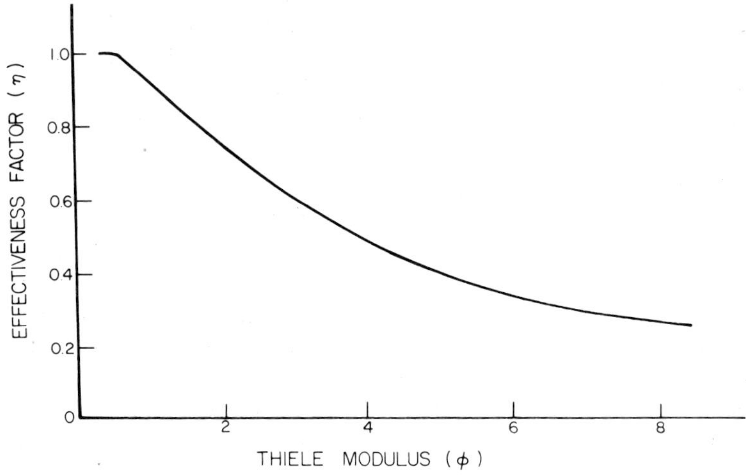

Fig. 2. Variation of effectiveness factor η with Thiele modulus ϕ for rectangular chips of immobilized cells (for pseudo-first-order reaction kinetics).

as substrate-inhibited kinetics. Lin (1979) has also presented a similar analysis for encapsulated-enzyme reaction systems. The general approach is to get a dimensionless reactor substrate balance of the following form that describes the concentration of the substrate at any point in the individual spherical particle in which the cells are entrapped.

$$\frac{d^2C_2}{dR^2} + \frac{2}{R}\frac{dC_2}{dR} - \frac{\phi^2 C_2}{C_2 + K_{m'}} = 0 \tag{31}$$

The overall substrate concentration change in a stirred-tank reactor (CSTR) can be described by

$$1 - C_1 = \frac{3\beta}{\phi^2}\left(\left.\frac{dC_2}{dR}\right|_{R=1}\right) \tag{32}$$

subject to the boundary conditions

$$R = 0; \qquad \frac{dC_2}{dR} = 0 \tag{33}$$

and

$$R = 1; \qquad \frac{dC_2}{dR} = N_{sh'}\,(C_1 - C_2) \tag{34}$$

where $C_1 = S/S_0$, $C_2 = S_2/S_0$, $K_{m'} = K_m/S_0$, $\beta = \alpha V_m V_R/QS_2$, $d = (V_m z_0^2/DS_0)^{1-2}$, $R = z/z_0$, and $N_{sh'} = k_L z_0/D_e$.

In the preceding equations, S_0 is inlet concentration, S is exit concentration, S_2 is the concentration at any point inside the pore, z_0 is the diameter of the particle, k_L is the mass-transfer coefficient, D_e is the diffusivity, K_m is the Michaelis constant, r is the radial coordinate, and $N_{sh'}$ is the modified Sherwood number. The equations are numerically integrated and the desired concentration profiles are generated for different conditions.

4. Combined External and Internal Diffusion Effects

Many practical reactor systems are likely to have both significant intra- and interparticle mass-transfer effects, which complicate even further the problem of reactor design. In its most general form, such a system will have to be modeled using Michaelis–Menten kinetics.

Vieth *et al.* (1976) have presented detailed solutions to this problem for the case of a spirally wound multipore membrane reactor for first-order (reversible and irreversible) as well as for Michaelis–Menten kinetics (numerical solution). The approach is to apply Eq. (26) to the specific problem. The equation is written in dimensionless form for the applicable rate expression in terms of rate constants and Thiele modulus. Appropriate boundary conditions are used, which normally involve a modified Sherwood number ($N_{sh'} = k_L\ell/D_e$). Numerical computation has to be resorted to in the case of Michaelis–Menten kinetics for the solution of the mass-transfer–kinetic model. Another approach to solve the combined external and internal mass transfer of substrate concomitant with the Michaelis–Menten reaction scheme has been advanced by Fink *et al.* (1973). They introduce a geometrical correction factor in the system equations in order to account for cylindrical, planar, or spherical geometries.

D. Other Considerations for Reactor Analysis

In the foregoing mathematical treatment, the fluid-flow pattern in the reactors has been idealized either as perfect plug flow or completely back-mixed. However, many reactor configurations fall in between these two patterns. Hence, the analysis of such reactors has to account for the nonideal flow characteristics as well. Such reactor configurations include fluidized-bed reactors and hollow-fiber reactors.

1. Fluidized-Bed Immobilized-Cell Reactors

Fluidized-bed reactors operate with complex flow patterns that fall somewhere between perfect plug flow or perfect back-mixing. Allen *et al.* (1979) have proposed an approximation to such a behavior by incorporating an additional term to account for back-mixing in the ideal plus flow. This term describes the nonideal effects in terms of a dispersion coefficient. In one dimension along the length of the reactor, this equation is

$$\frac{D'}{\varepsilon uL}\frac{d^2C}{dZ^2} - \frac{dC}{dZ} - \tau r = \frac{dC}{d\theta} \qquad (35)$$

where D' is the dispersion coefficient, ε the bed void fraction, u the superficial fluid velocity, L the reactor length or bed height, Z the normalized distance along the reactor length, r is the rate of chemical reaction per unit volume, and θ is the dimensionless time ($=t/\tau$). The authors have determined the dispersion coefficient values for fluidized-bed immobilized-enzyme reactors.

2. Hollow-Fiber Reactor Analysis

Webster and Shuler (1979) have modeled the diffusion and reaction processes that occur in an immobilized-whole-cell hollow-fiber reactor and presented the results in the form of effectiveness factor charts. Allowance was made for resistances in both hollow-fiber wall and the cell suspension. The relative rate of diffusion in these two regions was defined by a parameter Y—the ratio of permeation coefficient through the hollow-fiber wall to the effective diffusivity in the cell suspension. Redefined Thiele moduli are presented to account for the system-modeling geometry and order of the biochemical reaction.

Wandrey and Flaschel (1979) have analyzed a soluble-enzyme system in a cascade of CSTRs with a single ultrafiltration separation unit for recycle of the enzyme. The same analysis is applicable for whole cells (similar to a dialysis culture) provided that growth is restricted.

VI. IMMOBILIZED-LIVE-CELL REACTORS

A. Definitions and Assumptions

Throughout this analysis, it is assumed that the cells grow only on the surface of the catalyst. This assumption is valid to a large extent because oxygen limitation usually allows maximum growth near the surface of the catalyst only. The catalyst is therefore characterized by an available surface area, A (m^2/liter). This is a characteristic of the supporting material used. We then define a catalyst-loading capacity, X_s^* (g dry cell weight/m^2), which is the limit to which the supporting material can hold the biomass concerned. This will be a function of the support as well as the microorganism used. At any time, the cell concentration on the surface is defined by X_s(g/m^2), and in the bulk it is given by X (g/liter). The immobilized-cell concentration can be expressed in terms of reactor volume as $X_{im} = X_s A$ (g/liter).

As before, the reactor volume is $V_R(\ell)$ and substrate flow rate is Q (liters/min). The limiting substrate concentration is S in the reactor and S_0 at the inlet. The analysis that follows is mainly intended as a comparison to a conventional continuous-culture system, and hence a CSTR type of configuration is considered. This analysis is also applicable to relatively shallow fluidized beds.

We assume that the cells first grow on the surface with a specific growth rate μ_s where

$$\mu_s = \frac{1}{X_s}\frac{dX_s}{dt} = \frac{\mu_{ms} S}{K_S + S} \tag{36}$$

The surface growth continues until catalyst-loading capacity is reached. The data presented by Jirkú *et al.* (1981) suggest that μ_s is different (lower) from the bulk growth rate μ_b.

We further assume that once the catalyst is completely loaded, the cells grow into the bulk with specific growth rate μ_b, which depends on the substrate concentration, thus,

$$\mu_b = \frac{1}{X_{total}} \frac{dX_{total}}{dt} = \mu_m \frac{S}{K_S + S} \tag{37}$$

where $X_{total} = X_{im} + X$.

We also neglect the external and pore diffusional resistances in this analysis. It is further assumed that only live and growing cells remain attached to the catalyst while the dead cells wash away. We have recently obtained some experimental evidence that shows the preferential leaching of dead cells (Karkare *et al.*, 1981); however, this is a special case because others have shown cases where dead cells also remain attached.

B. Idealized Reactor Performance Equations

1. Steady-State Mass Balance

For steady-state mass balance, the catalyst is assumed to be fully loaded, that is, $X_{im} = X_s^*A$.

Taking inlet cell concentration to be zero, we have for a CSTR

$$\mu_b X V_R + \mu_b X_s^* A V_R = QX \tag{38}$$

Let the dilution rate $D = Q/V_R = 1/\tau$

$$DX = \mu_b(X + X_s^*A) \tag{39}$$

or

$$DX = \mu_b(X + X_{im}) \tag{40}$$

A balance on the limiting substrate gives

$$D(S_0 - S) = \frac{\mu_b}{y}(X + X_{im}) \tag{41}$$

where y is the biomass-yield coefficient.

Combining Eqs. (40) and (41) we get

$$X = y(S_0 - S) \tag{42}$$

Equation (42) is identical to the analogous continuous-culture equation. However, the value of S is always lower in the case of IMC reactors, as shown next.

Substituting Eq. (42) in (41) and Eq. (39) in (41), we get

$$\frac{\mu_{ms}}{K_s + S} = \frac{Dy(S_0 - S)}{y(S_0 - S) + X_{im}} \tag{43}$$

Equation (43) is a quadratic equation in S and has *one* meaningful root between zero and S_0. It is of interest to see that if $X_{im} \rightarrow 0$, we have the familiar continuous-culture relationship

$$\mu_b = D \tag{44}$$

Thus for the immobilized-cell process, we have $D > \mu_b$ at all values of D.

This implies that for any given D, the exit substrate concentration is lower for the IMC reactor. Hence, the IMC reactor is superior to the continuous-culture system in terms of conversion efficiencies. Consequently, the exit cell concentration for the immobilized-cell process is always higher. Figure 3 shows the effect of increasing X_{im} on the exit biomass concentration. The calculations are for growth of *Acetobacter*

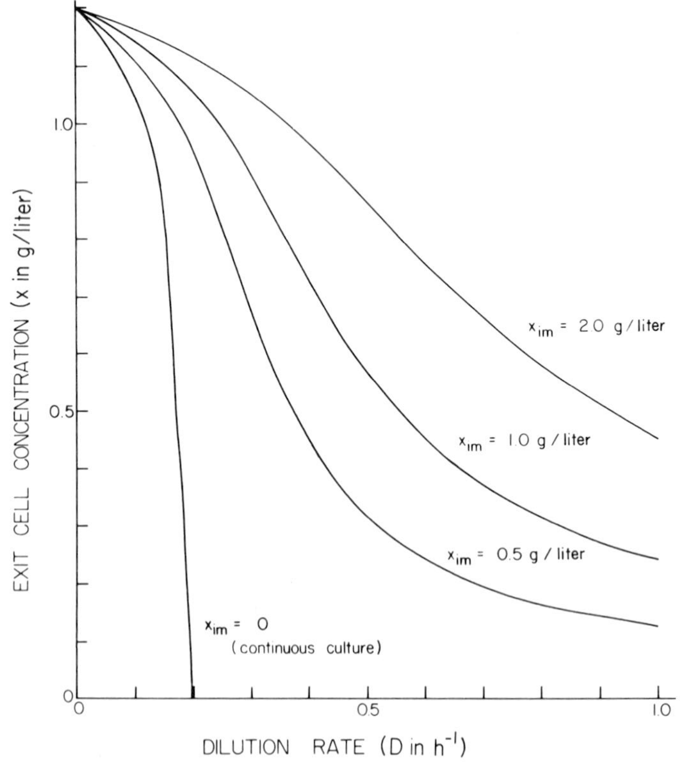

Fig. 3. Effect of dilution rate on effluent cell concentration in immobilized-live-cell CSTR with varying amounts of immobilized cells. Calculations were based on data of Vera-Solis (1976) for growth of *Acetobacter suboxydans* on ethanol. $\mu_m = 0.23\ h^{-1}$, $K_s =$ 6.5 g/liter, $Y = 0.0265$ g cells/g ethanol, and $S_0 = 45$ g/liter ethanol.

suboxydans. The relevant constants were obtained from the data of Vera-Solis (1976). It is apparent that the oft-dreaded washout conditioning in submerged continuous cultures is eliminated by using immobilized cells.

For the special case of $D = \mu_m$ we have a unique solution for Eq. (43).

$$S = \frac{K_s y S_0}{K_s y + X_{im}} \tag{45}$$

Again, the equation reduces to $S = S_0$ as $X_{im} \to 0$. Rearranging Eq. (43) we get

$$D = \frac{\mu_m S y (S_0 - S) + \mu_m S X_{im}}{y(K_s + S)(S_0 - S)} \tag{46}$$

Multiplying this by $X = y(S_0 - S)$, we get

$$DX = P_r = \frac{\mu_m S y (S_0 - S) + \mu_m S X_{im}}{K_s + S} \tag{47}$$

For maximum productivity with respect to substrate concentration

$$\frac{dP_r}{dS} = 0 \quad \text{and} \quad \left.\frac{d^2 P_r}{dS^2}\right|_{S_{opt}} < 0 \tag{48}$$

Differentiation and simplification lead to

$$y S_{opt}^2 + 2K_s y S_{opt} - K_s(y S_0 + X_{im}) = 0 \tag{49}$$

Using the meaningful positive root, we get

$$S_{opt} = \frac{\sqrt{K^2 y^2 + K_s y(y S_0 + X_{im})} - K_s y}{y} \tag{50}$$

and

$$D_{opt} = \frac{\mu_m S_{opt} y (S_0 - S_{opt}) + \mu_m S_{opt} X_{im}}{y(K_s + S_{opt})(S_0 - S_{opt})} \tag{51}$$

It can be easily verified that as $X_{im} \to 0$, Eqs. (50) and (51) reduce to their counterparts in continuous-culture systems.

For the example in Fig. 3, considering $X_{im} = 2$ g/liter, it can be readily calculated that

$$D_{opt} = 0.77 \text{ h}^{-1}$$

This is much greater than μ_m. The productivity at this dilution rate is 0.4633 g/liter/h, which is about 3.5 times the maximum productivity of a similar continuous culture. The exit cell concentrations for the two cases are quite comparable ($X = 0.6$ g/liter for IMC reactor and $X = 0.8$ g/liter for continuous culture). It is obvious that the cell concentration can be increased for the IMC reactor at the expense of a little productivity and the productivity levels will still be much higher than the continuous culture.

Figure 4 shows the change in biomass productivity with dilution rates for various immobilized-cell concentrations.

2. Concept of a Hybrid Reactor

The foregoing considerations lead us to the concept of a highly productive "biomass generator." If a large amount of cells can be immobilized in a small volume (by using highly porous supports), we can generate biomass at much higher rates than was possible before. We can thus conceptualize quantitatively the term "microbial generator" coined by Messing *et al.* (1981). Of particular interest here is single-cell protein production for food and/or feed use.

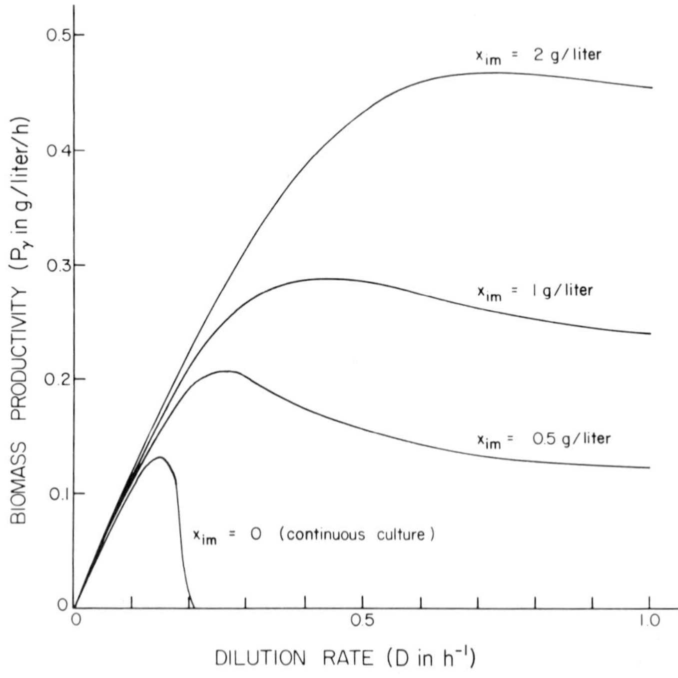

Fig. 4. Effect of dilution rate on biomass productivity of immobilized-live-cell CSTR with varying amounts of immobilized cells. Calculations based on data of Vera-Solis (1976) for growth of *Acetobacter suboxydans* on ethanol. Parameter values are the same as in Fig. 3.

The mathematical treatment of the immobilized-cell reactors presented here finds its roots in an article on wall growth in continuous culture by Topiwala and Hamer (1971). It should be noted, however, that the equations presented by these authors were of a very preliminary nature and dealt only in the deviations from the ideal chemostat. A tentative suggestion was made for use of "extended surfaces" in fermentation. Because for the production of biomass at enhanced rates, we are utilizing cell growth both in the immobilized phase and in the bulk phase, it is appropriate to refer to these reactors in particular as "hybrid reactors." Thus the hybrid reactors are a combination of the concepts of cell immobilization and continuous culturing of microorganisms. In general, wherever live immobilized cells are used for continuous fermentation (whether for growth-associated or non-growth-associated products), the term "structured-bed fermentation," first used by Vieth (1979), seems appropriate.

3. Incubation Time (Nonsteady-State Mass Balance)

As mentioned before, the cells first grow on the catalyst surface until its loading capacity is reached. This period is referred to as incubation time. In this period, the inlet and exit cell concentration is zero. A balance on the limiting substrate gives

$$D(S_0 - S) = \left(\frac{\mu_s}{y}\right) XA = \frac{\mu_{ms} S X_s A}{y(K_s + S)} \tag{52}$$

and a cell balance gives

$$\frac{dX_s}{dt} = \left(\frac{\mu_{ms} S}{K_s + S}\right) X_S \tag{53}$$

The equations just given can be numerically integrated between $X_S = X_{S0}$ (initial cell concentration on the surface) and X_S^* (the catalyst-loading capacity), to calculate the incubation time for a particular system. These equations can quantify the "conditioning" of the beads used by Chibata (1980) where the cells are being grown on or inside the beads in order to saturate the catalyst.

4. Metabolite Production

In general, metabolite production in fermentation systems can be described by the Leudeking–Piret model:

$$\frac{dP}{dt} = K_1 X + K_2 \frac{dX}{dt} \tag{54}$$

Where P is the metabolite concentration, if $K_2 \gg K_1$ we have a growth-associated product (or primary metabolite), and when $K_1 \gg K_2$ we have a secondary metabolite.

a. Primary (growth-associated) metabolites. In this case the rate of product formation is given by

$$\frac{dP}{dt} = K_2 \frac{dX}{dt} \tag{55}$$

and the metabolite productivity is described by

$$P_r = \frac{dP}{dt} = K_2 DX \tag{56}$$

Hence, maximizing the metabolite productivity is the same as maximizing biomass productivity. Therefore, the calculation of optimum dilution rate is the same as before.

b. Secondary metabolites. Rate of biosynthesis of secondary metabolites is dependent primarily on the cell concentration. Hence we can write this rate as

$$\frac{dP}{dt} = K_1 x \tag{57}$$

For an immobilized-cell process, this would become

$$\frac{dP}{dt} = K_1(X + X_{im}) \tag{58}$$

Hence, maximizing the productivity in this case involves maximizing X_{im}. This can be done by using a catalyst support with high loading capacities. It is also advantageous to keep X as high as possible. From Fig. 3 we can see that X does not change drastically until D becomes greater than μ_m. Thus any value of D would substantially yield the same productivity. Therefore, the choice of D would depend on the yield requirement of the process. The product mass balance in this case is given by

$$DP = K_1(X + X_{im})$$

or

$$P = \frac{K_1}{D}(X + X_{im}) \tag{59}$$

Because X is relatively constant, D can be chosen to suit the requirement of P (often dictated by the recovery process).

c. Metabolites with mixed growth model. When K_1 and K_2 are both significant, the product mass balance becomes

$$DP = K_1(X + X_{im}) + K_2\mu_b(X + X_{im}) = (K_1 + K_2\mu_b)(X + X_{im}) \tag{60}$$

Again, using the same techniques as in biomass productivity, we can calculate optimum dilution rate for maximum productivity. In this case,

$$S_{opt} = \frac{\sqrt{y^2K_S^2(K_1 + \mu_m K_2)^2 - yK_S(K_1 + \mu_m K_2)yK_SK_1 - \mu_m K_2(yS_0 + X_{im})} - yK_S(K_1 + \mu_m K_2)}{y(K_1 + \mu_m K_2)} \tag{61}$$

and

$$D_{opt} = \frac{\mu_m S_{opt} y(S_0 - S_{opt}) + \mu_m S_{opt} X_{im}}{y(K_S + S_{opt})(S_0 - S_{opt})} \tag{62}$$

Again, we can verify that as $K_1 \to 0$, we get S_{opt} identical to that of growth-associated products.

C. Mass-Transfer Considerations

Very little information is available in the literature on the analysis of the mass-transfer effects in immobilized-living-cell systems. However, the treatment of bulk and pore diffusional effects would be essentially the same as in the case of immobilized-single-enzyme-type reactor systems. Thus, the mathematical treatment of the substrate flow up to the surface of the immobilized cell would be based on the type of analysis covered in Section V,C. Next we have to consider the diffusion of the substrate and the product through the barrier imposed by the cell envelope (i.e., cell wall and cell membrane) itself. It behooves us to examine first the relative resistances presented by the carrier matrix and the cell envelope in this connection.

Even though the pore diffusion problem is often cited as a negative aspect of immobilized-cell reactors, if one considers the fact that most immobilized-cell processes are based on fixing the microbial cells in some form of hydrated gels (>90% water), it is not unreasonable to speculate that the diffusional resistance of the gel itself may not be that critical. This argument is further strengthened by the observation that the live cells tend to concentrate on or near the carrier surface.

In the case of monoenzyme reactions, it is often possible to increase the permeability of the cell envelope by specific treatments. One example in this connection is heat treatment of cells prior to immobilization in the case of cells containing glucose isomerase activity (Vieth and Venkatasubramanian, 1976). When concerned with more complex reaction sequences and pathways, this may not be feasible. The total cell structure needs to be retained intact in order to preserve the optimal arrangement of the enzymes and cofactor-generating machinery. Thus, the role of transport resistances through the cell envelope in the overall reaction assumes greater importance in the case of immobilized-living-cell systems.

1. Substrate Transport into the Immobilized Cell

It can generally be assumed with impunity that the substrate transport into the microbial cell is characterized by passive diffusion in the case of single-enzyme-type immobilized-cell systems. The precise mechanism of the substrate transport into an immobilized living cell has simply not been studied as yet. However, it is reasonable to suggest that active or facilitated transport mechanisms would prevail at least for those substrate molecules that are transported in such manner in equivalent submerged fermentations.

Thus, the type of diffusion mechanism would depend on the organism and the limiting substrate itself. Active transport seems to be a common mode of transport of sugars into cells. Hence, models based on this concept should be used to take into account the cell envelope resistance in the reactor performance calculation. Vieth *et al.* (1982) have described lactose transport through the cell membrane by an equation of the following type:

$$\frac{d(L_{\text{in}})}{dt} = \frac{A(t)L_0}{B(t) + L_0} = \frac{J_L}{V_c} \tag{63}$$

where L_{in} is the intracellular lactose concentration, L_0 is the extracellular concentration, and $A(t)$ and $B(t)$ are time-dependent constants. J_L/V_c is the molar flux. In the case of continuous systems, A and B are likely to be true constants. Little attention has been paid to this important concept and further work is necessary along the lines outlined previously in order to determine conditions to enhance rates of transport through the cell barrier.

Another important transport problem relates to oxygen transfer into the cells in the immobilized state in the case of aerobic organisms. Adequate oxygen supply is ensured in traditional fermentations through the use of properly designed aeration and agitation systems. It may not be feasible to extrapolate this approach to immobilized-cell reaction systems. Although a number of promising approaches are being investigated, rigorous mathematical analysis of this problem is yet to be undertaken.

Finally, it must be pointed out that the overall modeling problem is further complicated by the multitude of reactions that take place inside the cells in converting the substrate into additional cell mass, energy compounds, and desirable products. In order to be able to cope with these complexities, one has to resort to simplified engineering models of sufficient accuracy as discussed in the following section.

D. Engineering Models for Overall Reactions

It is obvious that the simple growth kinetics and product-formation models are not sufficient to describe the kinds of metabolic shifts that could occur due to nutrient changes and cell age differences. It is necessary to model the biochemistry in greater detail so that we can get a better insight into the kinds of nutrient changes necessary to bring about the desired metabolic shifts.

As mentioned in the previous section, one pragmatic approach to the aforementioned problem is to develop simplified practical engineering models describing the biochemistry of the metabolic pathways. The idea is to lump chains of reactions into a single reaction concept and focus mainly on certain key intermediates that control the channeling of the substrate into different pathways. Figure 5 shows a simplified version of the biochemical pathways involved in production of polyene macrolides (secondary metabolites) and cell maintenance. We can readily recognize acetyl-CoA as one of the key intermediates. Thus the fate of this intermediate can shift the metabolism of the cell. Therefore, the activity of the various enzymes acting on this intermediate is of particular interest. Another possible way of shifting the metabolism is to control the maintenance energy requirement of the cells by attacking the sources of maintenance energy requirement such as osmotic pressure, protein turnover rate, and pH.

Thus the immobilized-cell concept, which allows us to maintain a high cell density in the reactor at any dilution rate, opens up a host of nutrient feed options to impose more sophisticated controls on the cell metabolism than were possible before.

Control of Immobilized-Cell Metabolism

The effect of immobilization on the metabolism of microorganisms has not yet been studied in detail. Some evidence of changes in metabolism has been presented by Navarro and Durand (1977). However, it is not clear whether the increase in respiration rate consequent to immobilization is due to a shift in metabolism or to growth (i.e., production of additional biomass) of the cells, on the surface. The physiol-

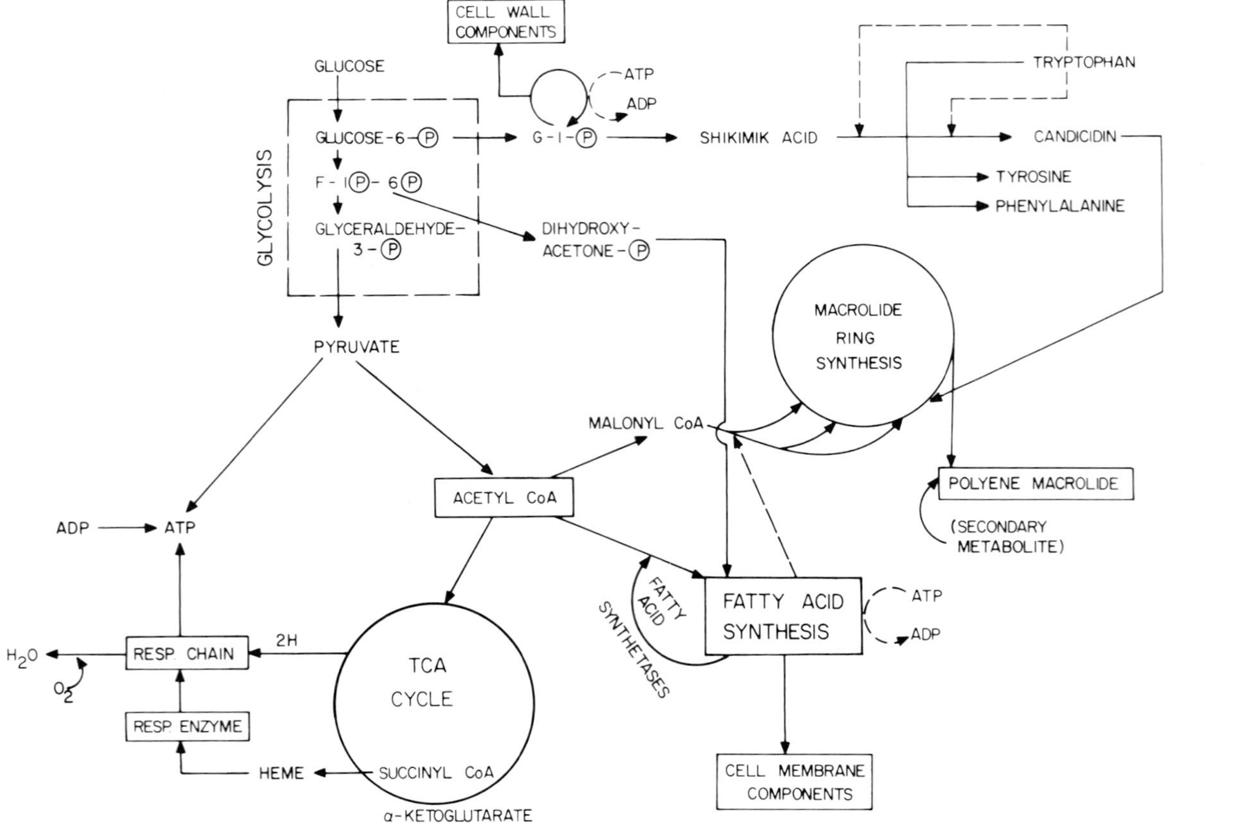

Fig. 5. Simplified scheme of biochemical pathways involved in the production of polyene macrolides (secondary metabolites) and cell maintenance.

ogy of the immobilized organisms will definitely play an important role in metabolite production. For example, in the case of secondary metabolite production, the cells may have maturity requirements (Martin and McDaniel, 1975) that necessitate a controlled growth of the organism. In other words, different metabolites would require different strategies for controlling the metabolism of the microbes in order to convert the substrate most efficiently into desired products.

One ready tool for this purpose is the nutrient concentration in the feed. By carefully controlling the growth factors and other product inducers in the feed, it would be possible to maintain an immobilized culture of appropriate cell density and physiology in order to increase the productivity of the desired metabolite. The aim would therefore be to shift the cell metabolism in such a way as to channel more substrate into the useful metabolite by reducing maintenance requirements and cell growth. This is of particular interest in secondary metabolite production.

The other variable at our disposal for controlling cell metabolism is the dilution rate (within the limits imposed by design considerations). The dilution rate has an impact on the average cell age in the reactor. Cell age has been shown to have an appreciable effect on immobilized-cell activity (Venkatasubramanian and Vieth, 1979). Hence, both the nutrient concentration and feed rate to the reactor are bound to have a great effect on the substrate conversion efficiency of the reactor.

VII. PRACTICAL CONSIDERATIONS FOR DESIGN AND OPERATION OF IMMOBILIZED-CELL SYSTEMS

Venkatasubramanian and Harrow (1979) have discussed some of the practical considerations involved in the design and operation of bound-cell reactors, especially when only a single enzyme activity is desired. However, the principles discussed therein are generally applicable to more complex IMC reactors as well. Because as of this time only single-enzyme-type IMC reactors have been scaled up and operated commercially—for example, the process for isomerization of glucose to fructose to produce high-fructose corn syrup (HFCS)—the following discussion is based on such systems.

A. Number of Reactor Columns and Flow Mode

In the case of single-enzyme-type IMC reactors, the activity of the bound enzyme decreases continuously. It is therefore necessary to have a number of packed reactor columns in contrast to one gigantic column. This would minimize production fluctuations with respect to capacity and conversion level. For a given plant capacity, the optimum number of columns can be estimated theoretically. The flow variation for a single-column operation would be too great to render it practical. Although increasing the number of reactors provides greater operational flexibility and minimum flow-rate variation, it must also be borne in mind that they result in higher cost for reactor piping, valves, and instrumentation. Also, more reactors imply more frequent enzyme changes, thereby increasing operational costs.

Both series and parallel operation of the reactor system are possible. In series operation there are fewer streams to control, and the upstream bed in a series can be nearly fully exhausted before removing from service. However, the latter point is not of great relevance because exhaustion of bed activity is very gradual. Furthermore, series operation suffers from one overriding disadvantage: fluid velocity. A three-column series will have three times the velocity of the same system with three columns in parallel. Therefore, the potential for pressure drop and compaction problems are greatly enhanced. Parallel operation, however, offers the greatest operational flexibility. Each reactor can be operated essentially independent of others. Each column can be brought into and taken out of service very readily.

Another consideration relates to upflow versus downflow operational mode. Upflow operation offers the advantages of good fluid–particle contact and constant minimum pressure drop. The downflow mode allows operation under essentially atmospheric pressure where the fluid flow through the bed is controlled readily by the hydraulic head above the bed. Therefore, the preferred reactor design appears to be a number of fixed-bed columns operating in the parallel mode in which fluid flow occurs in the downflow fashion.

B. Column Hydraulics

The hydraulic considerations dictate the maximum permissible bed depth. The immobilized-cell particles often form a compressible bed. Under normal downflow operating mode, compression is insignificant and the behavior of the bed is close to that of the rigid granule. However, under severe hydraulic force, the bed will compact and pressure drop will increase with time. The hydraulic force is greatest in the freshly formed bed when activity (hence throughput) is highest.

The pressure drop is shown as a function of loaded-bed height in Fig. 6 for a typical immobilized-cell reactor. For practical purposes, an allowable pressure drop of 3 psi is considered ideal. On the basis of Fig. 6 one can predict a maximum bed height of 15 ft corresponding to

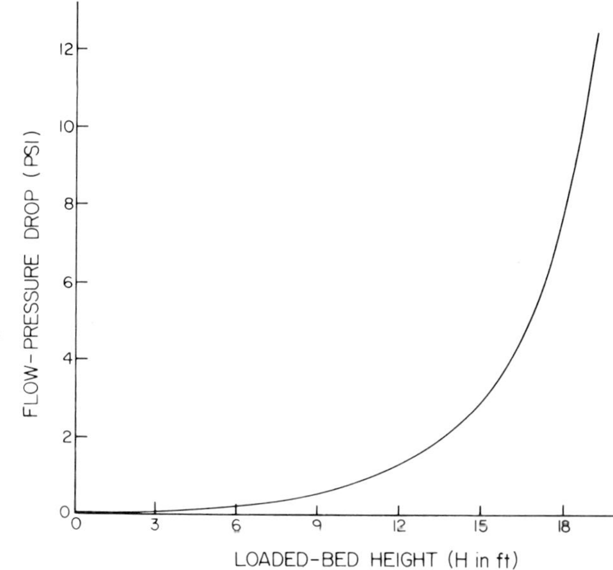

Fig. 6. Effect of loaded-bed height (ft) on the flow-pressure drop (psi) in a packed-bed immobilized-cell reactor.

this level. Above 3 psi pressure drop, there is a drastic reduction in bed permeability. Further, one can experience a time-dependent effect of the hydraulic stress on permeability. Should the bed go into compaction, it will be evident through increasing pressure drop and a concomitant reduction in the apparent activity of the enzyme. Under these conditions, the bed must be backwashed to reexpand and resettle. This must be done before the hydraulic pressure becomes too serious; otherwise, irreversible bed compaction would occur.

C. Operational Considerations and Process-Control Strategies

In downflow operation, the substrate percolates through the packed bed by gravity; to accomplish this, a sufficient liquid head above the bed must be maintained at all times. Feed inlet and outlet should also be designed so as to ensure that the packed bed is never allowed to become dry. Startup and shutdown procedures must be developed to obtain maximum efficiency of the enzyme as well as ease of column switchcover.

It is necessary to operate the reactor system in such a way as to obtain a finished product of uniform quality and constant conversion level. On-line measurements of substrate and product(s) levels and a priori knowledge of catalyst-decay profile are employed as feedback mechanisms to control flow rate through the reactors.

Typically, a single reactor column is operated for a period of three half-lives, after which the enzyme is replaced with a fresh batch. Practical flow-rate limitation and maximum allowable reactor residence time (mentioned earlier) dictate the cutoff point for a given reactor. Modest variations in operating temperature and conversion level are important process-control strategies. It is possible to control the entire operation on-line through a minicomputer or microprocessor.

D. Special Considerations for Live-Immobilized-Cell Systems

Due to the additional complexity of live-cell systems, many other considerations come into play in the design and operation of such systems.

First, for a single-enzyme system, the substrate is usually a clean liquid containing a single component that is easy to handle. For a living-cell system, the substrate is likely to be more complex in its composition; it is also likely to contain particulate matter. Some of the problems associated with such complex substrates are foaming, sterility maintenance, and pH control. The last two requirements arise directly due to the use of live cells in the reactor. Therefore, a practical design of the reactor should provide adequate control for these problems. These considerations favor a series of small reactors as they offer better control.

Particle size and compressibility of the particles is also an important consideration in live-cell systems as it affects the viability of the immobilized cells (because a compressed bed would lead to lower oxygen availability). Of particular concern is irreversible compression of the bed. If a fixed-bed system is used, we can also envisage plugging problems—especially in the case of growth-associated products. Mycelial fermentation processes are likely to present problems of non-Newtonian rheology and consequent problems of inadequate mixing. The foregoing considerations may impose a maximum allowable cell density in the reactor.

Another important consideration in a continuous process is the product concentration in the exit stream. Much of the published literature is overly concerned about the volumetric productivity per se. Although this is an important parameter, to achieve economical downstream processing and purification of the product, a certain minimum product concentration should be obtained in the reactor effluent.

For economic reasons, it is also necessary to have nearly complete substrate utilization in the reactor. However, the system should be designed and operated so as to channel the substrate toward the formation of the desirable end product rather than toward unwanted by-products. The question of by-product formation in IMC processes has received but cursory treatment in the research reported so far. In one instance, at least, we have found that it is possible to minimize by-product (isocitric acid) production during the course of citric acid biosynthesis by immobilized *Aspergillus niger* cells through proper nutrient control (Venkatasubramanian and Vieth, 1979). Again, this may be better achieved using a series of reactors introducing the substrate in between stages and optimizing the conditions in each reactor separately. The prospect of achieving higher productivities at high product concentrations would invariably necessitate the use of more sophisticated process controls advantageously. In traditional fermentations, the sophisticated control systems may be inordinately costly compared to the substrate utilization economy. With immobilized-cell systems, because the substrate economy can be realized to its full extent, it may be possible to think in terms of microprocessor control of the reactor by monitoring the effluent stream parameters such as cell concentration in the case of growth-associated products.

VIII. EPILOGUE

In the last few years, considerable progress has been made on development of immobilized-cell reactor systems using both nonliving and living cells. Theoretical reactor analysis of the former case is far simpler because it can be treated essentially as an extension of immobilized-monoenzyme reactors. In contrast to earlier approaches to modeling enzyme reactors that implicitly lump mass-transfer and kinetic effects together, recent mathematical analyses approach these effects separately and explicitly as a rational combination of elementary steps. This is indeed a welcome trend. However, such analyses have yet to be extended to immobilized-living-cell systems.

Most of the published reports on immobilized-living-cell systems are concerned with methodology development and feasibility demonstration rather than with rigorous reactor analysis. Indeed, reactor performance data are quite scanty. We have presented here a preliminary analysis of such systems and described further approaches to refine it. As pointed out earlier, much remains to be accomplished. Because fluidized-bed immobilized-cell reactors appear to be gaining in popularity, it is propitious to undertake rigorous analysis of these reactors. More work is also needed in characterizing the hydrodynamics, including analysis of residence-time distributions, and biomass and product synthesis kinetics of reactor systems of practical importance such as hybrid and hollow-fiber membrane reactors.

With the advent of immobilized-cell reactors mediating more and more complex biocatalytic processes, a thorough understanding of reactor dynamics and stability becomes even more important. Another area that warrants further work is the analysis of combined reaction–separation schemes.

Finally, there is a glaring lacuna in the area of reactor scale-up. Most laboratory data have been collected from small reactors. Even if these agree with theoretical predictions, their use in reactor scale-up is quite limited. Therefore, more effort should be devoted to shifting the scene from the laboratory to the pilot plant. This should provide a sound basis for formulating reliable scale-up procedures, as well as for developing realistic process economic evaluations.

IX. LIST OF SYMBOLS

A	Surface area available for immobilization (m^2/liter)
$A(t)$	Time-dependent constant (mol/liter/h)
a_m	Surface area for mass transfer (cm^2/g catalyst)
$B(t)$	Time-dependent constant (mol/liter)
C	Substrate concentration in reactor (g/liter)
C_1	Dimensionless exist substrate concentration ($= S/S_0$)
C_2	Dimensionless concentration of substrate inside pore ($= S_1/S_0$)
D	Dilution rate (h^{-1})
D'	Dispersion coefficient (cm^2/s)
D_e	Effective diffusivity (cm^2/s)
D_{opt}	Optimum dilution rate for maximum volumetric productivity (h^{-1})
E	Enzyme concentration in the reactor (g/liter)
J_L	Lactose flux into cells (mol/g cells/s)
K_1	Non growth-associated product constant (h^{-1})
K_2	Growth-associated product constant
K_m	Briggs–Haldane (Michaelis–Menten) constant (g/liter)
$K_{m'}$	Apparent Michaelis–Menten constant (g/liter)
K_p	Competitive product-inhibition constant (g/liter)
$K_{p'}$	Noncompetitive product-inhibition constant (g/liter)
K_s	Monod equation constant—"Half growth velocity constant" (g/liter)
k_2	Reaction-rate constant (min^{-1})
k_d	Enzyme-decay constant (min^{-1})
$k_{d'}$	Substrate-dependent enzyme-deactivation constant (g/liter/min)
k_f	Pseudo-first-order rate constant (min^{-1})
$k_{f'}$	Modified pseudo-first-order rate constant (min^{-1})
k_L	Mass-transfer coefficient (cm/s)

k_s	Substrate-inhibition constant (g/liter)
k_{true}	True kinetic constant (units depend on m)
L	Reactor length or bed height (cm)
L_{in}	Intracellular lactose concentration (mol/liter)
L_o	Extracellular lactose concentration (mol/liter)
ℓ	Characteristic dimension of catalyst particle (cm)
m	Order of reaction
$N_{sh'}$	Modified Sherwood number ($= k_L\ell/D_e$)
P	Product concentration (g/liter)
P_r	Reactor (volumetric) productivity (g/liter/h)
Q	Flow rate through the reactor (liters/min)
R	Dimensionless radial coordinate in catalyst particle
r	Reaction rate (g/liter/min)
r_m	Rate of mass transfer (g/min/g catalyst)
S	Substrate concentration in reactor (g/liter)
S_2	Substrate concentration at any point inside pore (g/liter)
S_F	Bulk substrate concentration (g/liter)
S_{opt}	Optimum substrate concentration for maximum productivity (g/liter)
S_0	Inlet substrate concentration (g/liter)
S_S	Substrate concentration at the catalyst surface (g/liter)
t	Time (min)
u	Superficial fluid velocity (cm/s)
V_c	Specific volume of cells (liters/g cell dry weight)
V_m	Maximum reaction rate (g/liter/min)
$V_{m'}$	Apparent maximum reaction rate (g/liter/min)
V_R	Reactor volume (liters)
X	Effluent cell concentration (g/liter)
X_{im}	Immobilized-cell concentration expressed in terms of reactor volume (g/liter)
X_s	Cell concentration on the surface of support (g/m^2)
X_s^*	Maximum possible cell concentration on the surface of support (g/m^2)
X_{total}	Total cell concentration in immobilized-cell reactor including free and immobilized cells (g/liter)
x	Fractional conversion of substrate
x_0	Conversion at time $t = 0$
x_t	Conversion at time t
Y	Ratio of permeation coefficient in hollow fiber to effective diffusivity in cell suspension
y	Biomass yield coefficient (g cell/g substrate)
Z	Normalized distance along reactor length
z	Distance from center of catalyst particle (cm)
z_0	Diameter of catalyst particle (cm)

Greek Letters

α	Fraction of reactor volume occupied by immobilized-cell particles ($= 1 - \varepsilon$)
β	Dimensionless parameter defined in Eq. (32)
ε	Fractional void volume of reactor
η	Effectiveness factor
θ	Dimensionless time ($= t/\tau$)
μ_b	Specific growth rate of cells in bulk (h^{-1})

μ_m	Maximum specific growth rate of cells in bulk (h^{-1})
μ_s	Specific growth rate of cells on the surface of the support (h^{-1})
μ_{ms}	Maximum specific growth rate of cells on the surface of the support (h^{-1})
ϕ	Thiele modulus
ϕ_m	Modified Thiele modulus ($= \ell[V_{m'}/K_{m'}D_e]^{0.5}$)
τ	Reactor spacetime (h)

ACKNOWLEDGMENT

The authors wish to thank Mrs. Terri Kumpa and Mrs. Gladys Dennison for typing and proofreading the manuscript. Parts of this work were supported by the H. J. Heinz Company and the National Science Foundation (Grant CBE-80-10865), for which the authors are grateful.

REFERENCES

Allen, B. R., Coughlin, R. W., and Charles, M. (1979). *Ann. N. Y. Acad. Sci.* **326,** 105.
Briffaud, J., and Engasser, J. M. (1979). *Biotechnol. Bioeng.* **21,** 2093.
Chibata, I. (1980). *Food Process Eng.* [*Proc. Int. Congr.*], *2nd, 1979* Vol. 2, p. 1.
Chibata, I., and Tosa, T. (1980). *Trends Biochem. Sci.* **4,** 88.
Chibata, I., Tosa, T., Sato, T., Mori, T., and Yamamoto, K. (1974). *Enzyme Eng.* **2,** 309.
Constantinides, A., Bhatia, D., and Vieth, W. R. (1981). *Biotechnol. Bioeng.* **23,** 899.
Denbigh, K. G. (1965). "Chemical Reactor Theory." Cambridge Univ. Press, London and New York.
Dutta, R., Armiger, W., and Ollis, D. (1973). *Biotechnol. Bioeng.* **15,** 993.
Fink, D. J., Na, T. Y., Schultz, J. S. (1973). *Biotechnol. Bioeng.* **15,** 879.
Halwachs, W. (1979). *Process Biochem.* **14,** 25.
Hamilton, B. K., Stockmeyer, L. J., and Colton, C. K. (1973). *J. Theor. Biol.* **41,** 547.
Jirkú, V., Turková, J., and Krumphanzl, V. (1981). *Biotechnol. Lett.* **3,** 509.
Kan, J. K., and Shuler, M. L. (1978). *Biotechnol. Bioeng.* **20,** 217.
Karkare, S. B., Chotani, G. K., and Venkatasubramanian, K. (1981). Rutgers University, New Brunswick, New Jersey (unpublished results).
Kennedy, J. F., Humphreys, J. D., and Barker, S. A. (1980). *Enzyme Microb. Technol.* **2,** 209.
Larreta Garde, V., Thomasset, B., and Barbotin, J. N. (1981). *Enzyme Microb. Technol.* **3,** 216.
Lin, S. H. (1978). *J. Appl. Chem. Biotechnol.* **28,** 677.
Lin, S. H. (1979). *Chem. Eng. J.* **17,** 55.
Lin, S. H., and Wei, C. K. (1979). *Chem. Eng. Sci.* **34,** 827.
Linko, Y. Y., and Linko, P. (1981). *Biotechnol. Lett.* **3,** 21.
Marconi, W., Galinelli, S., and Morisi, F. (1974). *Biotechnol. Bioeng.* **16,** 501.
Martin, J. F., and McDaniel, L. E. (1975). *Biotechnol. Bioeng.* **17,** 925.
Messing, R. A., Opperman, R. A., Simpson, L. B., and Takeguchi, M. (August 25, 1981). U.S. Patent 4,286,061.
Michaels, A. (1980). *Desalination* **35,** 337.
Navarro, J. M., and Durand, G. (1977). *Eur. J. Appl. Microbiol.* **4,** 243.

O'Neill, S. P., Wykes, J. R., Dunnill, P., and Lilly, M. D. (1971). *Biotechnol. Bioeng.* **13,** 319.
Rony, P. R. (1972). *J. Am. Chem. Soc.* **94,** 8247.
Smiley, K. L. (1971). *Biotechnol. Bioeng.* **13,** 309.
Tanaka, H., Uzman, S., and Dunn, I. J. (1981). *Biotechnol. Bioeng.* **23,** 1683.
Topiwala, H. H., and Hamer, G. (1971). *Biotechnol. Bioeng.* **13,** 919.
Venkatasubramanian, K., ed. (1979). "Immobilized Microbial Cells," ACS Symp. Ser. 106. Academic Press, New York.
Venkatasubramanian, K. (1980). *In* "Enzymes: The Interface Between Technology and Economics" (J. Danehy and B. Wolnak, eds.), p. 34. Plenum, New York.
Venkatasubramanian, K., and Harrow, L. S. (1979). *Ann. N.Y. Acad. Sci.* **326,** 141.
Venkatasubramanian, K., and Vieth, W. R. (1979). *Prog. Ind. Microbiol.* **15,** 61.
Vera-Solis, F. (1976). M.S. Thesis, Massachusetts Institute of Technology, Cambridge, Massachusetts.
Vieth, W. R. (1979). *Ann. N. Y. Acad. Sci.* **326,** 1.
Vieth, W. R., and Venkatasubramanian, K. (1976). *In* "Methods in Enzymology" (K. Mosbach, ed.), Vol. 44, p. 263. Academic Press, New York.
Vieth, W. R., Venkatasubramanian, K., Constantinides, A., and Davidson, B. (1976). *Appl. Biochem. Bioeng.* **1,** 221.
Vieth, W. R., Kaushik, K., and Venkatasubramanian, K. (1982). *Biotechnol. Bioeng.* **24,** 1455.
Wandrey, C., and Flaschel, E. (1979). *Adv. Biochem. Eng.* **12,** 147.
Weetal, H. H., and Havewala, N. B. (1972). *Biotechnol. Bioeng. Symp.* **3,** 221.

Index

K

L

M

N

O

P

R

S

T

U

V

W

KENT
LIBRARY
UNIVERSITY